Biology and Ecology of Eels

Biology and Ecology of Eels

Editor

Jose Martin Pujolar

MDPI

Basel • Beijing • Wuhan • Barcelona • Belgrade • Novi Sad • Cluj • Manchester

Editor
Jose Martin Pujolar
Technical University of Denmark
Kongens Lyngby
Denmark

Editorial Office
MDPI
St. Alban-Anlage 66
4052 Basel, Switzerland

This is a reprint of articles from the Special Issue published online in the open access journal *Fishes* (ISSN 2410-3888) (available at: https://www.mdpi.com/journal/fishes/special_issues/Biology_Ecology_Eels).

For citation purposes, cite each article independently as indicated on the article page online and as indicated below:

Lastname, A.A.; Lastname, B.B. Article Title. *Journal Name* **Year**, *Volume Number*, Page Range.

ISBN 978-3-7258-0227-2 (Hbk)
ISBN 978-3-7258-0228-9 (PDF)
doi.org/10.3390/books978-3-7258-0228-9

Contents

Preface

Eels have fascinated humankind since ancient times, and even the Greek philosopher Aristotle wondered about the life cycle and reproduction of eels and the reasons for their unpredictable presence and absence. Our knowledge on the biology and ecology of eels has advanced enormously, and the knowledge gap has been filled with information on the life cycle of many eel species, particularly European and Japanese eels, but there is still so much we do not know about these fascinating creatures that further research is certainly warranted. The purpose of this Special Issue is to gather the latest knowledge on the biology and ecology of eels in a wide variety of topics ranging from morphology and reproduction to genetics and stock assessment. This Special Issue is aimed at all researchers in general and the eel research community in particular. We would like to thank all authors that contributed papers, without them this issue would not have been possible.

Jose Martin Pujolar
Editor

fishes

Review

Footprints of Natural Selection in North Atlantic Eels: A Review

José Martin Pujolar [1,*], Francesca Bertolini [2,3] and Magnus W. Jacobsen [3]

1 Centre for Gelatinous Plankton Ecology and Evolution, National Institute of Aquatic Resources, Technical University of Denmark, 2800 Kongens Lyngby, Denmark
2 Department of Agricultural and Food Sciences, Division of Animal Sciences, University of Bologna, 40127 Bologna, Italy
3 Section for Marine Living Resources, National Institute of Aquatic Resources, Technical University of Denmark, 2800 Kongens Lyngby, Denmark
* Correspondence: jmapu@aqua.dtu.dk

Abstract: The study of natural selection and local adaptation is a thriving field of research. Local adaptation is driven by environment components and results in locally adapted phenotypes with higher fitness relative to other phenotypes from other locations in the species range. Tests of local adaptations have traditionally been done using transplant experiments, but the advent of next-generation sequencing methods have allowed the study of local adaptation to move from a phenotypic to a genomic approach. By using genome scans and state-of-the-art statistical tests, researchers can identify genes putatively under selection and study the genomic architecture of local adaptation, which often includes the observation of clustering of adaptive genes concentrated in fewer genomic regions known as "genomic islands of divergence". The two species of North Atlantic eels, the European and the American eel, are excellent species for studying selection since they are panmictic and present large population sizes, show a wide distribution range across extremely heterogenous environments, and are subject to high mortalities. We reviewed studies of natural selection and local adaptation in American eel, European eel, between life cycle stages, between European and American eel. Finally, we discussed genome architecture in relation to local adaptation in eels and the role of both genetic (i.e., local adaptation) and non-genetic (i.e., phenotypic plasticity) in the survival of eels across their distribution range.

Keywords: local adaptation; *Anguilla*; European eel; American eel; panmixia; genomic islands of divergence

Citation: Pujolar, J.M.; Bertolini, F.; Jacobsen, M.W. Footprints of Natural Selection in North Atlantic Eels: A Review. *Fishes* **2022**, *7*, 311. https://doi.org/10.3390/fishes7060311

Academic Editor: Joseph Michael Quattro

Received: 12 October 2022
Accepted: 27 October 2022
Published: 28 October 2022

1. Introduction

In order to understand the selective pressures acting upon natural populations, it is crucial to identify which regions of the genome are under selection and discriminate between neutral vs. adaptive genetic differentiation [1]. A growing number of studies have examined the factors driving historical and contemporary evolution in natural populations, searching for gene-environment interactions leading to local adaptation [2]. In particular, species with a wide geographic distribution across heterogeneous habitats in terms of environment drivers (e.g., temperature, productivity, depth, salinity, oxygen, photoperiod) may experience spatially varying selective pressures, which can result in local adaptation of ecologically important traits [3]. A meta-analysis on salmonids estimated the frequency of local adaptation to be ca. 55–70%, with local populations having, on average, a 20% fitness advantage relative to foreign populations [4].

Population genomic studies predict the observation of highly differentiated genomic regions referred to as genomic islands of divergence, which arise as a consequence of genome hitchhiking [5]. Under the hitchhiking model [6], when a selectively favoured beneficial mutation rises to fixation, the neutral variants located nearby the selected mutation will also rise to fixation, which is known as hitchhiking. Consequently, we can observe genomic regions with higher-than-average genetic differentiation together with a change in genetic variability. The loss or reduction in genetic variation in genomic regions adjacent to causative variants that may occur in response to directional selection within one or more populations is referred to as a selective sweep [7]. Genomic islands of divergence have been observed in many taxa from insects [8] to humans [9], including many classic studies in fish, such as stickleback [10,11], Atlantic cod [12,13] or whitefish [14].

2. Excellent Species for the Study of Footprints of Selection

The two species of North Atlantic eels, the European eel *Anguilla anguilla* and the American eel *Anguilla rostrata*, are optimal species for the study of footprints of natural selection. First, both species are panmictic [15–18] and present large effective population sizes despite the acknowledged stock declines. Using an RAD-sequencing approach, Pujolar et al. [17] estimated an effective population size (N_e) for European eel from 100,000 to 1 million individuals. A large Ne was also suggested by PSMC (pairwise sequentially Markovian coalescent) analysis of effective population size back in time, with estimates of >1 million individuals in both species [19]. Such high N_e is important as it renders natural selection the major evolutionary force determining the changes in genetic composition and suggests a negligible role of random genetic drift in the evolution of North Atlantic eels. Second, North Atlantic eels show a wide distribution range and are present across extremely heterogenous environments [20]; for instance, in terms of temperature, the European eel is distributed from subtropical habitats in the Mediterranean to subarctic habitats in Iceland and Scandinavia, while the American eel is distributed from Venezuela to Greenland [21]. Eels are extremely plastic regarding salinity and are regarded as facultative catadromous, with some eels being freshwater residents, some being brackish and marine water residents, and some shifting between habitats [20]. In such highly heterogenous environments, selective pressures vary from region to region so that signatures of selection and local adaptation are likely region-specific. Similarly, Indo-Pacific eel species would also be suitable for selection studies due to their panmictic status and wide geographic distribution in a variety of habitats.

Moreover, there is high potential for selective responses due to high mortalities in both early and late life stages. Bonhommeau et al. [22] estimated a 10% survival rate in European eel glass eels, while Åström and Dekker [23] estimated a natural mortality rate of $M = 0.14$ per year and a fishery mortality rate of $F = 0.54$ per year. Many sources of mortality in eels are anthropogenic, including fisheries, habitat loss, migration barriers and human-introduced parasites and viruses [24]. Anthropogenic mortality in the early continental phase notably includes fisheries targeting glass eels and upstream migration barriers, while in the final continental years of the eel's life cycle exploitation targets the last part of the yellow eel stage and the silver eel stage.

3. How Panmixia Affects the Detection of Signatures of Selection

North Atlantic eels are textbook examples of panmixia (i.e., the existence of one single randomly mating population), and together with the lack of larval homing, it has important implications for the type of signatures of selection we can detect in North Atlantic eel populations. Despite the wide distribution range of North Atlantic eels, there is conclusive evidence for panmixia in both species. In European eel, the comprehensive study of Als et al. [15] genotyped >1000 specimens collected across the entire distribution range in Europe at 21 microsatellites. A low nonsignificant genetic differentiation was found ($F_{ST} = 0.00024$), which was supported by the lack of substructuring found among larvae collected in the spawning site in the Sargasso Sea ($F_{ST} = 0.00076$) or when comparing Europe vs. Sargasso Sea samples ($F_{ST} = -0.00012$). Very low mean F_{ST} values were also reported in previous studies using low-density marker sets that included large numbers of sampling sites and individuals (e.g., $F_{ST} = 0.0017$, [25]; $F_{ST} = 0.0014$, [26]; $F_{ST} = 0.0099$, [27]; $F_{ST} = -0.00003$, [28]), the only exception being the study of Baltazar-Soares et al., [29] showing a microsatellite differentiation > 10 times more than reported in other studies

(F_{ST} = 0.02). Panmixia has also been confirmed at the genomic level using reduced representation sequencing [17], showing again no differentiation between geographic areas consistent with a single panmictic population (F_{ST} = 0.00007) after analyzing 259 RAD-sequenced juvenile (glass eel) individuals from eight locations between 34 and 64° N at >450,000 SNPs. A recent paper analyzing full genome data reached the same conclusion of panmixia in European eel [18]. Similarly, Côté et al. [16] conducted the most comprehensive study on American eel, including a total of 2142 eels from 32 sampling locations genotyped at 18 microsatellite loci. Data showed all measures of genetic differentiation to be practically zero, providing decisive evidence for panmixia in American eel.

The existence of single randomly mating panmictic populations for both North American eel species suggests there is no larval homing and larvae do not return to the parental original freshwater habitats. Despite panmixia, there have been suggestions of cryptic female philopatric behavior in this species based on analysis of mitochondrial DNA [29]. However, the conclusion is not supported as it has not been verified in other studies utilizing the same marker [30]. If larvae showed philopatry (i.e., if larvae from Mediterranean parents always returned to the Mediterranean), genetic differences would be expected to accumulate across regions over time. Hence, the lack of genetic differentiation found between geographic areas consistent with a single panmictic population suggests that larval migratory routes are random.

One direct consequence of panmixia and random dispersal of larvae is that heritable local adaptation is not possible in North Atlantic eels, despite the high potential for selection. If for instance, an individual thrives in the Mediterranean because its genetic composition makes it better adapted to survive the particular environmental conditions of the Mediterranean. However, due to lack of homing, its progeny might end up randomly in Iceland or Scandinavia, hence not benefiting from the pre-adapted alleles in their genetic composition. Therefore, all signatures of spatially varying selection in a given generation are expected to be lost in the next generation, which prevents heritable trans-generational local adaptation [31]. However, single-generation signatures of local selection should still be detectable [17,31].

We present now a summary of all selection studies to date on North Atlantic eels (Table 1), including studies on American eel, on European eel, between life cycle (juvenile vs. adult) stages, between the two species and finally studies discussing the role of genome architecture in relation with local adaptation, including epigenetic studies.

Table 1. Summary of all selection studies on North Atlantic eels including species studied, genetic markers used, sampling details and main results.

Study	Species	Marker	Sampling Details	Main Results
Gagnaire et al. [31]	American eel	100 candidate SNPs	A total of 992 individuals from 16 sampling sites from Florida to Quebec	Positive correlations at 8 loci with environmental variables, mostly related to energy production and metabolism
Ulrik et al. [32]	European eel	80 candidate SNPs	A total of 321 glass eels collected at 8 sampling sites from Iceland to the Mediterranean	Positive associations at 11 loci with environmental variables, mostly related to metabolism
Pujolar et al. [17]	European eel	RAD sequencing	A total of 259 glass eels from 8 sampling sites ranging from Iceland to Morocco	A total of 754 loci under selection with a variety of functions including calcium signaling and circadian rhythms
Jacobsen et al. [33]	European and American eel	RAD sequencing	A total of 30 American eel and 30 European eel individuals either glass or yellow eels	A total of 3757 highly differentiated candidate SNPs, located in genes mostly related to development and energy production

Table 1. *Cont.*

Study	Species	Marker	Sampling Details	Main Results
Pavey et al. [34]	American eel	RAD sequencing	A total of 379 individuals from 8 locations each of freshwater and brackish/marine habitats from the Atlantic Canada and St. Lawrence river region	A total of 331 loci under selection associated with environmental variables showing differences between ecotypes (freshwater vs. brackish/marine)
Pujolar et al. [35]	European eel	80 candidate SNPs + RAD sequencing	A total of 123 glass eels and 113 silver eels collected in Iceland, Ireland and Spain	A total of 1413 potentially selected SNPs, located in genes related to growth among other functions
Enbody et al. [18]	European eel	Whole-genome sequencing	A total of 417 Europeam eel samples from 10 locations across Europe	When comparing Baltic vs. Mid-Atlantic samples only a small region under selection located on chromosome 1 and two other regions on chromosomes 13 and 15 were found
Pujolar et al. [36]	European and American eel	RAD sequencing	Re-analysis of 359 samples retrieved from Genbank, including 254 European eel and 105 American eel	Islands of divergence were detected at 7 chromosomes, with candidate genes involved in energy production, development and regulation
Liu et al. [37]	European eel	DNA methylation	A total of 50 individuals were analysed representing 7 localities in Europe and northern Africa	Differentially methylated regions were reported including genes involved in development, in particular Hox genes.

4. Studies of Selection in American Eel

Using a candidate gene approach, Gagnaire et al. [31] studied the evolutionary effects of spatially varying selection in American eel. A panel of 100 candidate single nucleotide polymorphisms (SNPs) were genotyped in 992 individuals from 16 sampling sites at different life stages of the same cohort as well as in glass eels of the following cohort. Evidence for spatially varying selection was suggested at 13 coding genes in American eel showing significant correlations with environmental variables (latitude, longitude, and temperature) across the entire species range. Within glass eels, associations with environmental variables were found at eight loci using generalized linear models. Most loci under selection represented key metabolic genes involved in lipid metabolism (*ACP*, acyl carrier activity; *ANX2*, inhibition of phospholipase A2; *GPX4*, phospholipid–hydroperoxide glutathione peroxidase activity), saccharide metabolism (*MDH*, malate dehydrogenase activity; *UGP2*, UDP-glucose pyrophosphorylase activity) and protein biosynthesis (*PRP-40*, pre-mRNA–processing activity). The environmental variable showing the highest associations was temperature, arguably a key factor influencing enzymatic activities and metabolic pathways [38]. In fact, a decreased metabolism has been observed below certain threshold temperatures in both North Atlantic eels [39,40] and early-life history traits such as glass eel upstream migration have been shown to be temperature-related [41]. Using RAD sequencing, Pavey et al. [34] studied natural selection between American eels inhabiting freshwater vs. brackish/saltwater habitats. Out of 42,424 SNPs analyzed, 331 were associated with habitat, located in genes representing vascular and morphological development, calcium ion regulation, growth and transcription factors, and olfactory receptors. This points to the existence of differential selective pressures in American eel in the two distinct habitats.

5. Studies of Selection in European Eel

Early studies of adaptive evolution in European eel focused on the detection of signatures of local selection in glass eels. Ulrik et al. [32] used a panel of 80 coding-gene SNPs previously analyzed in American eel [31] to genotype individuals collected from eight locations across Europe. Signatures of selection were found at 11 coding-gene SNPs, four from outlier tests and seven from environmental correlations. Most genes were involved in major metabolic functions, including *GAPDH* (Glyceraldehyde 3-phosphate dehydrogenase, glycolysis pathway, catalyzes the conversion of glyceraldehyde 3-phosphate to D-glycerate 1,3-bisphosphate), *ALDH2* (Aldehyde dehydrogenase 2, major oxidative pathway of alcohol metabolism, catalyzes acetaldehyde to acetic), and *ALD_R* (Aldose reductase, polyol pathway of glucose metabolism, catalyzes the reduction of glucose to sorbitol). None of the above genes linked to metabolic pathways was correlated with temperature, despite its known importance for enzymatic activities and metabolism.

When comparing the results from Ulrik et al. [32] on European eel with Gagnaire et al. [31] on American eel using the same SNP panel, no genes putatively under selection were shared across studies. The contrasting pattern found suggested no apparent parallel footprints of selection in North Atlantic eels. It also suggested no common genetic-by-environment associations between European and American eel.

As an alternative to the candidate gene approach, Pujolar et al. [17] tested for footprints of natural selection in glass eels at the genome level using a total of 50,354 SNPs generated by RAD sequencing. A total of 754 potentially locally selected SNPs were identified using F_{ST}-based outlier tests and significant correlations with environmental variables. Candidate genes for local selection constituted a wide array of functions, including calcium signalling, neuroactive ligand-receptor interaction, and circadian rhythm. One of the candidate genes identified was the circadian clock gene *Period*, possibly related to differences in local photoperiod associated with the >30° difference in latitude between localities (34 and 64° N).

6. Signature of Selection between Life-Cycle Stages

Eels present a complex life cycle organized into morphologically distinct phases separated by abrupt metamorphic transitions (metamorphosis). Life stages represent alternative adaptations for optimal food and niche exploitation (e.g., growth, feeding, growth, dispersal) as well as specific tasks (e.g., reproduction) [42]. North Atlantic eels present a particularly complex life cycle that includes two metamorphoses. After spawning in partial sympatry in the Sargasso Sea, larvae are transported by currents to the coasts of Europe (European eel) and North America (American eel). On reaching the continental shelf, eels undergo a first metamorphosis from larvae into glass eels (juvenile stage). After an extensive period of feeding and growth as yellow eels, eels undergo a second metamorphosis into silver eels (adult stage). The latter encompasses modifications both at the morphological (skin colour, eye size, body length and weight) and physiological level (loss of digestive tract, development of gonads). The second metamorphosis prepares animals for the spawning migration back to the Sargasso Sea, where eels reproduce once and die [21]. Given the drastic changes associated with metamorphic transitions, selective pressures should differ before and after metamorphosis and different genes and pathways should be under selection at different life stages. Hence North Atlantic eels provide an excellent opportunity to study the genetic associations between life cycle stages.

Pujolar et al. [35] compared juvenile glass eels vs. adult silver eels in European eel using two different sets of markers to test for selection: the same panel of functional genes developed by Gagnaire et al. [31] for American eel and a new set of ca. 150,000 SNPs generated by RAD sequencing. A total of 2413 (1.57%) candidate SNPs were identified with signal transduction pathway as the most over-represented group of genes, including MAPK signalling, calcium signalling and GnRH (gonadotropin-releasing hormone) signalling. The majority of the over-represented pathways were related to growth, while others could result from the different conditions that eels inhabit during their life cycle.

The observation of many different genes and pathways under selection when comparing juvenile and adult eels supports the adaptive decoupling hypothesis for the benefits of metamorphosis [43]. The hypothesis states that partitioning the life cycle into discrete morphological phases may be overall beneficial as it allows the different life stages to respond independently to their unique selection pressures. In turn, this might translate into a more effective use of niche resources and a better performance of phase-specific tasks (e.g., feeding and growth in juveniles, migration, and reproduction in adults).

7. Signatures of Selection between European and American Eel

The European and American eel overlap in many morphological parameters and only differ in vertebral counts (Figure 1) [44–46] but are considered as sister species based on mitogenomic differences [47]. Divergence time between the two species is estimated to range between 1.3–2.4 Mya [19], based on the joint allele frequency spectrum (JAFS) and PSMC plots, and 3.8 Mya [30], based on the mitogenome. Surprisingly, genetic differentiation between the two species is low/moderate as shown by SNP data, with an overall $F_{ST} = 0.087$ [36], which confirmed earlier studies using AFLP and microsatellites [15,25,48,49]. This is a likely result of gene flow between species, which are known to breed together in partial sympatry in the Sargasso Sea [50]. Several genetic studies have detected hybrid individuals in larvae, juvenile and adult samples, with the highest signatures of admixture detected in Iceland on the basis of vertebrae counts (Figure 1) and especially molecular data [15,44,49,51–53].

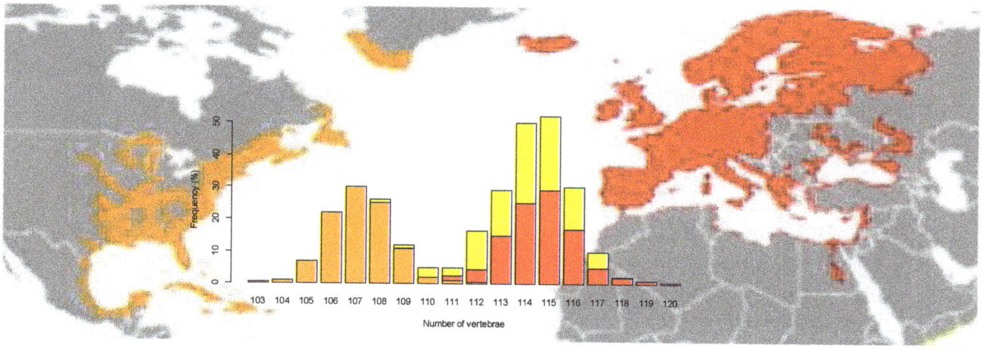

Figure 1. Vertebrae counts in European eel (red), American eel (orange) and Icelandic individuals (yellow) after re-examining data from Avise et al. [44]. Geographic distribution for each species is adapted from Jacoby et al. [54].

When testing for positive selection between European and American eel, Jacobsen et al. [33] found that candidate SNPs were located within genes related to development and phosphorylation, consistent with the hypothesis that larval phase duration and migration loops play a key role in the speciation of North Atlantic eels, which was also confirmed recently in the re-analysis of Pujolar et al. [36]. The different footprints of selection between species could be due to distinct selection pressures associated with the much longer larval migration for European eel (from seven months to two years) relative to American eel (from six to 12 months) [22]. The potential extra year spent in the open sea could impose a stronger selective pressure on European eel larvae relative to American eel. Similarly, the migration loop of adults returning to the Sargasso Sea for spawning is ca. 5000 km for European eel vs. ca. 2000 km for American eel [21]. Hence the distinct signatures of selection between North American eels could be attributable to the different metabolic and energetic requirements in larval and adult migration between species. The same argument was also used to explain the contrasting pattern of spatially varying selection in American eel [31] and European eel [32] using the same panel of candidate SNPs.

8. Genomic Islands of Divergence

Early studies aiming at finding signatures of selection in eels had the disadvantage of utilizing a European eel draft genome assembled into a large number of small contigs and scaffolds [55], which was the only eel genome available at the time. This meant that the existence of regions of elevated differentiation across the genome (or genomic islands of divergence) could not be properly tested. However, a new high-quality reference European eel genome has been recently released, assembled at the chromosome level [56].

Taking advantage of the newly assembled and annotated European eel genome, Enbody et al. [18] searched for genetic footprints of differentiation within European eel using whole genome resequencing data. The study compared samples collected in the Baltic and Mid-Atlantic (England, Ireland, France) and reported little evidence for islands of selection, finding only a small region under selection located on chromosome 1 (covering ca. 6 kb around 81.2 Mbp) and two other regions on chromosomes 13 and 15.

Similarly, Pujolar et al. [36] used genome-wide data from a total of 359 RAD-sequenced individuals retrieved from the Sequence Nucleotide Archive (SNA) to identify genomic islands of selection between North Atlantic eels using the new European eel genome as reference. State-of-the-art statistic tests were used, including methods based on higher population differentiation than under neutral expectation (F_{ST} value) and measures of linkage disequilibrium (iHS, XP-EHH) [57]. First, two between-population methods (F_{ST} and XP-EHH) were used to identify islands of selection between North Atlantic eels. Larger regions or islands were observed in a total of seven chromosomes (2, 4, 5, 6, 7, 9 and 10), most recognizable on chromosome 6 with a 3.6 Mbp region from 31.45 to 35.10 Mbp and on chromosome 10 with a 950 kb region from 28.80 to 29.75 Mbp (Figure 2). The two between-population methods are best at detecting complete or nearly complete signatures of selection. Due to the long time required until fixation is reached, F_{ST} and XP-EHH are expected to identify older selection events between populations in the more distant past [58]. In this sense, both FST and XP-EHH are powerful tools to detect "hard selective sweeps", which occur when a new mutation arises and spreads quickly to fixation due to natural selection [6]. Other scenarios might be more difficult to detect, especially when selection leads to changes in allele frequencies without reaching fixation. Genes included in the islands of selection detected in the study showed significant enrichment for terms related mainly to ATP phosphorylation and development. This is in accordance with previous studies comparing European vs. American eel [31,33], suggesting different selective pressures in relation to metabolism and energetics.

Second, shared signatures of selection within European eel and American eel were detected using iHS at a total of 11 chromosomes. Regions were generally small (100–300 kb) except for two large regions, a region of 800 kb on chromosome 8 from 52.35 to 52.75 Mbp and a region of 850 kb on chromosome 16 from 2.45 to 3.30 Mbp. Unlike the two between-population methods (F_{ST} and XP-EHH), the iHS test has higher statistical power when selected alleles are at intermediate frequencies that have not yet reached fixation [57]. Hence, it can detect signatures of recent and even ongoing selective sweeps [59]. Those scenarios include "soft selective sweeps", in which multiple haplotypes harboring advantageous mutations are all favoured [60]. Several hypotheses might account for the detection of shared islands of selection in North Atlantic eels, including parallel evolution due to adaptation to similar habitats and introgression.

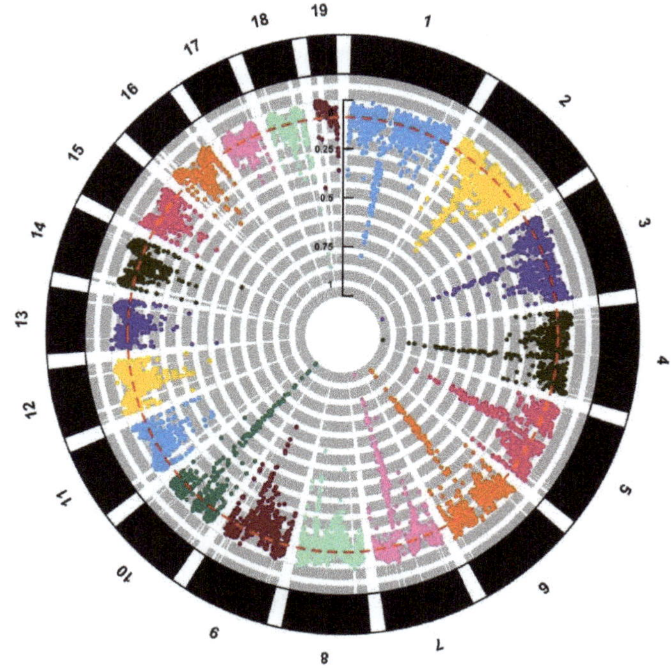

Figure 2. Circular Manhattan Plot showing sliding-window F_{ST} between European and American eel at 19 chromosomes after re-examining data from Pujolar et al. [36]. Average F_{ST} (red dotted line) is reported.

9. Making Sense of Genomic Islands of Divergence

Islands of genomic divergence are regions in which differentiation between populations or species is the highest and can in some cases also constitute "islands of speciation" [61]. Such islands represent regions in the genome in which a selective sweep leads to the increase and fixation of adaptive mutations, resulting in a reduction of genetic variability in the region nearby the favorable allele due to genetic hitchhiking [62].

While many studies on genomic regions of elevated differentiation focus on searching for speciation genes, there is an increasing realization that selection might be acting mainly on regulation and expression rather than functional changes. In this sense, it is interesting that Jacobsen et al. [33] found upstream regions, likely involved in regulation, to include a lower percentage of outlier SNPs (F_{ST} = 1) compared to the rest of the genome. This suggests a conserved and important role of these regions in both species [33]. The recent paper of Pujolar et al. [36] looked at the genomic location and effects of each candidate SNP under selection when comparing North Atlantic eels using genome-wide RAD sequencing data. Only 1.5% of SNPs were in coding regions (exons), with most variants found in introns (66.6%) and intergenic regions (25.6%). Mutations in the exons were mostly synonymous and only two mutations had a moderate effect producing a different amino acid. Given that most SNPs putatively under selection in the study were found in noncoding regions, this possibly reflects regulatory differences between North Atlantic eels. This is also supported by gene expression analysis performed on American and European eel leptocephali larvae collected in the Sargasso Sea, which suggests differential timing of gene expression regulation during early development [63]. Similarly, when comparing marine and freshwater stickleback populations, Jones et al. [11] reported up to 83% of SNPs under selection located in noncoding regions with an assumed regulatory role. It should also be taken into account that within an island of genomic differentiation, most mutations are likely to be hitchhiking and not the direct target of selection, making it difficult to distinguish between neutral and adaptive variation. On this account, searching for candidate genes within genomic islands of differentiation can only be used as an indication of putative selection and functional validation (i.e., gene expression analyses or QTL mapping from genetic crosses) would be required to demonstrate causality.

Finally, epigenetic variation as a mechanism of adaptive plasticity could play a role in local adaptation of North Atlantic eels to the heterogenous habitat conditions they experience throughout their life cycle. Epigenetics mechanisms, including DNA methylation, are defined as DNA modifications affecting gene expression without changing the DNA sequence [64]. Emerging evidence suggests significant methylation differences of functional importance associated with environmental variation [65]. When studying methylation variation in European eel, Liu et al. [37] reported differentially methylated regions including genes involved in developmental processes, particularly Hox genes. Overall, methylation results highlight the importance of epigenetics in the adaptation and resilience of eels and suggest interactions between habitat, development, and epigenetic variation. The importance of epigenetic variation as a mechanism of adaptive plasticity in eels merits further research.

Author Contributions: J.M.P. wrote the manuscript with contribution from F.B. and M.W.J. All authors have read and agreed to the published version of the manuscript.

Funding: This research received no external funding.

Conflicts of Interest: The authors declare no conflict of interest.

References

1. Stapley, J.; Reger, J.; Feulner, P.G.D.; Smadja, C.; Galindo, J.; Ekblom, R.; Bennison, C.; Ball, A.D.; Beckerman, A.P.; Slate, J. Adaptation genomics: The next generation. *Trends Ecol. Evol.* **2010**, *25*, 705–712. [CrossRef] [PubMed]
2. Bourret, V.; Dionne, M.; Kent, M.P.; Lien, S.; Bernatchez, L. Landscape genomics in Atlantic salmon (*Salmo salar*): Searching for gene-environment interactions driving local adaptation. *Evolution* **2013**, *67*, 3469–3487. [CrossRef]
3. Kawecki, T.J.; Ebert, D. Conceptual issues in local adaptation. *Ecol. Lett.* **2004**, *7*, 1225–1241. [CrossRef]
4. Fraser, D.J.; Weir, L.K.; Bernatchez, L.; Hansen, M.M.; Taylor, E.B. Extend and scale of local adaptation in salmonid fishes: Review and meta-analysis. *Heredity* **2011**, *106*, 404–420. [CrossRef] [PubMed]
5. Feder, J.L.; Nosil, P. The efficacy of divergence hitchhiking in generating genomic islands during ecological speciation. *Evolution* **2010**, *64*, 1729–1747. [CrossRef] [PubMed]
6. Maynard Smith, J.; Haigh, J. The hitch-hiking effect of a favourable gene. *Genet. Res.* **1974**, *23*, 23–35. [CrossRef]
7. Stephan, W. Selective sweeps. *Genetics* **2019**, *211*, 5–13. [CrossRef]
8. Nosil, P.; Egan, S.P.; Funk, D.J. Heterogenous genomic differentiation between walking-stick ecotypes: "Isolation by adaptation" and multiple roles for divergent selection. *Evolution* **2008**, *62*, 316–336. [CrossRef]
9. Hoffer, T.; Foll, M.; Excoffier, L. Evolutionary forces shaping genomic islands of population differentiation in humans. *BMC Genomics* **2012**, *13*, 107. [CrossRef]
10. Hohenlohe, P.A.; Basshan, S.; Etter, P.D.; Stiffler, N.; Johnson, E.A.; Cresko, W.A. Population genomics of parallel adaptation in threespine stickleback using sequenced RAD tags. *PLoS Genet.* **2010**, *6*, e1000862. [CrossRef]
11. Jones, F.C.; Grabherr, M.G.; Chan, Y.F.; Russell, P.; Mauceli, E.; Johnson, J.; Swofford, R.; Pirun, M.; Zody, M.C.; White, S.; et al. The genomic basis of adaptive evolution in threespine sticklebacks. *Nature* **2012**, *484*, 55–61. [CrossRef] [PubMed]
12. Hemmer-Hansen, J.; Nielsen, E.E.; Therkildsen, N.J.; Taylor, M.I.; Ogden, R.; Geffen, A.J.; Bekkevold, D.; Helyar, S.; Pampoulie, C.; Johansen, T.; et al. A genomic island linked to ecotype divergence in Atlantic cod. *Mol. Ecol.* **2013**, *22*, 2653–2667. [CrossRef]
13. Rodríguez-Ramilo, S.T.; Baranski, M.; Moghadam, H.; Grove, H.; Lien, S.; Goddard, M.E.; Meuwissen, T.H.E.; Sonesson, A.K. Strong selection pressures maintain divergence on genomic islands in Atlantic cod (*Gadus morhua*) populations. *Genet. Sel. Evol.* **2019**, *51*, 61. [CrossRef] [PubMed]
14. Gagnaire, P.A.; Pavey, S.A.; Normandeau, E.; Bernatchez, L. The genetic architecture of reproductive isolation during speciation-with-gene-flow in lake whitefish species pairs assessed by RAD sequencing. *Evolution* **2013**, *67*, 2483–2497. [CrossRef] [PubMed]
15. Als, T.D.; Hansen, M.M.; Maes, G.E.; Castonguay, M.; Riemann, L.; Aerestrup, K.; Munk, P.; Sparholt, T.; Hanel, R.; Bernatchez, L. All roads lead to home: Panmixia of European eel in the Sargasso Sea. *Mol. Ecol.* **2011**, *20*, 1333–1346. [CrossRef] [PubMed]
16. Côté, C.; Gagnaire, P.A.; Bourret, V.; Verrault, G.; Castonguay, M.; Bernatchez, L. Population genetics of the American eel (*Anguilla rostrata*): $F_{ST} = 0$ and North Atlantic Oscillation effects on demographic fluctuations of a panmictic species. *Mol. Ecol.* **2013**, *22*, 1763–1776. [CrossRef]

17. Pujolar, J.M.; Jacobsen, M.W.; Als, T.D.; Frydenberg, J.; Munch, K.; Jónsson, B.; Jiang, X.; Cheng, L.; Maes, G.E.; Bernatchez, L.; et al. Genome-wide signatures of within-generation local selection in the panmictic European eel. *Mol. Ecol.* **2014**, *23*, 2514–2528. [CrossRef]
18. Enbody, E.D.; Petterson, M.E.; Sprehn, C.G.; Palm, S.; Wickstrom, H.; Andersson, L. Ecological adaptation in European eels is based on phenotypic plasticity. *Proc. Natl. Acad. Sci. USA* **2021**, *118*, e2022620118. [CrossRef]
19. Nikolic, N.; Liu, S.; Jacobsen, M.W.; Jónsson, B.; Bernatchez, L.; Gagnaire, P.A.; Hansen, M.M. Speciation history of European (*Anguilla Anguilla*) and American eel (*A. rostrata*), analysed using genomic data. *Mol. Ecol.* **2020**, *29*, 565–577. [CrossRef]
20. Daverat, F.; Limburg, K.E.; Thibaut, I.; Shiao, J.C.; Dodson, J.J.; Caron, F.; Tzeng, W.N.; Iizuka, Y.; Wickstrom, H. Phenotypic plasticity of habitat use by three temperate eel species *Anguilla anguilla, A. japonica* and *A. rostrata. Mar. Ecol. Progr. Ser.* **2006**, *308*, 231–241. [CrossRef]
21. Tesch, F. *The Eel*; Blackwell Science Ltd.: Oxford, UK, 2003.
22. Bonhommeau, S.; Blanke, B.; Tréguier, A.M.; Grima, N.; Rivot, E.; Vermand, Y.; Greiner, E.; Le Pape, O. How fast can the European eel (*Anguilla anguilla*) larvae cross the Atlantic Ocean? *Fish. Oceanogr.* **2009**, *18*, 371–385. [CrossRef]
23. Åström, M.; Dekker, W. When will the eel recover? A full life cycle model. *ICES J. Mar. Sci.* **2007**, *64*, 1491–1498. [CrossRef]
24. Van den Thillart, G.; Rankin, J.C.; Dufour, S. *Spawning Migration of the European Eel: Reproduction Index, a Useful Tool for Conservation Management*; Springer: Dordecht, The Netherlands, 2009.
25. Wirth, T.; Bernatchez, L. Decline of Atlantic eels: A fatal synergy? *Proc. Biol. Sci.* **2003**, *270*, 681–688. [CrossRef] [PubMed]
26. Dannewitz, J.; Maes, G.E.; Johansson, L.; Wickström, H.; Volckaert, F.A.M.; Jarvi, T. Panmixia in the European eel: A matter of time. *Proc. Biol. Sci.* **2005**, *272*, 1129–1137. [CrossRef]
27. Maes, G.E.; Pujolar, J.M.; Hellemans, B.; Volckaert, F.A.M. Evidence for isolation by time in the European eel (*Anguilla Anguilla*). *Mol. Ecol.* **2006**, *15*, 2095–2107. [CrossRef]
28. Palm, S.; Dannewitz, J.; Prestegaard, T.; Wickström, H. Panmixia in European eel revisited: No genetic difference between maturing adults from southern and northern Europe. *Heredity* **2009**, *103*, 82–89. [CrossRef]
29. Baltazar-Soares, M.; Biastoch, A.; Harrod, C.; Hanel, R.; Marohn, L.; Prigge, E.; Evans, D.; Bodles, K.; Behrens, E.; Böning, C.W.; et al. Recruitment collapse and population structure of the European eel shaped by local ocean current dynamics. *Curr. Biol.* **2014**, *24*, 104–108. [CrossRef]
30. Jacobsen, M.W.; Pujolar, J.M.; Gilbert, M.T.P.; Moreno-Mayar, J.V.; Bernatchez, L.; Als, T.D.; Lobón-Cervià, J.; Hansen, M.M. Speciation and demographic history of Atlantic eels (*Anguilla anguilla* and *A. rostrata*) revealed by mitogenome sequencing. *Heredity* **2014**, *113*, 432–442. [CrossRef]
31. Gagnaire, P.A.; Normandeau, E.; Côté, C.; Hansen, M.M.; Bernatchez, L. The genetic consequences of spatially varying selection in the panmictic American eel (*Anguilla rostrata*). *Genetics* **2012**, *190*, 725–736. [CrossRef]
32. Ulrik, M.G.; Pujolar, J.M.; Ferchaud, A.L.; Jacobsen, M.W.; Als, T.D.; Gagnaire, P.A.; Frydenberg, J.; Bøcher, P.K.; Jónsson, B.; Bernatchez, L.; et al. Do North Atlantic eels show parallel patterns of spatially varying selection? *BMC Evol. Biol.* **2014**, *14*, 138. [CrossRef]
33. Jacobsen, M.W.; Pujolar, J.M.; Bernatchez, L.; Munch, K.; Jian, J.; Niu, Y.; Hansen, M.M. Genomic footprints of speciation in Atlantic eels *Anguilla anguilla* and *A. rostrata. Mol. Ecol.* **2014**, *23*, 4785–4798. [CrossRef] [PubMed]
34. Pavey, S.A.; Gaudin, J.; Normandeau, E.; Dionne, M.; Castonguay, M.; Audet, C.; Bernatchez, L. RAD sequencing highlights polygenic discrimination of habitat ecotypes in the panmictic American eel. *Curr. Biol.* **2015**, *25*, 1666–1671. [CrossRef] [PubMed]
35. Pujolar, J.M.; Jacobsen, M.W.; Bekkevold, D.; Lobon-Cervià, J.; Jónsson, B.; Bernatchez, L.; Hansen, M.M. Signatures of natural selection between life cycles stages separated by metamorphosis in European eel. *BMC Genom.* **2015**, *16*, 600. [CrossRef] [PubMed]
36. Pujolar, J.M.; Jacobsen, M.W.; Bertolini, F. Comparative genomics and signatures of selection in North Atlantic eels. *Mar. Genom.* **2022**, *62*, 100933. [CrossRef]
37. Ling, S.; Tengstedt, A.N.B.; Jacobsen, M.W.; Pujolar, J.M.; Jónsson, B.; Lobón-Cervià, J.; Bernatchez, L.; Hansen, M.M. Genome-wide methylation in the panmictic European eel (*Anguilla anguilla*). *Mol. Ecol.* **2022**, *31*, 4286–4306. [CrossRef]
38. Somero, G.N. Adaptation of enzymes to temperature: Searching for basic "strategies". *Comp. Biochem. Physiol. B Biochem. Mol. Biol.* **2004**, *139*, 321–333. [CrossRef]
39. Nyman, L. Some effects of temperature on eel (*Anguilla*) behaviour. *Rep. Inst. Freshw. Res. Drottingholm* **1972**, *52*, 90–102.
40. Walsh, P.J.; Foster, G.D.; Moon, T.W. The effects of temperature and metabolism of the American eel *Anguilla rostrata*: Compensation in the summer and torpor in the winter. *Physiol. Zool.* **1983**, *56*, 532–540. [CrossRef]
41. Linton, E.D.; Jónsson, B.; Noakes, D.L.G. Effects of water temperature on the swimming and climbing behavior of glass eels *Anguilla* spp. *Environ. Biol. Fishes* **2007**, *78*, 189–192. [CrossRef]
42. Heyland, A.; Moroz, L.L. Signalling mechanisms underlying metamorphic transitions in animals. *Integr. Comp. Biol.* **2006**, *46*, 743–759. [CrossRef]
43. Moran, N.A. Adaptation and constraint in the complex life cycles of animals. *Annu. Rev. Ecol. Syst.* **1994**, *25*, 573–600. [CrossRef]
44. Avise, J.C.; Nelson, W.S.; Arnold, J.; Koehn, R.K.; Williams, G.C.; Thorsteinsson, V. The evolutionary genetic status of Icelandic eels. *Evolution* **1990**, *44*, 1254–1262. [CrossRef] [PubMed]
45. Lecomte-Finiger, R. The genus *Anguilla*: Current state of knowledge and questions. *Rev. Fish. Biol. Fish.* **2003**, *13*, 265–279. [CrossRef]

46. Van Ginneken, V.J.T.; Maes, G.E. The European eel (*Anguilla Anguilla*), its life cycle, evolution and reproduction: A literature review. *Rev. Fish. Biol. Fish.* **2005**, *15*, 367–398. [CrossRef]
47. Mineguishi, Y.; Aoyama, J.; Inoue, J.G.; Miya, M.; Nishida, M.; Tsukamoto, K. Molecular phylogeny and evolution of the freshwater eels genus *Anguilla* based on the whole mitochondrial genome sequences. *Mol. Phylogenet. Evol.* **2005**, *34*, 134–146. [CrossRef]
48. Mank, J.E.; Avise, J.C. Microsatellite variation and differentiation in North Atlantic eels. *J. Hered.* **2003**, *94*, 30–34. [CrossRef]
49. Gagnaire, P.A.; Albert, V.; Jónsson, B.; Bernatchez, L. Natural selection influences AFLP intraspecific variability and introgression patterns in Atlantic eels. *Mol. Ecol.* **2009**, *18*, 1678–1691. [CrossRef]
50. Miller, M.J.; Bonhommeau, S.; Munk, P.; Castonguay, M.; Hanel, R.; McCleave, J.D. A century of research on the larval distribution of the Atlantic eels: A re-examination of the data. *Biol. Rev. Camb. Philos. Soc.* **2015**, *90*, 1035–1064. [CrossRef]
51. Albert, V.; Jónsson, B.; Bernatchez, L. Natural hybrids in Atlantic eels (*Anguilla anguilla, A. rostrata*): Evidence for successful reproduction and fluctuating abundance in space and time. *Mol. Ecol.* **2006**, *15*, 1903–1916. [CrossRef]
52. Pujolar, J.M.; Jacobsen, M.W.; Als, T.D.; Frydenberg, J.; Magnussen, E.; Jónsson, B.; Jiang, X.; Cheng, L.; Bekkevold, D.; Maes, G.E.; et al. Assessing patterns of hybridization between North Atlantic eels using diagnostic single nucleotide polymorphisms. *Heredity* **2014**, *112*, 627–637. [CrossRef]
53. Jacobsen, M.; Smedegaard, L.; Sørensen, S.; Pujolar, J.M.; Munk, P.; Jónsson, B.; Magnussen, E.; Hansen, M.M. Assessing pre- and post-zygotic barriers between North Atlantic eels (*Anguilla anguilla* and *A. rostrata*). *Heredity* **2017**, *118*, 266–275. [CrossRef] [PubMed]
54. Jacoby, D.M.P.; Casselman, J.M.; Crooks, V.; Delucia, M.B.; Ahn, H.; Kaifu, K.; Kurwie, T.; Sasal, P.; Silfvergrip, A.M.C.; Smith, K.; et al. Synergistic patterns of threat and the challenges facing global anguillid eel conservation. *Glob. Ecol. Conserv.* **2015**, *4*, 321–333. [CrossRef]
55. Henkel, C.V.; Burgerhout, E.; de Wijze, D.L.; Dirks, R.P.; Minegishi, Y.; Jansen, H.J.; Spaink, H.P.; Dufour, S.; Weltzien, F.A.; Tsukamoto, K.; et al. Primitive duplicate Hox clusters in the European eel's genome. *PLoS ONE* **2012**, *7*, e32231. [CrossRef] [PubMed]
56. Rhie, A.; McCarthy, S.A.; Fedrigo, O.; Damas, J.; Formenti, G.; Koren, S.; Uliano-Silva, M.; Chow, W.; Fungtammasan, A.; Kim, J.; et al. Towards complete and error-free genome assemblies of all vertebrate species. *Nature* **2021**, *592*, 737–746. [CrossRef] [PubMed]
57. Ma, Y.; Ding, X.; Qanbari, S.; Weigend, S.; Zhang, Q.; Simianer, H. Properties of different selection signature statistics and a new strategy to combine them. *Heredity* **2015**, *115*, 426–436. [CrossRef] [PubMed]
58. Cadzow, M.; Boocock, J.; Nguyen, H.T.; Wilcox, P.; Merriman, T.R.; Black, M.A. A bioinformatics workflow for detecting signatures of selection in genomic data. *Front. Genet.* **2014**, *5*, 293. [CrossRef]
59. Voight, B.F.; Kudaravalli, S.; Wen, X.; Pritchard, J.K. A map of recent positive selection in the human genome. *PLoS Biol.* **2006**, *4*, e72. [CrossRef]
60. Hermisson, J.; Pennings, P.S. Soft sweeps: Molecular population genetics of adaptation from standing genetic variation. *Genetics* **2005**, *169*, 2335–2352. [CrossRef]
61. Via, S.; West, J. The genetic mosaic suggests a new role for hitchhiking in ecological speciation. *Mol. Ecol.* **2008**, *17*, 4334–4345. [CrossRef]
62. Storz, J.F. Using genome scans of DNA polymorphism to infer adaptive population divergence. *Mol. Ecol.* **2005**, *14*, 671–688. [CrossRef]
63. Bernatchez, L.; St-Cyr, J.; Normandeau, E.; Maes, G.E.; Als, T.D.; Kalujnaia, S.; Cramb, G.; Castonguay, M.; Hansen, M.M. Differential timing of gene expression between leptocephali of the two *Anguilla* eel species in the Sargasso Sea. *Ecol. Evol.* **2011**, *1*, 459–467. [CrossRef] [PubMed]
64. Dupond, C.; Armant, D.R.; Brenner, C.A. Epigenetics: Definition, mechanisms and clinical perspective. *Semin. Reprod. Med.* **2009**, *27*, 351–357. [CrossRef]
65. Srikant, T.; Drost, H.H. How stress facilitates phenotypic innovation through epigenetic diversity. *Front. Plant Sci.* **2020**, *11*, 606800. [CrossRef] [PubMed]

fishes

MDPI

Review

Swimbladder Function in the European Eel *Anguilla anguilla*

Bernd Pelster [1,2]

[1] Institute for Zoology, Universität Innsbruck, 6020 Innsbruck, Austria; bernd.pelster@uibk.ac.at; Tel.: +43-512-5075-1860; Fax: +43-512-5075-1899

[2] Center for Molecular Biosciences, Universität Innsbruck, 6020 Innsbruck, Austria

Abstract: Eels use the swimbladder for buoyancy control. The ductus pneumaticus connecting the esophagus with the swimbladder is closed soon after initial opening of the swimbladder in the glass eel stage, so that eels are functionally physoclist. Subsequent filling of the swimbladder is achieved by activity of gas gland cells in the swimbladder epithelium and countercurrent concentration in the rete mirabile. Gas gland cells produce and release lactic acid and CO_2. In blood, acidification induces a release of oxygen from the hemoglobin (Root effect). The resulting increases in PO_2 and PCO_2 provide diffusion gradients for the diffusion of oxygen and CO_2 into the swimbladder, the main gases secreted into the swimbladder. In addition, the partial pressure of these two gases remains elevated in venous blood leaving the swimbladder epithelium and returning to the rete mirabile. Back-diffusion from venous to arterial capillaries in the rete results in countercurrent concentration, allowing for the generation of high gas partial pressures, required for filling the swimbladder under elevated hydrostatic pressure. The transition of the yellow eel to the silver eel stage (silvering) is accompanied by a significant improvement in swimbladder function, but swimbladder volume cannot be kept constant during the daily vertical migrations silver eels perform during their spawning migration back to the spawning grounds in the Sargasso Sea. Infection of the swimbladder with the nematode *Anguillicola crassus* significantly impairs the function of the swimbladder as a buoyancy organ.

Keywords: gas gland; countercurrent exchange; metabolism; Root effect; *Anguilla anguilla; Anguillicola crassus*; rete mirabile; hemoglobin; oxygen transport

Citation: Pelster, B. Swimbladder Function in the European Eel *Anguilla anguilla*. *Fishes* **2023**, *8*, 125. https://doi.org/10.3390/fishes8030125

Academic Editor: Jose Martin Pujolar

Received: 27 January 2023
Revised: 17 February 2023
Accepted: 20 February 2023
Published: 22 February 2023

1. Introduction

Eels are usually considered catadromous fish spawning in the marine habitat, and spending most of their life cycle in freshwater systems. It has been shown, however, that some eels may skip the freshwater phase completely and stay in coastal water, or move between brackish water and freshwater (semi-catadromous behavior) [1]. The spawning area of the European eel *Anguilla anguilla* (L. 1758) is the Sargasso Sea. In a recent tracking experiment, it was shown for the first time that adult eels released near the Azores swim to the Sargasso Sea [2]. Tagged eels released near the European coast so far could not be tracked all the way down to the Sargasso Sea [3–6]. Fertilized eggs develop into a larvae named Leptocephalus [7], and later it was realized that the Leptocephali are the larvae of the European eel [8]. The Leptocephali drift with the Gulf stream [9,10] and reach the European or North African continental slope after a journey of about 7 months to 2 years [10]. Before entering the European freshwater system, Leptocephali metamorphose into glass eels. The translucent Leptocephali do not have a swimbladder and appear to be positively buoyant with overall densities of 1.028–1.043 $g \cdot mL^{-1}$ [11]. This low density value appears to be due to a high concentration of glycosaminoglycans in the translucent extracellular matrix. The swimbladder develops and is first inflated in glass eels. The glass eels then develop into so-called yellow eels, which typically spend 5–25 years in the European freshwater system [9,12].

The onset of the spawning migration requires preparation for the transition from freshwater to sea water. This occurs with a process called silvering, originally considered to be a second metamorphosis, but based on endocrine activity, it appears to be more like a puberty [13]. Silvering includes a number of physiological and morphological changes, including changes in ventral color from yellow to silver, and a significant darkening of the dorsal side. Fat stores and eyes are enlarged [12,14,15], and in the swimbladder, the length of the rete mirabile is increased [16–18]. An increased guanine incrustation has been observed in the American eel [19].

This review summarizes our current knowledge on swimbladder function in the different developmental stages of the European eel. The eel swimbladder has attracted the attention of scientists for over 100 years. Because of the bipolar rete mirabile, which allows taking blood samples in front of the rete and between the rete and the gas gland cells, eels have become a model for swimbladder function. Much information has been gained in a remarkable number of studies, but eels remain a mystery, with a large number of open questions.

2. Opening of the Swimbladder in Glass Eel

The swimbladder develops as a dorsal outgrowth of the esophagus. Leptocephali do not have a swimbladder. The first sign of swimbladder development is detected in early metamorphic stages to the glass eel [20]. In many glass eels caught on arrival along the European Coast, the swimbladder does not yet contain any gas, but is filled with surfactant [20]. Surfactant is required to reduce surface tension. Its presence has also been shown in yellow eel swimbladder, and it is produced and secreted by gas gland cells [21]. In early glass eels, the epithelial gas gland cells do not show the extensive basolateral labyrinth, characteristic for gas gland cells of yellow and silver eels (Figure 1). The intimate connection between gas gland cells and blood capillaries also is not yet visible [20]. The histological appearance of the gas gland cells therefore suggested that the initial filling of the swimbladder with gas, observed in the early glass eel stage, is not achieved by secretory activity of these cells, but by gulping air or by taking up small gas bubbles from the water [20]. Within 3 to 4 months of development after the initial inflation, the connective tissue of the swimbladder and the gas gland cells differentiate to the adult state present in yellow eels with an extensive basolateral labyrinth, which is in close contact with swimbladder capillaries [20]. The connection between esophagus and the swimbladder, the ductus pneumaticus, is functionally closed so that the eel cannot gulp air at the water surface. The ductus pneumaticus is used as the resorbing section of the swimbladder, separated from the secretory section via a sphincter muscle, located between the two retia mirabilia [22].

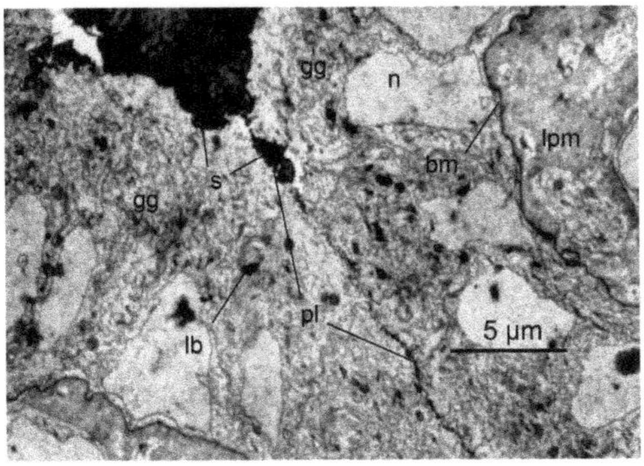

Figure 1. Histology of glass eel swimbladder. In Leptocephali developing into glass eels, the swimbladder initially is filled with surfactant; opposing gas gland cells appear to be separated by a thin layer of surfactant. Gas gland cells do not yet express an extensive basolateral labyrinth and the connection between gas gland cells and blood capillaries is not as tight as in yellow eels. gg, gas gland cell; s, surfactant; n, nucleus; pl, plasma membrane; lb, lamellar body; bm, basal membrane; lpm, lamina propria mucosae. Modified after [20].

3. Swimbladder Function in Yellow Eels

In the European freshwater system consisting of rivers and lakes, eels experience a limited depth range. Hydrostatic pressure increases by one atmosphere for every ten meters of water depth, so the range of hydrostatic pressures experienced in freshwater is limited. Nevertheless, the swimbladder wall is flexible, and therefore, according to Boyle's law, in a fish descending from the water surface to a water depth of 10 m, it results in a 50% decrease in swimbladder volume. However, neutral buoyancy, a status in which a fish can stay at a certain water depth without any swimming movement, requires a constant swimbladder volume, exactly compensating for the high density of other fish tissues, which exceeds water density [23,24]. The compression of the swimbladder encountered when descending into deeper water therefore must be compensated by gas secretion to keep the swimbladder volume constant. Ascending closer to the water surface, in turn, requires the removal of gas to avoid an increase in swimbladder volume.

Although eels are anatomically physostome, the ductus pneumaticus, the connection between the esophagus and the swimbladder, is functionally closed soon after the original opening of the swimbladder [9,22]. The yellow eel cannot gulp air at the surface and is functionally a physoclist fish. The ductus pneumaticus is used as a resorbing part of the swimbladder. It is a thin-walled, almost translucent gas cavity, highly vascularized, draining into the main venous system. Via a sphincter muscle located between the two retia mirabilia, gas from the secretory section can be transferred to the resorbing part. Along partial pressure gradients, gases then diffuse from this resorbing section to the blood and the venous circulatory system.

The wall of the secretory swimbladder has a silvery appearance due to guanine incrustation in connective tissues, and the epithelium consists of gas gland cells. Arterial blood supply to the secretory swimbladder passes a remarkable countercurrent system, the rete mirabile or red body. In the rete mirabile, the swimbladder artery gives rise to several tens of thousands of capillaries, running in parallel for a distance of several millimeters [25]. The capillaries reunify, forming two or three larger arterial vessels supplying the gas gland cells. Venous return to the rete again forms several tens of thousands of venous capillaries running in parallel and surrounding the arterial capillaries in the rete. The diffusion distance between arterial and venous capillaries is in the range of only one to two micrometer [25]. This arrangement of blood vessels allows taking blood samples in front of the countercurrent system, i.e., at the heart pole, but also between the rete and the gas gland cells, i.e., at the swimbladder pole. Thus, the role of the rete mirabile and the function of gas gland cells can be analyzed separately. This explains why the eel became a model species for the analysis of swimbladder function in physoclist fish [26]. In many other species, this separation of rete and gas gland cell function is not possible because of the close and intimate connection between the two structures [27,28].

In situ studies and experiments with primary cultured gas gland cells confirmed that gas gland cells are specialized for the production of acidic metabolites with a low fraction of aerobic metabolism, although they are chronically exposed to high oxygen partial pressures. A large fraction of glucose taken up from the blood is converted into lactic acid [29–32]. Some glucose is also shifted to the pentose phosphate shunt, where the activity of 6-phosphogluconate dehydrogenase results in the production of CO_2, by far exceeding the amount of CO_2 produced in the aerobic metabolism [31,33,34] (Figure 2). The activity of the pentose phosphate shunt also results in the generation of NADPH, an important reduction equivalent involved in the degradation of reactive oxygen species.

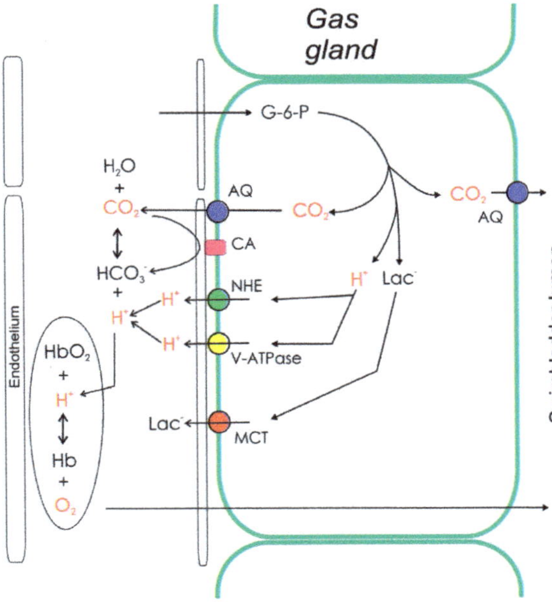

Figure 2. Secretory activity of gas gland cells. The metabolism of gas gland cells of the European eel depends on blood glucose, which is converted to lactic acid (lactate anion and a proton), and used for the production of CO_2. CO_2 production in the pentose phosphate shunt by far exceeds the amount of CO_2 produced in the aerobic metabolism. Protons, CO_2, and lactate are released into the blood, where the acidification switches on the Root effect. The resulting increase in PO_2 provides the partial pressure gradient for the diffusion of oxygen into the swimbladder. The production of CO_2 in gas gland cells also generates a partial pressure gradient towards the swimbladder lumen, allowing for the secretion of CO_2 into the swimbladder. AQ, aquaporin; CA, carbonic anhydrase; G-6-P, glucose-6-phosphate; Hb, hemoglobin; Lac, lactate; NHE, sodium proton exchanger; MCT, monocarboxylate carrier.

Pharmacological studies indicated that gas gland cells produce and release lactate into the blood stream by a monocarboxylate carrier [35], and analysis of the transcriptome as well as of the proteome of these cells confirmed the presence of monocarboxylate carriers [36,37]. Proton secretion is achieved via sodium proton exchange proteins (NHE-proteins) and a proton ATPase (V-ATPase). V-ATPase subunits and several sodium proton exchange proteins, including NHE1 and NHE2, have been identified in the proteome and in the transcriptome [36–39], and the inhibition of NHE proteins and of V-ATPase both resulted in a significant decrease in acid secretion in isolated gas gland cells [35,40]. Figure 2 summarizes the activity of gas gland cells based on data obtained from the European eel.

Due to the production of CO_2 in gas gland cells, the highest PCO_2 is expected in these cells, providing a driving force for the diffusion of CO_2 into the swimbladder lumen as well as into the blood stream (Figure 2). A membrane-bound carbonic anhydrase has been detected in gas gland cells, establishing a rapid equilibrium between CO_2 and bicarbonate in swimbladder blood, and the inhibition of carbonic anhydrase activity reduced the rate of acid secretion [35,41]. In a recent study, the presence of aquaporin 1 was demonstrated in apical and basolateral membranes of gas gland cells, and also in endothelial cells of the swimbladder [42]. In a physoclist swimbladder, water movements between the gas cavity and the surrounding tissue do not make sense. Aquaporin 1 is known to be a water channel, but to be also permeable to CO_2 [43–45]. Therefore, it was suggested that these aquaporins act as a CO_2 channel [42], facilitating the diffusion of CO_2 into the swimbladder as well as into the blood.

Acidification of the blood during the passage of the gas gland cells causes a decrease in the oxygen carrying capacity of the hemoglobin, switching on the Root effect [46–49]. The acid-induced partial deoxygenation of hemoglobin results in a significant increase in PO_2 in blood passing the gas gland cells. Thus, a partial pressure gradient is established, driving the diffusion of oxygen into the swimbladder. It also assures that the PO_2 in venous blood returning to the rete mirabile exceeds the PO_2 of arterial blood leaving the rete.

The activity of the gas gland cells thus induces an initial increase in gas partial pressures in swimbladder capillaries, the so-called single concentrating effect [50]. In a second step, this initial increase in gas partial pressure is multiplied by back-diffusion and countercurrent concentration in the rete mirabile [46,50]. Back-diffusion of oxygen and CO_2 from venous to arterial capillaries in the rete mirabile of the European eel resulted in a 7-fold increase in PO_2 and an 8-fold increase in PCO_2 during arterial passage of the rete mirabile [51].

Gas molecules diffuse through membranes, and the rete mirabile was considered to be a passive exchange system [50]. However, analysis of water and lactate movements in the rete provided the first evidence that not only oxygen and CO_2, but also lactate diffuses back to the arterial side in the eel rete [52]. A detailed study of the transcriptome and the proteome of rete capillaries revealed the expression of a large number of membrane transport proteins and several membrane ATPases, including Ca^{2+}-ATPase, V-ATPase, and Na^+/K^+-ATPase [53]. This suggests that ion transport and proton transport mechanisms indeed contribute to the countercurrent exchange and support the back-diffusion of lactate and protons to the arterial side of the rete in order to switch on the Root effect during the arterial passage of the rete. The data also revealed aquaporin expression in the rete. Because no water shift was detected in the rete, this suggests that aquaporin facilitates the diffusion of CO_2 not only in gas gland cells but also in the rete, supporting acidification of the blood in order to switch on the Root effect [53].

4. The Effect of Silvering

Silvering, which has been considered to be a puberty-like transition [13], is characterized by remarkable morphological and physiological changes. While the dorsal skin darkens, the ventral skin becomes silvery in preparation for the migration in the open ocean. This counter-shading is known to occur in many pelagic fish [54]. Eyes are significantly enlarged, and the diameter of the eye is among other parameters used to describe the silvering status of eels [55]. Neuromasts develop near the lateral line and fat stores increase by up to 28%. Eels do not feed during their spawning migration, and the alimentary tract degenerates [3,9,12,14,15].

An elongation of the rete mirabile has been described for the American eel [16], and a reduction in the gas permeability of the swimbladder wall occurs by increasing wall thickness and guanine deposition [18]. Cholesterol in cell membranes also contributes to a low gas permeability, but compared to yellow eels, in silver eels the cholesterol content was not increased [42]. Comparison of the transcriptome of yellow and silver European eels indicated significant expression changes for genes related to the extracellular matrix, supporting the conclusion that gas permeability of the swimbladder wall is reduced in silver eels [37]. Moreover, genes related to the degradation of reactive oxygen species (ROS) were over-expressed in silver eels. Due to greater depth encountered in the open ocean, higher oxygen pressures must occur in the swimbladder, probably coinciding with an increased level of ROS production [37]. Genes related to glucose transport were elevated in their transcription level in silver eels, while the expression of glycolytic enzymes was not enhanced [37].

An elevated expression of glucose transport and of monocarboxylate transport proteins was confirmed by analysis of the proteome [36]. The proteome also revealed expression of glucose-6-phosphate dehydrogenase, a crucial enzyme of the pentose phosphate shunt. A large number of enzymes related to ROS defense, including catalase, superoxide dismutase, and peroxiredoxin, were detected in the proteome, and glutathione peroxidase (GPX-7) was upregulated [36]. Measurement of enzyme activities confirmed an elevated ROS defense capacity in silver eels [56]. A summary of the basic functional characteristics of a yellow eel swimbladder and of the improvements observed during silvering is presented in Figure 3.

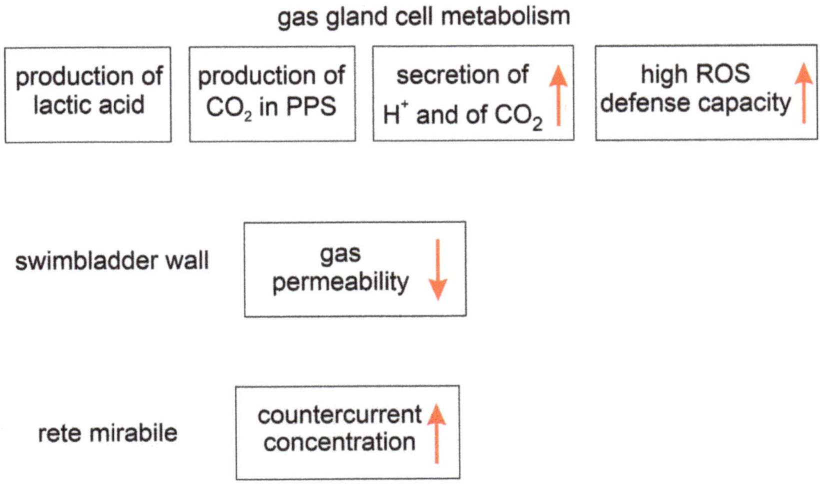

Figure 3. Functional characteristics of eel swimbladder and improvements related to silvering, indicated by a red arrow. Silvering results in an improvement of the secretory activity of gas gland cells, and in an improvement of the ROS defense capacity. Gas permeability of the swimbladder wall is reduced; the countercurrent concentrating capacity of the rete mirabile is enhanced. PPS, pentose phosphate shunt, ROS, reactive oxygen species.

5. Swimbladder Function during the Spawning Migration

Many studies have attempted to track tagged eels during their spawning migration to the Sargasso Sea. Depending on the starting point in Europe, the spawning migration covers a distance of 5000 to 6000 km, but eels could successfully be tracked for up 1000 km, and only occasionally for up to 2000 km [4–6]. Only eels released at the Azores, i.e., about halfway between Western Europe and the Sargasso Sea, could be tracked to the Sargasso Sea [2]. An unexpected result of all tracking studies was the observation that eels perform diurnal vertical migrations. While in the Baltic or in the North Sea, the depth range is below 100 m, beyond the continental shelf, eels swim at depths of about 200–300 m at nighttime, and at depths of 600–800 m during daytime. Eels have even been recorded at depths below 1000 m [5,6,57,58]. Eels do not feed during their spawning journey [9,59]; therefore, these diurnal excursions cannot be related to feeding. Thermoregulation and predator avoidance have been discussed as possible reasons [3,58,60], and a recent study suggested that eels follow an isolume with these vertical movements [4]. It is generally expected that the journey takes about 4 to 6 months, but recent considerations indicate that it may also take more than a year [4,59,61]. Experiments indeed revealed that eels are able to swim for several months and can cover a distance of up to 5500 km in a swim tunnel without feeding [62,63]. Body composition of eels after six months in a swim tunnel was not different from control eels, indicating that eels used fat, protein, and carbohydrates in the same proportion [63]. Based on the decrease in energy reserves and on oxygen consumption, optimal swimming speed and the cost of transport have been assessed, and all studies consistently show that eels swim very efficiently, with the cost of transport being significantly lower than in salmonids [59,61,63,64].

These studies have been performed under atmospheric pressure. The measuring of oxygen uptake of eels kept under elevated hydrostatic pressure for prolonged periods revealed decreased oxygen consumption as compared to values recorded under atmospheric pressure, suggesting that the actual costs of transport during the spawning migration may even be lower than calculated based on swim tunnel experiments under atmospheric pressure [65]. These data clearly show that eels are able to successfully complete the spawning migration, and preserve sufficient energy for gamete production and successful reproduction.

With the diurnal vertical migration, eels experience significant changes in hydrostatic pressure. While at a depth of 300 m, hydrostatic pressure amounts to 31 atm, at 800 m it increases to 81 atm, resulting in proportional compression of the flexible walled swimbladder according to Boyle's law. With a reduced swimbladder volume, eels would become negatively buoyant, if the eels would not be able to fully compensate for the increase in hydrostatic pressure by secreting gas in order to keep the swimbladder volume constant. The depth records obtained from tagged eels reveal that the descent from the depths at night to greater depths in the daytime takes no more than 1 to 2 h, and the same time is required for the return to the depth range at the beginning of the night [4,6]. Accordingly, at constant temperature, about 2.500 mL of gas would be required for a 1 kg eel to keep the swimbladder volume constant during a descent from 300 m to 800 m [66]. In the European yellow eel, gas secretion rates did not exceed 1–2 $mL \cdot h^{-1}$ [51,67]. Values below 1 $mL \cdot h^{-1}$ have also been reported for the American yellow eel *Anguilla rostrata*, and in silver eel, the rate of gas secretion increased to up to 3 $mL \cdot h^{-1}$ [18]. These values are far too low to keep the swimbladder volume constant during the descent from 300 m to 800 m in less than 2 h. Taking into account the oxygen transport capacity of European eel blood and the possible glucose turnover in gas gland cells, it seems unlikely that the required amount of gas can be supplied to keep the volume constant [66].

If the swimbladder would provide neutral buoyancy at a depth of 800 m, the same amount of gas would have to be resorbed and transported in the blood within 1 to 2 h during ascent to avoid increasing positive buoyancy. This also appears unlikely based on gas transport capacities determined in eel blood. Therefore, it can be assumed that the swimbladder may provide neutral buoyancy at the depths encountered at night time, and is compressed during descent in the daytime. In this case, the lack of buoyancy during descent must be compensated by hydrodynamic lift [66].

The changes in swimbladder and rete mirabile structure and in swimbladder metabolism outlined above, however, clearly indicate an improvement in swimbladder function. Additionally, although the rate of gas secretion is not sufficient to keep the swimbladder volume constant during these vertical migrations, there must be some secretory activity. Gas partial pressure in the water hardly increases with depth, so that the partial pressure gradient between swimbladder lumen and the surrounding water increases with depth [68]. Although the gas permeability of the swimbladder wall is low, it certainly is not impermeable to gases, and the diffusional loss of gas through the wall increases with depth along the increasing partial pressure gradient between the lumen and the surrounding tissues. This loss must be compensated for by gas secretion. Moreover, even if a full compensation is not possible, this does not mean that there is no secretory activity at all to achieve at least a partial compensation. Silver eels traveling in the open ocean experience a much larger depth range than yellow eels in freshwater, and this may explain the improved secretory activity detected in silver eels.

6. The Nematode *Anguillicola crassus* Impairs Swimbladder Function

In the early 1980s, the swimbladder nematode *Anguillicola crassus* was accidentally introduced to Europe, and within about a decade, a large fraction of the European eel population was infected with this nematode [69–75]. The eel is infected by feeding on intermediate hosts of the nematode (copepods) or diverse paratenic hosts, such as smaller fish species, amphibians, insect larvae, crustacea, or mollusks [73,76]. Eels are infected with L3-stage nematode larvae, which enter the swimbladder wall. Experiments with glass eels revealed that, even at this early stage, eels can be infected when fed with the L3-stage-nematode-infected copepods [77]. In the swimbladder, the histophagic L3 larvae develop into L4-stage larvae and the preadult stage, which enters the swimbladder lumen. Adult nematodes live in the swimbladder lumen and feed on blood from swimbladder capillaries [78,79]. Reductions in hematocrit and plasma protein content have been observed as a result of this hematophagy [79]. Eggs of the nematode hatch in the swimbladder and leave the swimbladder via the ductus pneumaticus and the gut [73,77,80].

A comparison of the oxygen content of infected and non-infected yellow eels revealed a significant reduction in oxygen content, suggesting a significant impairment of swimbladder function [41] (Figure 4). A heavily infected swimbladder of yellow eels may be almost completely filled with nematodes and dark fluid, containing almost no gas at all. In this situation, the buoyancy function of the swimbladder is completely lost, and hydrodynamic lift will be required to prevent sinking [23,24]. An infection of the swimbladder not only changes the gas composition and the amount of gas in the bladder, but it also results in significant changes in the swimbladder wall. Swimbladder epithelial cells proliferate and the single-layered epithelium thickens, forming a multilayered epithelium. The extent of the basolateral labyrinth is reduced, and the intimate contact to blood capillaries and the polarity of the cells is partially lost [74,80,81]. This results in a significant increase in diffusion distances between blood capillaries and the swimbladder lumen, and also between gas gland cells and the blood, explaining the decrease in oxygen content observed in infected yellow eel swimbladders [41]. In addition, the elasticity of the swimbladder wall is significantly reduced [82].

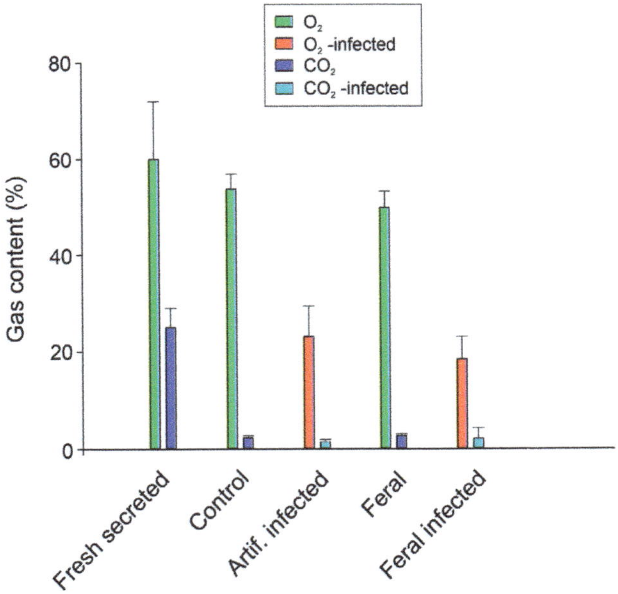

Figure 4. Oxygen and CO_2 content of healthy and infected eel swimbladder. Oxygen makes up the largest fraction of freshly secreted gas in the eel swimbladder, under steady-state conditions and in the swimbladder of feral eels collected in the wild. The fraction of CO_2 is also high in freshly secreted gas, but under steady-state conditions, it is much lower because of the preferential resorption due to the high physical solubility of CO_2. In eels artificially infected with the nematode *Anguillicola crassus* and in feral eels with infected swimbladders, the oxygen content is largely reduced, and the CO_2 content is also reduced. Data from [67,77].

Infection of the swimbladder in yellow eels resulted in a modification of the mRNA level of more than 1600 genes [38]. Functional annotation based on GO terms revealed that a large number of genes related to metabolism were affected in infected eels. Several transcripts coding for proteins involved in glucose and monocarboxylate transport were elevated in their expression level. The highest number of affected genes, however, were related to the term 'immune response', indicating that yellow eels tried to defend the nematode. Inflammatory components, complement proteins, and also immunoglobulins were significantly elevated in their mRNA expression level [38]. An increased activity of the immune system has also been detected by testing the antibody response following an artificial infection in both European and Japanese eel [73]. In addition, an elevated non-specific immune response in response to the nematode infection has been observed in wild and farmed European eels [83]. The presence of macrophages in infected swimbladders underlines the initiation of a strong immune response by the nematode [72,84,85].

While a large number of genes were differentially expressed at the mRNA level in infected yellow eel swimbladders, parasite infestation caused only minor expression changes in silver eels, with no apparent changes in the mRNA content of genes connected to glycolytic acid production or ROS defense, both crucial for swimbladder function [38]. Silvering includes remarkable physiological modifications, including the preparation for the transition from freshwater to sea water with appropriate changes in ion and osmoregulation. It also marks the onset of sexual maturation [86,87], although sexual maturity is only achieved at some point during the spawning migration. These changes require a lot of energy, and this may explain the significantly diminished defense response of silver eel swimbladder tissue to the nematode infection.

Comparison of transcriptional changes in non-infected vs. infected swimbladders in silver eels exercising in a swim tunnel under elevated hydrostatic pressure (8 atm) [88] showed a 3-fold greater number of differentially expressed genes in infected eels. In healthy eels, genes with an elevated transcription level were related to glycolytic activity, glucose transport, and carbonic anhydrase expression, suggesting an improvement of the gas secretory activity. Elevated levels of angiopoietin transcripts suggested an improvement of the gas gland cell blood connection, facilitating acid release and switching on the Root effect in swimbladder capillaries [88]. In infected eels, elevated transcriptional levels of glycolytic enzymes including hexokinase 2 point to the activation of glycolysis due to tissue thickening and elevated diffusion distances. Transcriptional changes in infected swimbladders also showed an elevated transcription of immune response genes [88], confirming the results obtained with unexercised yellow and silver eels [37].

Focusing on the physiology and the performance of the swimbladder, a comparison of oxygen consumption of glass eels infected and non-infected with nematodes revealed no significant difference in oxygen consumption [77]. A comparison of the swimming capacity of female silver eels with infected and non-infected swimbladders revealed a reduced swimming capacity in infected eels [89]. Eels with nematode-infected swimbladder had lower cruising speeds, and 27 out of 74 eels swam unsteadily and stopped swimming at comparatively low aerobic swimming speeds. These 27 eels had a significantly higher oxygen consumption in the swim tunnel, and in eels with infected swimbladder, the costs of transport were elevated. In endurance trials, eels with heavily infected swimbladders failed to swim 1000 km [89]. In eels up to 45 cm body length with more than 10 nematodes in the swimbladder, a significant reduction in swimming speed has been observed [90], and silver eels with several adult nematodes in their swimbladder have been shown to avoid accelerating water velocity [91]. This could result in a delayed passage of downstream rapids and thus in a delayed start in the spawning migration.

The study of Simon et al. [92] suggested that nematode infestation does not influence diurnal diving behavior. However, given that the study was conducted only with one single individual infected with three nematodes and a limited depth (60 m), this observation does not exclude the possibility of a negative impact with more parasites or on longer and deeper dives. On the basis of tracking studies, Wysujack et al. [5] speculated that nematode infestation did not preclude daily vertical migrations, although the study did not include the actual infection rate of the silver eels.

Taken together, these data clearly show the tremendous negative impact of the nematode *Anguillicola crassus* on the swimming capacity of eels and on swimbladder function. The nematode infection, therefore, is expected to be a serious threat to successful spawning migration to the Sargasso Sea, and it appears to be one of the parameters contributing to the current decline in the eel population.

7. Future Directions

Although the study of the swimbladder has attracted the attention of scientists for many decades, it remains a mystery. We have learned much about the basic concept of gas secretion, including the secretory activity of gas gland cells, the Root effect, and countercurrent multiplication of the single concentrating effect. However, we know little about the exact transport characteristics of the countercurrent system, the rete mirabile, which, as recently detected, is not only a passive exchange system—transport proteins and especially ATPases may be controlled in its activity. Therefore, it will be interesting to assess possible regulatory mechanisms for the back-diffusion of ions and metabolites in the rete mirabile. Although repeatedly assessed and important, because gas gland cell metabolism and the rate of acid secretion determine the rate of gas deposition [67], the mechanisms controlling gas gland cell activity remain undefined. Still unclear are the role and the performance of the swimbladder during diurnal vertical migrations. It appears unlikely that the swimbladder can provide neutral buoyancy during the rapid descent and ascent associated with these migrations, but the improvements connected with silvering of the eel clearly demonstrate that the swimbladder is of importance during these migrations.

Funding: This research was funded by Austrian Fonds zur Förderung der wissenschaftlichen Forschung (FWF), P26363; I2984.

Conflicts of Interest: The author declares no conflict of interest.

References

1. Tzeng, W.N.; Wang, C.H.; Wickström, H.; Reizenstein, M. Occurrence of the semi-catadromous European eel *Anguilla anguilla* in the Baltic Sea. *Marine Biol.* **2000**, *137*, 93–98. [CrossRef]
2. Wright, R.; Piper, A.; Aarestrup, K.; Azevedo, J.; Cowan, G.; Don, A.; Gollock, M.; Rodriguez Ramallo, S.; Velterop, R.; Walker, A.; et al. First direct evidence of adult European eels migrating to their breeding place in the Sargasso Sea. *Sci. Rep.* **2022**, *12*, 15362. [CrossRef]
3. Righton, D.; Aarestrup, K.; Jellyman, D.; Sebert, P.; Van den Thillart, G.; Tsukamoto, K. The *Anguilla* spp. migration problem: 40 million years of evolution and two millennia of speculation. *J. Fish Biol.* **2012**, *81*, 365–386. [CrossRef]
4. Righton, D.; Westerberg, H.; Feunteun, E.; Okland, F.; Gargan, P.; Amilhart, E.; Metcalfe, J.; Lobon-Cervia, J.; Sjöberg, J.; Acou, A.; et al. Empirical observations of the spawning migration of European eels: The long and dangerous road to the Sargasso Sea. *Sci. Adv.* **2016**, *2*, e1501694. [CrossRef]
5. Wysujack, K.; Westerberg, H.; Aarestrup, K.; Trautner, J.; Kurwie, T.; Nagel, F.; Hanel, R. The migration behaviour of European silver eels (*Anguilla anguilla*) released in open ocean conditions. *Mar. Freshw. Res.* **2015**, *66*, 145–157. [CrossRef]
6. Aarestrup, K.; Okland, F.; Hansen, M.; Righton, D.; Gargan, P.; Castonguay, M.; Bernatchez, L.; Howey, P.; Sparholt, H.; Pedersen, M.; et al. Oceanic spawning migration of the european eel (*Anguilla anguilla*). *Science* **2009**, *325*, 1660. [CrossRef] [PubMed]
7. Kaup, J.J. *Catalogue of Apodal Fish, in the Collection of the British Museum*; British Museum: London, UK, 1856.
8. Grassi, G.B.; Calandruccio, S. Fortpflanzung und Metamorphose des Aales. *Allg. Fisch Zeitung.* **1897**, 22, 402–408.
9. Tesch, F.-W. *Der Aal*; Paul Parey Verlag: Singhofen, Germany, 1999.
10. Bonhommeau, S.; Castonguay, M.; Rivot, E.; Sabatié, R.; Le Pape, O. The duration of migration of Atlantic Anguilla larvae. *Fish Fish.* **2010**, *11*, 289–306. [CrossRef]
11. Tsukamoto, K.; Yamada, Y.; Okamura, A.; Kaneko, T.; Tanaka, H.; Miller, M.; Horie, N.; Mikawa, N.; Utoh, T.; Tanaka, S. Positive buoyancy in eel leptocephali: An adaptation for life in the ocean surface layer. *Mar. Biol.* **2009**, *156*, 835–846. [CrossRef]
12. Durif, C.M.F.; van Ginneken, V.; Dufour, S.; Müller, T.; Elie, P. Seasonal Evolution and Individual Differences in Silvering Eels from Different Locations. In *Spawning Migration of the European Eel*; Springer: Dordrecht, The Netherlands, 2009; pp. 13–38.
13. Rousseau, K.; Aroua, S.; Schmitz, M.; Elie, P.; Dufour, S. Silvering: Metamorphosis or Puberty? In *Spawning Migration of the European Eel*; Springer: Dordrecht, The Netherlands, 2009; pp. 39–63.
14. van Ginneken, A.C.G.; Maes, G.E. The European eel (*Anguilla anguilla*, Linnaeus), its lifecycle, evolution and reproduction: A literature review. *Rev. Fish Biol. Fish.* **2005**, *15*, 367–398. [CrossRef]
15. van Ginneken, V.; Durif, C.; Balm, S.; Boot, R.; Verstegen, M.; Antonissen, E.; van den Thillart, G. Silvering of European eel (*Anguilla anguilla* L.): Seasonal changes of morphological and metabolic parameters. *Anim. Biol.* **2007**, *57*, 63–77. [CrossRef]
16. Kleckner, R.C.; Krueger, W.H. Changes in swimbladder retial morphology in *Anguilla rostrata* during premigration metamorphosis. *J. Fish Biol.* **1981**, *18*, 569–577. [CrossRef]

17. Yamada, Y.; Zhang, H.; Okamura, A.; Tanaka, S.; Horie, N.; Mikawa, N.; Utoh, T.; Oka, H. Morphological and histological changes in the swim bladder during maturation of the Japanese eel. *J. Fish Biol.* **2001**, *58*, 804–814. [CrossRef]
18. Kleckner, R.C. Swim bladder volume maintenance related to migratory depth in silver phase *Anguilla rostrata*. *Science* **1980**, *208*, 1481–1482. [CrossRef] [PubMed]
19. Kleckner, R.C. Swimbladder wall guanine enhancement related to migratory depth in silver phase *Anguilla rostrata*. *Comp. Biochem. Physiol. Part A* **1980**, *65*, 351–354. [CrossRef]
20. Zwerger, P.; Nimeth, K.; Würtz, J.; Salvenmoser, W.; Pelster, B. Development of the swimbladder in the European eel (*Anguilla anguilla*). *Cell Tissue Res.* **2002**, *307*, 155–164. [CrossRef] [PubMed]
21. Prem, C.; Salvenmoser, W.; Würtz, J.; Pelster, B. Swimbladder gas gland cells produce surfactant: In vivo and in culture. *Am. J. Physiol. Regul. Integr. Comp. Physiol.* **2000**, *279*, 2336–2343. [CrossRef] [PubMed]
22. Dorn, E. Über den Feinbau der Schwimmblase von *Anguilla vulgaris* L. Licht- und Elektronenmikroskopische Untersuchungen. *Zeitschrift für Zellforsch.* **1961**, *55*, 849–912. [CrossRef]
23. Alexander, R.M. Physical aspects of swimbladder function. *Biol. Rev.* **1966**, *41*, 141–176. [CrossRef]
24. Pelster, B. Buoyancy control in aquatic vertebrates. In *Cardio-Respiratory Control in Vertebrates*; Glass, M.L., Wood, S.C., Eds.; Springer: Berlin/Heidelberg, Germany, 2009; pp. 65–98.
25. Wagner, R.C.; Froehlich, R.; Hossler, F.E.; Andrews, S.B. Ultrastructure of capillaries in the red body (rete mirabile) of the eel swim bladder. *Microvasc. Res.* **1987**, *34*, 349–362. [CrossRef]
26. Pelster, B. The swimbladder. In *Eel Physiology*; Trischitta, F., Takei, Y., Sebert, P., Eds.; CRC Press: Enfield, FL, USA, 2013; pp. 44–67.
27. Fänge, R. The mechanisms of gas transport in the euphysoclist swimbladder. *Acta Physiol. Scand. Suppl.* **1953**, *30*, 1–133. [PubMed]
28. Jones, F.R.H.; Marshall, N.B. The structure and functions of the teleostean swimbladder. *Biol. Bull.* **1953**, *28*, 16–83.
29. Steen, J.B. The physiology of the swimbladder in the eel *Anguilla vulgaris*. III. The mechanism of gas secretion. *Acta Physiol. Scand.* **1963**, *59*, 221–241. [CrossRef] [PubMed]
30. Pelster, B.; Scheid, P. Glucose Metabolism of the Swimbladder Tissue of the European Eel *Anguilla anguilla*. *J. Exp. Biol.* **1993**, *185*, 169–178. [CrossRef]
31. Pelster, B. Metabolism of the swimbladder tissue. *Biochem. Mol. Biol. Fishes* **1995**, *4*, 101–118.
32. Drechsel, V.; Schneebauer, G.; Sandbichler, A.M.; Fiechtner, B.; Pelster, B. Oxygen consumption and acid secretion in isolated gas gland cells of the European eel *Anguilla anguilla*. *J. Comp. Physiol. B* **2022**, *192*, 447–457. [CrossRef] [PubMed]
33. Walsh, P.J.; Milligan, C.L. Roles of buffering capacity and pentose phosphate pathway activity in the gas gland of the gulf toadfish *Opsanus beta*. *J. Exp. Biol.* **1993**, *176*, 311–316. [CrossRef]
34. Pelster, B.; Hicks, J.; Driedzic, W.R. Contribution of the Pentose Phosphate Shunt to the formation of CO_2 in swimbladder tissue of the eel. *J. Exp. Biol.* **1994**, *197*, 119–128. [CrossRef]
35. Pelster, B. Mechanisms of acid release in isolated gas gland cells of the European eel *Anguilla anguilla*. *Am. J. Physiol.* **1995**, *269*, R793–R799. [CrossRef]
36. Sialana, F.J.; Schneebauer, G.; Paunkov, A.; Pelster, B.; Lubec, G. Proteomic Studies on the Swim Bladder of the European Eel (*Anguilla anguilla*). *Proteomics* **2018**, *18*, 1700445. [CrossRef]
37. Pelster, B.; Schneebauer, G.; Dirks, R.P. *Anguillicola crassus* infection significantly affects the silvering related modifications in steady state mRNA levels in gas gland tissue of the European eel. *Front. Physiol.* **2016**, *7*, 175. [CrossRef]
38. Schneebauer, G.; Dirks, R.P.; Pelster, B. *Anguillicola crassus* infection affects mRNA expression levels in gas gland tissue of European yellow and silver eel. *PLoS ONE* **2017**, *12*, e0183128. [CrossRef] [PubMed]
39. Niederstätter, H.; Pelster, B. Expression of two vacuolar-type ATPase B subunit isoforms in swimbladder gas gland cells of the European eel: Nucleotide sequences and deduced amino acid sequences. *Bba. Gene Struct. Express* **2000**, *1491*, 133–142. [CrossRef]
40. Pelster, B. pH regulation and swimbladder function in fish. *Respir. Physiol. Neurobiol.* **2004**, *144*, 179–790. [CrossRef]
41. Würtz, J.; Taraschewski, H.; Pelster, B. Changes in gas composition in the swimbladder of the European eel (*Anguilla anguilla*) infected with *Anguillicola crassus* (Nematoda). *Parasitology* **1996**, *112*, 233–238. [CrossRef] [PubMed]
42. Drechsel, V.; Schneebauer, G.; Fiechtner, B.; Cutler, C.P.; Pelster, B. Aquaporin expression and cholesterol content in eel swimbladder tissue. *J. Fish Biol.* **2022**, *100*, 609–618. [CrossRef]
43. Chen, L.M.; Zhao, J.; Musa-Aziz, R.; Pelletier, M.F.; Drummond, I.A.; Boron, W.F. Cloning and characterization of a zebrafish homologue of human AQP1: A bifunctional water and gas channel. *AJP-Regul. Integr. Comp. Physiol.* **2010**, *299*, R1163–R1174. [CrossRef]
44. Perry, S.F.; Braun, M.H.; Noland, M.; Dawdy, J.; Walsh, P.J. Do zebrafish Rh proteins act as dual ammonia-CO_2 channels? *J. Exp. Zool. Part A Ecol. Genet. Physiol.* **2010**, *313*, 618–621. [CrossRef]
45. Talbot, K.; Kwong, R.W.M.; Gilmour, K.M.; Perry, S.F. The water channel aquaporin-1a1 facilitates movement of CO_2 and ammonia in zebrafish (*Danio rerio*) larvae. *J. Exp. Biol.* **2015**, *218*, 3931–3940. [CrossRef] [PubMed]
46. Pelster, B. The Generation of Hyperbaric Oxygen Tensions in Fish. *Physiology* **2001**, *16*, 287–291. [CrossRef]
47. Root, R.W. The respiratory function of the blood of marine fishes. *Biol. Bull.* **1931**, *61*, 427–456. [CrossRef]
48. Berenbrink, M.; Koldkjaer, P.; Kepp, O.; Cossins, A.R. Evolution of Oxygen Secretion in Fishes and the Emergence of a Complex Physiological System. *Science* **2005**, *307*, 1752–1757. [CrossRef]
49. Berenbrink, M.; Koldkaer, P.; Hannah Wright, E.; Kepp, O.; Jose da Silva, A. Magnitude of the Root effect in red blood cells and haemoglobin solutions of fishes: A tribute to August Krogh. *Acta Physiol.* **2011**, *202*, 583–592. [CrossRef] [PubMed]

50. Kuhn, W.; Ramel, A.; Kuhn, H.J.; Marti, E. The filling mechanism of the swimbladder. Generation of high gas pressures through hairpin countercurrent multiplication. *Experientia* **1963**, *19*, 497–511. [CrossRef]
51. Kobayashi, H.; Pelster, B.; Scheid, P. CO_2 back-diffusion in the rete aids O_2 secretion in the swimbladder of the eel. *Respir. Physiol.* **1990**, *79*, 231–242. [CrossRef]
52. Kobayashi, H.; Pelster, B.; Scheid, P. Water and lactate movement in the swimbladder of the eel, *Anguilla anguilla*. *Respir. Physiol.* **1989**, *78*, 45–57. [CrossRef] [PubMed]
53. Schneebauer, G.; Drechsel, V.; Dirks, R.; Faserl, K.; Sarg, B.; Pelster, B. Expression of transport proteins in the rete mirabile of european silver and yellow eel. *BMC Genom.* **2021**, *22*, 866. [CrossRef]
54. Kelley, J.L.; Taylor, I.; Hart, N.S.; Partridge, J.C. Aquatic prey use countershading camouflage to match the visual background. *Behav. Ecol.* **2017**, *28*, 1314–1322. [CrossRef]
55. Pankhurst, N.W. Relation of visual changes to the onset of sexual maturation in the European eel *Anguilla anguilla* (L.). *J. Fish Biol.* **1982**, *21*, 127–140. [CrossRef]
56. Schneebauer, G.; Hanel, R.; Pelster, B. *Anguillicola crassus* impairs the silvering-related enhancements of the ROS defense capacity in swimbladder tissue of the European eel (*Anguilla anguilla*). *J. Comp. Physiol. B* **2016**, *186*, 867–877. [CrossRef]
57. Westerberg, H.; Lagenfelt, I.; Svedang, H. Silver eel migration behaviour in the Baltic. *ICES J. Mar. Sci.* **2007**, *64*, 1457–1462. [CrossRef]
58. Wahlberg, M.; Westerberg, H.; Aarestrup, K.; Feunteun, E.; Gargan, P.; Righton, D. Evidence of marine mammal predation of the European eel (*Anguilla anguilla* L.) on its marine migration. *Deep Sea Res. Part I Oceanogr. Res. Pap.* **2014**, *86*, 32–38. [CrossRef]
59. Palstra, A.; van Ginneken, V.; van den Thillart, G. Cost of transport and optimal swimming speed in farmed and wild European silver eels (*Anguilla anguilla*). *Comp. Biochem. Physiol. Part A Mol. Integr. Physiol.* **2008**, *151*, 37–44. [CrossRef]
60. Trancart, T.; Tudorache, C.; van den Thillart, G.; Acou, A.; Carpentier, A.; Boinet, C.; Gouchet, G.; Feunteun, E. The effect of thermal shock during diel vertical migration on the energy required for oceanic migration of the European silver eel. *J. Exp. Mar. Bio. Ecol.* **2015**, *463*, 168–172. [CrossRef]
61. Palstra, A.P.; van den Thillart, G. Swimming physiology of European silver eels (*Anguilla anguilla* L.): Energetic costs and effects on sexual maturation and reproduction. *Fish Physiol. Biochem.* **2010**, *36*, 297–322. [CrossRef]
62. van den Thillart, G.; van Ginneken, V.; Korner, F.; Heijmans, R.; van der Linden, R.; Gluvers, A. Endurance swimming of European eel. *J. Fish Biol.* **2004**, *65*, 312–318. [CrossRef]
63. van Ginneken, V.; Antonissen, E.; Müller, U.K.; Booms, R.; Eding, E.; Verreth, J.; van den Thillart, G. Eel migration to the Sargasso: Remarkably high swimming efficiency and low energy costs. *J. Exp. Biol.* **2005**, *208*, 1329–1335. [CrossRef] [PubMed]
64. Tudorache, C.; Burgerhout, E.; Brittijn, S.; van den Thillart, G. Comparison of swimming capacity and energetics of migratory European eel (*Anguilla anguilla*) and New Zealand short-finned eel (*A. australis*). *Front. Physiol.* **2015**, *6*, 256. [CrossRef] [PubMed]
65. Sébert, P.; Scaion, D.; Belhomme, M. High hydrostatic pressure improves the swimming efficiency of European migrating silver eel. *Respir. Physiol. Neurobiol.* **2009**, *165*, 112–114. [CrossRef]
66. Pelster, B. Swimbladder function and the spawning migration of the European eel *Anguilla anguilla*. *Front. Physiol.* **2015**, *5*, 486. [CrossRef]
67. Pelster, B.; Scheid, P. The influence of gas gland metabolism and blood flow on gas deposition into the swimbladder of the European eel *Anguilla anguilla*. *J. Exp. Biol.* **1992**, *173*, 205–216. [CrossRef]
68. Pelster, B. Buoyancy at Depth. In *Fish Physiology*; Randall, C.D.J., Farrel, A.P., Eds.; Academic Press: San Diego, CA, USA, 1997; Volume 15, pp. 195–237.
69. Haenen, O.L.M.; van Banning, P. Experimental transmission of *Anguillicola crassus* (Nematoda, Dracunculoidea) larvae from infected prey fish to the eel *Anguilla anguilla*. *Aquaculture* **1991**, *92*, 115–119. [CrossRef]
70. Lefebvre, F.; Wielgoss, S.; Nagasawa, K.; Moravec, F. On the origin of *Anguillicoloides crassus*, the invasive nematode of anguillid eels. *Aquat. Invasions* **2012**, *7*, 443–453. [CrossRef]
71. Schabuss, M.; Kennedy, C.R.; Konecny, R.; Grillitsch, B.; Reckendorfer, W.; Schiemer, F.; Herzig, A. Dynamics and predicted decline of *Anguillicola crassus* infection in European eels, *Anguilla anguilla*, in Neusiedler See, Austria. *J. Helminthol.* **2005**, *79*, 159–167. [CrossRef] [PubMed]
72. Würtz, J.; Knopf, K.; Taraschewski, H. Distribution and prevalence of *Anguillicola crassus* (Nematoda) in eels *Anguilla anguilla* of the rivers Rhine and Naab, Germany. *Dis. Aquat. Organ.* **1998**, *32*, 137–143. [CrossRef] [PubMed]
73. Knopf, K. The swimbladder nematode *Anguillicola crassus* in the European eel *Anguilla anguilla* and the Japanese eel *Anguilla japonica*: Differences in susceptibility and immunity between a recently colonized host and the original host. *J. Helminthol.* **2006**, *80*, 129–136. [CrossRef] [PubMed]
74. Kirk, R.S. The impact of *Anguillicola crassus* on European eels. *Fish. Manag. Ecol.* **2003**, *10*, 385–394. [CrossRef]
75. Lefebvre, F.; Fazio, G.; Mounaix, B.; Crivelli, A.J. Is the continental life of the European eel *Anguilla anguilla* affected by the parasitic invader *Anguillicoloides crassus*? *Proc. R. Soc. B Biol. Sci.* **2013**, *280*, 20122916. [CrossRef]
76. Moravec, F.; Skoríkková, B. Amphibians and larvae of aquatic insects as new paratenic hosts of *Anguillicola crassus* (Nematoda: Dracunculoidea), a swimbladder parasite of eels. *Dis. Aquat. Organ.* **1998**, *34*, 217–222. [CrossRef]
77. Nimeth, K.; Zwerger, P.; Würtz, J.; Salvenmoser, W.; Pelster, B. Infection of the glass-eel swimbladder with the nematode *Anguillicola crassus*. *Parasitology* **2000**, *121*, 75–83. [CrossRef]

78. Polzer, M.; Taraschewski, H. Identification and characterization of the proteolytic enzymes in the developmental stages of the eel-pathogenic nematode *Anguillicola crassus*. *Parasitol. Res.* **1993**, *79*, 24–27. [CrossRef]
79. Boon, J.H.; Cannaerts, V.; Augustijn, H.; Machiels, M.; De Charleroy, D.; Ollevier, F. The effect of different infection levels with infective larvae of *Anguillicola crassus* on haematological parameters of European eel (*Anguilla anguilla*). *Aquaculture* **1990**, *87*, 243–253. [CrossRef]
80. Würtz, J.; Taraschewski, H. Histopathological changes in the swimbladder wall of the European eel *Anguilla anguilla* due to infections with *Anguillicola crassus*. *Dis. Aquat.Org.* **2000**, *39*, 121–134. [CrossRef] [PubMed]
81. Molnár, K.; Szakolczai, J.; Vetési, F. Histological changes in the swimbladder wall of eels due to abnormal location of adults and second stage larvae of *Anguillicola crassus*. *Acta Vet. Hung.* **1995**, *43*, 125–137.
82. Barry, J.; McLeish, J.; Dodd, J.A.; Turnbull, J.F.; Boylan, P.; Adams, C.E. Introduced parasite *Anguillicola crassus* infection significantly impedes swim bladder function in the European eel *Anguilla anguilla* (L.). *J. Fish Dis.* **2014**, *37*, 921–924. [CrossRef] [PubMed]
83. van Banning, P.; and Haenen, O. Effects of the swimbladder nematode *Anguillicola crassus* in wild and farmed eel, *Anguilla anguilla*. In *Pathology in Marine Science;* Elsevier: Amsterdam, The Netherlands, 1990; pp. 317–330.
84. Knopf, K.; Madriles Helm, A.; Lucius, R.; Bleiss, W.; Taraschewski, H. Migratory response of European eel (*Anguilla anguilla*) phagocytes to the eel swimbladder nematode *Anguillicola crassus*. *Parasitol. Res.* **2008**, *102*, 1311–1316. [CrossRef]
85. Beregi, A.; Molnár, K.; Békési, L.; Székely, C. Radiodiagnostic method for studying swimbladder inflammation caused by *Anguillicola crassus* (Nematoda:Dracunculoidea). *Dis. Aquat. Organ.* **1998**, *34*, 155–160. [CrossRef] [PubMed]
86. Burgerhout, E.; Minegishi, Y.; Brittijn, S.; de Wijze, D.; Henkel, C.; Jansen, H.; Spaink, H.; Dirks, R.; van den Thillart, G. Changes in ovarian gene expression profiles and plasma hormone levels in maturing European eel (*Anguilla anguilla*); Biomarkers for broodstock selection. *Gen. Comp. Endocrinol.* **2016**, *225*, 185–196. [CrossRef] [PubMed]
87. Dufour, S.; Burzawa-Gerard, E.; Le Belle, N.; Sbaihi, M.; Vidal, B. Reproductive Endocrinology of the European Eel, *Anguilla anguilla*. In *Eel Biology;* Springer: Tokyo, Japan, 2003; pp. 373–383.
88. Schneebauer, G.; Lindemann, C.; Drechsel, V.; Marohn, L.; Wysujack, K.; Santidrian, E.; Dirks, R.; Hanel, R.; Pelster, B. Swimming under elevated hydrostatic pressure increases glycolytic activity in gas gland cells of the European eel. *PLoS ONE* **2020**, *15*, e0239627. [CrossRef]
89. Palstra, A.P.; Heppener, D.; van Ginneken, V.; Székely, C.; van den Thillart, G. Swimming performance of silver eels is severely impaired by the swim-bladder parasite *Anguillicola crassus*. *J. Exp. Mar. Bio. Ecol.* **2007**, *352*, 244–256. [CrossRef]
90. Sprengel, G.; Luchtenberg, H. Infection by endoparasites reduces maximum swimming speed of European smelt *Osmerus eperlanus* and European eel *Anguilla anguilla*. *Dis. Aquat. Organ.* **1991**, *11*, 31–35. [CrossRef]
91. Newbold, L.R.; Hockley, F.A.; Williams, C.; Cable, J.; Reading, A.; Auchterlonie, N.; Kemp, P. Relationship between European eel *Anguilla anguilla* infection with non-native parasites and swimming behaviour on encountering accelerating flow. *J. Fish Biol.* **2015**, *86*, 1519–1533. [CrossRef] [PubMed]
92. Simon, J.; Westerberg, H.; Righton, D.; Sjöberg, N.; Dorow, M. Diving activity of migrating silver eel with and without *Anguillicola crassus* infection. *J. Appl. Ichthyol.* **2018**, *34*, 659–668. [CrossRef]

fishes MDPI

Review

Side Effects of Human Drug Use: An Overview of the Consequences of Eels' Exposure to Cocaine

Luigi Rosati [1,2], Ivana Caputo [3,4], Lillà Lionetti [3,4], Mayana Karoline Fontes [5,6], Camilo Dias Seabra Pereira [6] and Anna Capaldo [1,2,*]

1 Department of Biology, University of Naples Federico II, Via Cinthia Edifice 7, 80126 Naples, Italy
2 Center for Studies on Bioinspired Agro-Environmental Technology (BAT Center), 80055 Portici, Italy
3 Department of Chemistry and Biology "A. Zambelli", University of Salerno, Via Giovanni Paolo II 132, 84084 Fisciano, Italy
4 ELFID (European Laboratory for Food-Induced Diseases), University of Naples Federico II, Via Pansini 5, 80131 Naples, Italy
5 Institute of Biosciences, Litoral Paulista Campus, São Paulo State University "Júlio de Mesquita Filho", Praça Infante Dom Henrique s/n, São Vicente 11330-900, Brazil
6 Department of Marine Sciences, Federal University of São Paulo, Rua Maria Máximo 168, Santos 11030-100, Brazil
* Correspondence: anna.capaldo@unina.it or nevotta16@libero.it

Abstract: The widespread use of drugs is a global problem which affects not only humans but also the environment around them, as research is showing the presence of these substances in different environmental matrices, like air, water, and soil. Above all, due to the remarkable pharmacological properties of drugs, it is discovered that organisms accidentally exposed to them, as aquatic organisms, undergo behavioral and physiological changes that can compromise their health, survival, and reproduction ability. In addition to this, we must consider the ability of some drugs to accumulate within these organisms, thus entering the food chain, and the possible interactions that drugs in water can establish with each other and with other possible pollutants, making the final effects on exposed organisms unpredictable. This article is an overview of the effects of one of these drugs, cocaine, one of the drugs commonly found in the aquatic environment, on European eel, an endangered species and known biomonitor of aquatic contamination.

Keywords: *Anguilla anguilla*; cocaine; endocrine system; histopathological changes; illicit drugs; neuroendocrine effects; water pollution

Citation: Rosati, L.; Caputo, I.; Lionetti, L.; Fontes, M.K.; Pereira, C.D.S.; Capaldo, A. Side Effects of Human Drug Use: An Overview of the Consequences of Eels' Exposure to Cocaine. *Fishes* **2023**, *8*, 166. https://doi.org/10.3390/fishes8030166

Academic Editor: Jose Martin Pujolar

Received: 27 February 2023
Revised: 14 March 2023
Accepted: 15 March 2023
Published: 17 March 2023

1. Introduction

Illicit drug use is a growing phenomenon, which, as numerous studies have shown, poses a threat not only to human health but also to the environment. In 2020, an estimated 284 million people worldwide, aged 15–64, had used a drug of abuse within the last 12 months; of these users, approximately 13.6 per cent are estimated to suffer from drug use disorders [1]. At the same time, the environmental impact of the drugs of abuse appears increasingly evident; cultivation and production of plant-based and synthetic drugs, and drug use, have many consequences like energy use, deforestation, soil and water pollution and depletion, air pollution, food chain effects and biodiversity loss. Although the global effects of the activities related to the cultivation and production of illicit drugs are less significant than those of the pharmaceutical industry and agriculture, they can be important at local or community level [2].

Cocaine, for example, is estimated to have been consumed at least once by about 21.5 million people (0.4 percent of the global population aged 15–64) in 2020 [1]. It is estimated that 1982 tons of pure cocaine were produced in 2020, an increase of 11% over the previous year. The carbon footprint (a measure that expresses the total emissions of greenhouse gases, represented as carbon equivalents) of cocaine, related to cultivation of coca plants, processing of cocaine, disposal of waste generated in the manufacturing process and land-use change, is 4500 kg CO_2e per kg of cocaine produced. Therefore, referring to the 2020 data, we obtain a mean value of the total emissions per year of 1.9 million tons of CO_2e, a value significantly higher than that of other crops, as sugar cane or cocoa beans [2].

Another important environmental effect of drugs of abuse, as well as human and veterinary medicines, is their ability to contaminate the aquatic environment and influence the organisms living there [3–7]. Indeed, the intake of drugs of abuse is followed by their metabolization, usually partial, and the excretion of the parent drug and its metabolites especially by the renal way, that is the primary way in which these substances are removed from the body, though they can also be excreted through stool, sweat, tears and saliva [8]. Moreover, another possible source of contamination is the occasional discharge of drugs of abuse by clandestine laboratory wastes into sewage systems [3]. Therefore, illicit drugs and their metabolites reach treatment plants that do not always remove these substances, mainly because they are not always designed to do so. As a result, both drugs of abuse and their metabolites can be found in treated wastewater effluents and surface water, groundwater and drinking water [9–14]. Although the concentrations of drugs of abuse and their metabolites, found in surface waters around the world, are rather low (between $ng \cdot L^{-1}$ and $\mu g \cdot L^{-1}$), the continuous exposure and the remarkable pharmacological properties of drugs of abuse and their metabolites, sometimes greater than the parent drug, raise concerns about the fate of aquatic species living in contaminated sites. Indeed, scientific evidence is increasingly showing that different animal and plant aquatic organisms can bioaccumulate the different drugs of abuse and their metabolites, often suffering toxic effects [4–7]. In this paper, particular consideration will be given to cocaine present in the aquatic environment, and its effects on aquatic fauna and especially on European eels.

2. Cocaine

In nature, cocaine is present as an alkaloid in plants belonging to the Erythroxylaceae family, and in greater quantities in *Eritroxylum coca* and *Eritroxylum novogranatense*, two shrubs that grow spontaneously in South America. Cocaine can exist in several formulations; the predominantly used, and historically best known, cocaine formulation is the hydrochloride form. It is highly soluble in water and taken mainly intranasally (snorted), but may also be injected (subcutaneously, intramuscularly, or intravenously) and swallowed. The intranasal route is the most frequently used by regular users and ensures a quick onset of effects due to the high vascularization of the nasal mucosa. The subcutaneous and intramuscular pathway, due to the vasoconstrictor effect, involve slower absorption and therefore the effects are less rapid than the intravenous route. Intravenous bioavailability is 100%. Another widespread form is cocaine free base (basic form of cocaine hydrochloride), which is the transformation of water-soluble cocaine into the alkaloid base. It is mostly smoked but can also be injected; the effects appear within 5–10 s giving a very short "high" but very intense. Given the remarkable absorption surface of pulmonary alveoli, the inhalation route ensures the absorption of doses particularly high in the short term, which may explain the danger of cocaine in this form. It is relatively ineffective when administered intranasally or in a vein. Finally, "crack" cocaine is the name given to cocaine crystals obtained by processing cocaine to turn it into a substance that you can smoke. It's mostly smoked or injected, but it can also be ingested; its effects are like freebase [8]. Cocaine production is mainly concentrated in the Andean area of South America with the three largest producing countries (Colombia, Peru, and Bolivia), as *Erythroxylon coca* grows spontaneously in the warm and humid tropical climates of South America at an altitude of between 700 and 2000 metres. The neighbouring countries (Brazil, Venezuela, Argentina, and the Caribbean) play an important role as storage areas and transit zones for exports to Europe and the United States of America [2].

Cocaine belongs to the class of psychomotor stimulants, a group of drugs including nicotine, caffeine, and related substances, also known as minor stimulants, and amphetamines, cocaine, methylphenidate, and several dopamine agonists, also known as major stimulants. These stimulants, at low to moderate doses, increase vigilance, resistance to sleep and alertness; within a certain dose range, they also increase the performance in different tasks and locomotor activity, due to their sympathomimetic effects, as increased blood pressure and heart rate, constriction of blood vessels to the viscera, increased body temperature and muscle tension, and bronchial dilation. However, also direct adverse consequences on the body can be observed, as in the case of cocaine, which can cause neurological and cardiovascular problems like cerebral hemorrhage, convulsion induction, lethal cardiac arrhythmias, and acute myocardial infarction, even at low doses [8].

The main effect of cocaine on the nervous system stems from its ability to bind to monoamine transporters on nerve endings: serotonin transporter (SERT), dopamine transporter (DAT) and norepinephrine transporter (NET). This binding inhibits the reuptake of monoamines from the synaptic cleft, where the concentration of monoamines, and the consequent stimulation of the respective receptors, is increased. Although the effect of cocaine is exerted on all monoamines, the reinforcing properties of cocaine are mainly due to its effects on dopamine (DA), the extracellular levels of which increase dramatically following chronic cocaine exposure [15]. This increase, however, is followed by a rapid return to normal DA levels, for the activation of homeostatic mechanisms such as the activation of DA autoreceptors, which causes the user to increase the dose of cocaine administered. In time, the users come to the depletion of intraneuronal DA, due to different mechanisms: reduced DA synthesis via autoreceptor activity; enhanced extraneuronal catabolization via catecholamine o-methyl transferase (COMT) enzyme, or intraneuronal catabolization via monoamine oxidase (MAO) enzyme; decreased postsynaptic DA receptor availability due to excessive and prolonged exposure to DA. All these processes result in decreased dopaminergic activity with chronic cocaine use and contribute to cocaine dependence [8].

While the rewarding effects of cocaine are basically linked to DAT transporters, cocaine toxicity is especially related to both DAT and SERT transporters. In addition, cocaine is also able to bind to sigma and muscarinic acetylcholine receptors. Sigma receptors can be divided into two subtypes: sigma-1 and sigma-2; these receptors contribute to the toxicity of cocaine although it is not known which of the two subtypes is involved. As regards muscarinic receptors, cocaine probably behaves as a partial agonist, but the exact mechanism of action is not known. Cocaine can also bind to different voltage-gated ion channels, a binding responsible for the properties of local anesthetic of cocaine and for cocaine cardiotoxicity. Moreover, cocaine can bind, in a pH dependent manner, to two serum proteins, involved in cocaine toxicity: albumin and alpha 1 acid glycoprotein; the extent of binding to these proteins affects the concentration of cocaine free to exert its effects. Although the toxic effects of cocaine are mainly related to its binding with monoamine transporters, sigma and muscarinic acetylcholine receptors and cardiac ion channels, it is believed that two other neurotransmitters, gamma-aminobutyric acid (GABA) and glutamate (GLU), the most common inhibitory and excitatory neurotransmitters, respectively, may be involved in them [15].

After intake, cocaine is distributed throughout the body and is rapidly metabolized by different enzymatic pathways and non-enzymatic hydrolysis, also according to route of administration [16,17] in several metabolites, as benzoylcgonine, ecgonine methyl ester and norcocaine, the latter responsible for the toxic effects on the liver. A small amount of cocaine is excreted intact in the urine; the half-life of cocaine is between 30 and 90 min, whereas that of its metabolites is longer.

Following the excretion of cocaine and its metabolites with the urine, and occasional discharge of drugs of abuse by clandestine laboratory wastes into sewage systems [3], cocaine and its metabolites reach fresh and marine surface waters, where cocaine can be found in concentrations ranging from about 0.13 ng L^{-1} to 5896 ng L^{-1} in freshwaters [18–20] and from 2.4 ng L^{-1} to 537 ng L^{-1} in coastal zones [10,21]. The pharmacological characteristics of cocaine, and continuous exposure to it, make this substance potentially harmful to fish living in contaminated waters, such as eels.

3. European Eel

The European eel (*Anguilla anguilla*) is a migratory, catadromous fish species, consisting of a single stock, distributed throughout the European continent as well as in the Mediterranean basin, which reproduces in the Atlantic Ocean and for which the panmixia hypothesis is currently accepted. Eel fishing is carried out throughout the distribution area of the species and concerns the juvenile and preadult stages, but the conservation of the stock depends on the recruitment and emigration of breeding animals at sea. Eel farming is practiced in many countries, for a European level of around 8000 tons; the production completely depends on the wild seed, since the reproduction artificial, although implemented experimentally, does not go beyond the larval stage, at least in the European eel. Adult eels can survive in both air and water thanks to the fact that respiratory exchanges can occur both through the gills and through the skin. Eels are present in a wide range of aquatic habitats (rivers, canals, estuaries, lakes, ponds, and lagoons), in relation to their great adaptability to different environmental conditions; moreover, being a euryhaline species, eels can adapt to both fresh and sea water, and well tolerate variations in oxygen concentrations. As a primarily bottom-dwelling fish, the eels rely in their feeding on the prey population that is present there and prefer to eat at night [22,23].

The biological cycle of the eel is considered unique in relation to the nature and extent of reproductive migration. This life cycle takes place mainly in continental fresh waters, for most of the life of the eel, and in open ocean during the reproductive phase. Moreover, the life cycle involves a succession of metamorphic stages: leptocephalus larvae, hatching from eggs in the Sargasso Sea and heading for continental waters; glass eels that move into continental waters; yellow eels that remain in continental waters until sexual maturity is reached; silver eels that migrate back to reproductive areas [22,23].

This complex life cycle exposes the eels to different and numerous stressors, and many studies have shown that overfishing, habitat loss, pest attack and water contamination pose a serious threat to eel survival. In particular, the presence of fat in the eels, their long stay in the same area and their ability to accumulate contaminants especially lipophilic, make this species particularly susceptible to aquatic contamination. Indeed, to date, the European eel is included, according to IUCN, among the endangered species and considered at risk of extinction [22,24–26]. In addition to the different types of contaminants present in the water, illicit drugs are proving to be a new, widespread class of contaminants with remarkable pharmacological properties, that are likely to have effects on aquatic fauna [3–7].

Therefore, our research group has begun to study the effects of one of these illicit drugs, cocaine, on European eel, to understand how this substance could affect the physiology and reproductive capacity of this species. The experiments were carried out on silver eels, which, after a month's acclimatization, were exposed to an environmental concentration of cocaine (20 ng/L^{-1}), among those measured in surface water. The eels were housed in aquariums with a capacity of 300 L, in dechlorinated and well-aerated tap water, exposed to natural photoperiod and not fed, as in the silver stage eels do not normally feed. Moreover, the following parameters of the water were established and monitored: dissolved oxygen 8.1 $\pm$ 0.5 mg/L; salinity 0, temperature 15 $\pm$ 1 °C, pH 7.3 $\pm$ 0.2, ammonia $<$ 0.1 mg/L. For the exposure, a stock solution of 3 mg/500 mL of cocaine free-base in ethanol was prepared and kept in the refrigerator. During the exposure, 1 mL of the stock solution was administered every day, after the change of the water, directly in each aquarium for 30–50 days. At the same time control groups, exposed to tap water only, and carrier groups, exposed to ethanol only, in the same concentrations as eels receiving cocaine, were set up. At the end of exposure, some of the exposed eels were deprived of cocaine and exposed to tap water only, for 3–10 days, to verify their recovery ability. At the end of the exposure, or the recovery period, histological, histochemical, biochemical, and molecular biology analyses were performed [27]. All these studies were carried out in accordance with EU

Directive 2010/63/EU for animal experimentation and institutional guidelines for care and use of laboratory animals and were authorized by the Italian Ministry of Health's General Directorate of Animal Health and Veterinary Drugs.

4. The Effects of Cocaine on the European Eel

4.1. Accumulation of Cocaine in Eel Tissues

Cocaine is able to accumulate in biological tissues. From early studies, which showed the presence of cocaine in tissues of victims of drug-related deaths [28] and in rats exposed to cocaine [29], there is increasing evidence that aquatic organisms can accumulate cocaine [6,30–32] and its metabolites, for example benzoylecgonine [33], present in the aquatic environment. Consistent with literature data, eels chronically exposed to cocaine accumulated this drug in almost all their tissues, albeit to a different degree, with highest concentrations in brain, muscle, liver and kidney; lower concentrations in digestive tract, gills and skin, and lowest concentrations in spleen and gonads. After a recovery period of three days, cocaine was still present, but at much lower concentrations, indicating that recovery is possible, but takes longer [34]. The differential accumulation of cocaine in different organs of the eel agrees with the tissue-specific accumulation of methamphetamine and ketamine in zebrafish (*Danio rerio*) [35], and antidepressants [36] in several species of fish. The presence of high levels of cocaine in the brain of the eel reflects the affinity of this drug for nervous tissue and agrees with the accumulation of antidepressants in brain of white suckers [37] and fathead minnows [38]. The presence of high levels of cocaine in muscle can be explained by the fat solubility of cocaine and its tendency to deposit in adipose tissue [29], of which the eel muscle is rich (35.06%) [39]. In contrast, the low methamphetamine and ketamine level found in the muscle of zebrafish may depend on the low lipid content typical of the lean fish group, to which zebrafish belongs [40]. Finally, liver and kidney are involved in drug metabolization, and this may explain the high level of cocaine accumulation found in eel. Digestive tract, skin and gills, the first interfaces exposed to the water and its pollutants, represent the main routes of entry of cocaine, since the eels during the exposure were not feed, and showed low levels of cocaine, whereas spleen and gonads had the lowest levels, a result in accordance with what was observed for other drugs [35,36] and perhaps related to a poor metabolization capacity.

4.2. Neuroendocrine Effects of Cocaine in the Eels

4.2.1. Nervous Tissue

It is well known that main effects of cocaine stem from its ability to bind to DAT transporters and increase DA concentration in the synaptic cleft [16]. In fish, DA plays a key role in nervous system physiology, regulating numerous activities as hypothalamic and pituitary functions, locomotor activity, thermoregulation, and food intake. Moreover, in eels, DA was found to be involved (1) in sexual maturation, since it inhibits the synthesis and release of gonadotropins, and gonadal development, and (2) in last steps of eel reproduction and in reproductive migration [41,42]. Therefore, a variation in the levels of cerebral DA could seriously affect the physiology of this species. Our results showed that eels chronically exposed to cocaine had increased brain DA levels, which increased even more in recovery specimens [43]. This observation agrees with the increases in brain dopamine content found in *Danio rerio* after 72 h withdrawal from repeated cocaine administration, probably due to overall decrease in the expression of mRNA for DA transporter [44], a hypothesis that could also be valid for eels.

Cocaine is also a sympathomimetic drug, that facilitates norepinephrine (NE) transmission, and produces autonomic effects that reflect increases in sympathetic activity such as increased heart rate or blood pressure or dilation of bronchioles in the lungs [8]. Many studies, indeed, showed that cocaine activates the sympathoadrenal system, increasing the levels of plasma catecholamines (CA): NE and epinephrine (E) produced by the adrenal chromaffin cells [45]. Consistent with this data, eels chronically exposed to cocaine, and even more recovery eels, showed elevated plasma levels of DA, NE, and E [43]. It is well known that, in fish, physiological functions during stress are mainly regulated by the humoral release of CA, which also perform key functions such as increase in glucose synthesis and lipid metabolization, control of breathing, gas transfer and cardiovascular functions [46]. Therefore, it is reasonable to expect that these functions may be altered by increased levels of circulating catecholamines.

4.2.2. Endocrine System

While there is a great deal of information on the effects of cocaine on the nervous system, the available data on the effects of cocaine on the endocrine system are not numerous and relate mainly to research in humans and mammals, as rats and monkeys. Most studies concern the hypothalamus-pituitary-adrenal (HPA) axis, which seems to be the main target of the action of cocaine; in our study, also the effects of cocaine on hypothalamus-pituitary-thyroid (HPT) axis, prolactin (PRL), and gonadotropins: follicle-stimulating hormone (FSH) and luteinizing hormone (LH), were studied.

- HPA axis

The HPA axis, also called stress axis, because involved in the stress response, is activated by the release of corticotropin-releasing-hormone (CRH). CRH is produced in one of the numerous nuclei of the hypothalamus, the paraventricular nucleus (PVN), although it is also present in other brain regions. CRH is released into the portal circulation of the median eminence, from which it reaches the pituitary corticotropic cells and induces the production of a polypeptide precursor, pro-opiomelanocortin (POMC). However, CRH is also a stimulator of growth hormone (GH) release and an inhibitor of LH release. The corticotropic cells of the pars distalis cut POMC into several end products as adrenocorticotropic hormone (ACTH), melanocyte-stimulating hormone (α-, β-, and γ-MSH), β-lipotropin (LPH), β-endorphin and enkephalins, endogenous opioid peptides. However, POMC and related peptides, have been found also in many peripheral tissues, as skin and osteoarticular system [47]. ACTH, in turn, stimulates the fasciculata and reticularis areas of the adrenal cortex to release glucocorticoid hormones, basically cortisol in primates, along with a small amount of corticosterone, and corticosterone in rats. These hormones affect energy metabolism (1) inhibiting glucose utilization by peripheral tissues; (2) stimulating the liver to convert amino acids into glucose, and to store glucose in glycogen; (3) increasing mobilization of fat stores in non-neural tissues [48]. In turn, glucocorticoids regulate the activity of both corticotropic cells as well as PVN and other brain areas through a negative feed-back mechanism. In the brain, glucocorticoids bind to glucocorticoid receptors (GRs) and mineralocorticoid receptors (MRs), exerting effects through genomic and non-genomic mechanisms, that ultimately modify the behaviour and physiology of the organism [49].

Also, in fish there is a hypothalamus-pituitary-interrenal (HPI) axis, diversified between the different groups, and with anatomical and functional differences compared to mammals. Hypothalamic control of pituitary functions is mainly based on two regions that contain most neurosecretory neurons, the preoptic area (POA) and the hypothalamus proper, which in teleosts provide extensive innervation to adenohypophysis. Since in most teleosts there is a lack of a true median eminence and a portal system, the control of pituitary hormones is neuroglandular, by direct aminergic or peptidergic innervation [48]. In the Japanese eel, it has been observed that CRH has a high homology with that of mammals, and that CRH-immunoreactive (CRH-ir) fibers are very close and sometimes in contact with gonadotropin-releasing hormone immunoreactive (GnRH-ir) neurons, suggesting that CRH can regulate the activity of GnRH neurons [50]. In European eel, three crh paralogs genes were found (crh1a, crh1b, crh2), of which crh1b, mainly expressed in the brain, is considered the main regulator of corticotropic axis, whereas crh1a is mainly expressed in peripheral tissues as muscle and heart, and crh2 is weakly expressed in brain and peripheral tissues [51]. As in mammals, following CRH stimulation, in pituitary corticotropic cells, ACTH is obtained from a POMC precursor; ACTH, in turn, stimulates in teleosts the adrenocortical cells, present in the head kidney, to produce and release corticosteroids, mainly cortisol, and in lesser quantities corticosterone, aldosterone and some others. Cortisol is the major corticosteroid, having many roles as regulation of sodium fluxes and gluconeogenesis, and response to stress [48].

Many bibliographical data indicate that cocaine affects HPA axis increasing the levels of plasma ACTH and cortisol/corticosterone in man, non-human primates and rodents, [52] and it is considered that also oxytocin (OXY) and arginine-vasopressin (AVP), together with CRH, are also involved in cocaine-induced ACTH release [53]. In eels, after a chronic exposure to cocaine, decreased levels of ACTH and corticosterone and high levels of cortisol were observed. After discontinuation of cocaine exposure, ACTH levels returned to normal, while corticosterone levels remained low and cortisol levels continued to be high. The increase in cortisol, induced by cocaine, is consistent with what has been observed in humans and mammals, whereas the decrease in ACTH levels could be explained with a negative feedback mechanism [43]. In fish, an increase in cortisol plasma levels is considered a marker of HPI axis activation and, therefore, of stress in the animals [54]. In the eels, the reproductive migration and gonadal maturation require the presence of adequate amounts of lipids as an energy source. Therefore, the increase in cortisol induced by cocaine could induce an excessive lipid use and pre-migratory energy deficit, with the possible consequence of delaying or preventing or blocking reproductive migration, as has already been suggested for many contaminants [55,56].

- HPT axis

The HPT axis is activated by the release of thyrotropin-releasing-hormone (TRH), a hypothalamic tripeptide hormone, produced by many nuclei but mainly in the PVN, and present also in other brain areas. TRH release is stimulated by NE secreting neurons and inhibited by DA secreting neurons, which both innervate the PVN. TRH is released into the portal circulation of the median eminence and stimulates pituitary thyrotropic cells of the adenohypophysis to release thyroid-stimulating hormone (TSH). It should be noted that while TRH stimulates TSH secretion, DA, and somatostatin (SST) inhibit it. TSH, in turn, stimulates the thyroid gland to synthesize and release the thyroid hormones, triiodothyronine (T_3) and tetraiodothyronine or thyroxine (T_4) and to increase iodide uptake by the thyroid cells. Normally, the circulating levels of T_4 are higher than levels of T_3, but T_3 has a higher biological activity, and most of T_3 is derived from peripheral T_4 deiodination. The thyroid hormones influence numerous processes as general metabolism, growth, differentiation, and reproduction; moreover, they regulate the synthesis of TRH and TSH through a mechanism of negative feedback [48].

Also in fish, an HPT axis exists, but with several differences than mammals, and between different groups of fish. For example, TRH in many species does not stimulate TSH release, and often the stimulating action of CRH on TSH release exceeds that of TRH. Moreover, TRH can induce prolactin (PRL), ACTH, α-MSH and growth hormone (GH) release. Thyroid hormones are T_3, the active form, and T_4, and in teleosts, they are essential for development, growth, and metamorphosis, and are also involved in osmoregulation, whereas their relationship with reproductive cycles may vary in different species. Moreover, thyroid hormones influence nutrient metabolism of carbohydrates, lipids, and proteins: in *A. anguilla* at glass stage, thyroid hormones decrease body protein content, and increase plasma glucose levels [57]. Unlike mammals, where thyroid hormones regulate the activity of the HPT axis through a negative feed-back on TRH and TSH release, it does not appear that in fish thyroid hormones can inhibit the release of TRH, but only of TSH at the pituitary level, as in the case of European eel [48,58]. Finally, numerous evidence shows that there is a close interaction between the HPI axis and the HPT axis in fish, and that pituitary and hypothalamic hormones of HPI and HPT axes may interact with each other, regulating the effects of thyroid hormones and cortisol. This evidence suggests that thyroid hormones, in addition to classical hormones, cortisol and adrenaline, also play an important role in responding to stress in fish [59]. Finally, in the European eel, thyroid hormones inhibit GH synthesis and release, acting directly at pituitary level [60].

There is little information about the effects of cocaine on the PT axis, which in rats [61] and humans [62] is not affected by chronic or acute cocaine administration; in humans recent data indicate that cocaine use should be considered as a trigger for thyroid storm [63]. In our eels chronically exposed to cocaine, an increase in TSH levels and a decrease in T_3 levels were found, whereas T_4 levels appeared unchanged. After discontinuation of cocaine exposure, TSH values increased to normal levels, T_3 values increased but remained lower than control levels, and T_4 levels increased [43]. In the European eel, cortisol suppresses plasma thyroid hormone concentrations and increases plasma T_3, but not T_4 clearance [64], and in brook trout, cortisol stimulates the hepatic conversion of T_4 to T_3 [65]. In our experiment, increasing plasma cortisol levels, induced by cocaine exposure, is likely to lead to a reduction in T_3 levels, which in turn would increase TSH levels. We did not find a reduction in T_4 hormone levels, but the difference could be due to different exposure (chronic versus acute) and/or different dosage, or also to an increase of conversion of T_4 to T_3, that may be due to chronic, not acute, cocaine exposure.

- Prolactin

Prolactin (PRL) is produced in pituitary lactotrophs cells; the protein has a single chain of 199 amino acids, having many isoforms. The main PRL effects are associated with reproduction, growth, osmoregulation and integument, and the action of PRL often requires the interaction with other hormones. For example, glucocorticoids potentiate PRL actions, whereas progesterone inhibit PRL effects on the mammary gland competing for glucocorticoid binding sites and/or blocking gene activation by glucocorticoids. In lactating mammals, a neuroendocrine reflex triggered by teat stimulation induces PRL secretion, but many other factors regulate PRL release. An inhibitory control is provided by DA produced by neurons located in the arcuate nucleus of the hypothalamus; the regulation of the secretion of PRL also happens indirectly through many factors that act on the neurons DA stimulating or inhibiting them. For example, serotonin (5-HT), NE, histamine (H), somatostatin (SST) and many others inhibit DA neurons, stimulating PRL release, whereas TRH, acetylcholine (ACh), OXY, vasoactive intestinal polypeptide (VIP), and many others stimulate DA neurons, inhibiting prolactin release. Moreover, PRL itself feeds back on hypothalamic DA neurons regulating its own secretion [48,66].

In teleost, prolactin controls osmotic regulation in freshwater, regulating ion and water fluxes in urinary bladder, kidney, skin, gills, and intestine. Moreover, PRL stimulates mucus secretion by skin, intestine, and gills [67] and is involved in epithelial cell proliferation, together with cortisol [68].

In our experiment, after chronic cocaine exposure, the plasma PRL levels of eels were increased, but gradually decreased after discontinuation of cocaine exposure, to become lower than the control values ten days after cessation of exposure [69]. The increase in eel PRL levels agrees with the increase in PRL levels, which has been observed in rhesus monkeys after cocaine exposure [70]. In the eels, DA is a powerful PRL inhibitor [71]; therefore, the increase in plasma PRL level observed, despite the increase in cocaine-induced DA levels, can be explained by a change in dopaminergic regulation of PRL, as hypothesized in rhesus monkey [70].

- Gonadotropins

The gonadotropins, LH and FSH, are glycoproteins synthetized by the pituitary gland under the control of hypothalamic factors GnRH and gonadotropin-inhibiting hormone (GnIH), as well as activins and inhibins, produced by the gonads. LH induces in both sexes the synthesis of androgen hormones, and gamete release; FSH regulates in both sexes gamete development, and mainly in females, the conversion of androgens into estrogens by inducing the enzyme P450 aromatase [48].

Also, in fish LH and FSH, together with cortisol, are involved in the regulation of reproduction, in the development and maturation of gonads. In teleost, gametogenesis and the synthesis of androgens and estrogens is regulated by FSH, whereas final gamete maturation and release is regulated by LH, that also induces the production of a progesterone-like hormone [48]. In many teleosts, GnRH stimulates gonadotropins synthesis, whereas DA acts as an inhibitor. DA may also inhibit in some species the early steps of gametogenesis and interact with GnRH in the control of puberty. In turn, sex steroids are part of a negative feedback circuit that affects DA synthesis and the expression of DA D_2 receptors, involved in DA inhibition of gonadotropins [72]. Moreover, in the European eel, cortisol stimulates LH synthesis [73].

In our experiment, after chronic cocaine exposure, eels had lower levels of LH and FSH, compared to control, not exposed eels [74]. These results are different from those obtained in humans [75], where gonadotropins increase, following acute cocaine administration, or in female rhesus monkeys, where increased LH levels in the presence of low basal E2 levels were found after cocaine administration [76]. Although the difference may be due to differences in the doses administered, the type of treatment (acute vs. chronic) or the different species considered, it is likely that the increase in DA levels had an important role. DA inhibits the release of gonadotropins, decreasing therefore FSH and LH levels. Cocaine also induced an increase in cortisol levels, that stimulates in the eel LH synthesis [76], but evidently the inhibitory effect of dopamine must have exceeded the stimulatory effect of cortisol.

4.3. *Histopathological Changes Induced by Cocaine*

4.3.1. Skin, Intestine, and Gills

In fish, skin, intestine, and gills are organs immediately exposed to water and its contaminants. In fact, they possess a thin epithelial layer that is characterized by a large surface area, which makes them particularly susceptible to aquatic contaminants [77]; moreover, skin, intestine and gills can accumulate cocaine [34]. The skin has a mechanical protection function and protects the animal from the invasion of chemical, physical or biological agents; moreover, skin also regulates osmotic exchanges between the body and the external environment. In the eels, the skin has a surface layer composed of stratified epithelium (epidermis) and a deep layer, of connective nature, separated by a thin basal lamina. The stratified epithelium has a deep stratum germinativum, a middle layer with many large club cells, marked by an apical vacuole, and a superficial layer with flattened epithelial cells and mucous cells (Figure 1). The club cells are involved in alarm reactions, as in their vacuoles are stored substances that are released into the water in dangerous situations and warn the conspecifics of the presence of a danger. The mucous cells are born in the germinative layer and migrate to the superficial one, where they release mucus, playing a key role in osmoregulation, defense, and social interactions. The turnover of mucous cells is continuous since they die when release mucus [23,78,79].

The intestine is involved in osmoregulation and feeding; in the eel, four layers can be observed: mucosa, submucosa, a muscular layer, and serosa. The mucosa has many irregular folds and consists of a simple cylindrical epithelium with caliciform mucous cells and absorbent enterocytes, and of a lamina propria of loose connective tissue. Under the submucosa, of connective nature, there are two muscular layers of smooth muscle, an inner, thick, circular layer, and an outer, thin, longitudinal layer. Finally, serosa has a connective nature too. However, the morphology of the alimentary tract varies in relation to the life cycle and metamorphic stage of eels, and at the silver stage, which precedes reproductive migration, when the eels stop feeding, the intestine shows signs of degeneration, as disappearance of folds, decrease in number of mucous cells, epithelial histolysis and thinning of the muscle layer (Figure 2) [23].

The gills play many roles in fish, as gas exchange, ionic and osmotic regulation, acid-base balance, and excretion of nitrogenous waste products. The gills are formed by a branchial filament delimited by a stratified epithelium in which flattened epithelial cells, mucous cells, chloride cells, neuroepithelial cells and undifferentiated cells are present. From each filament, perpendicular to it, originates a double row of secondary lamellae, which have externally flattened cells and, inside, undifferentiated cells. Below the epithelium are evident pillar cells, which separate the lamellar venous sinuses (Figure 3) [23].

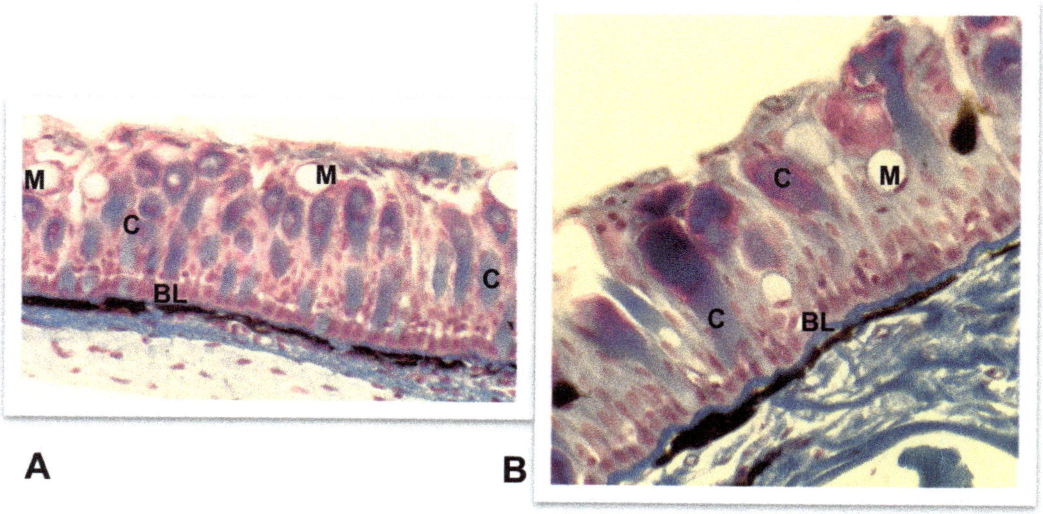

Figure 1. Light micrographs of the skin of *Anguilla anguilla*. Mallory staining. (**A**) Control and (**B**) exposed specimen. Cocaine exposure (**B**) caused the thickening and folding of the basal lamina (BL), the loss of apical vacuoles in club cells (C) and the decrease of mucous cells (M). Magnification: 400×.

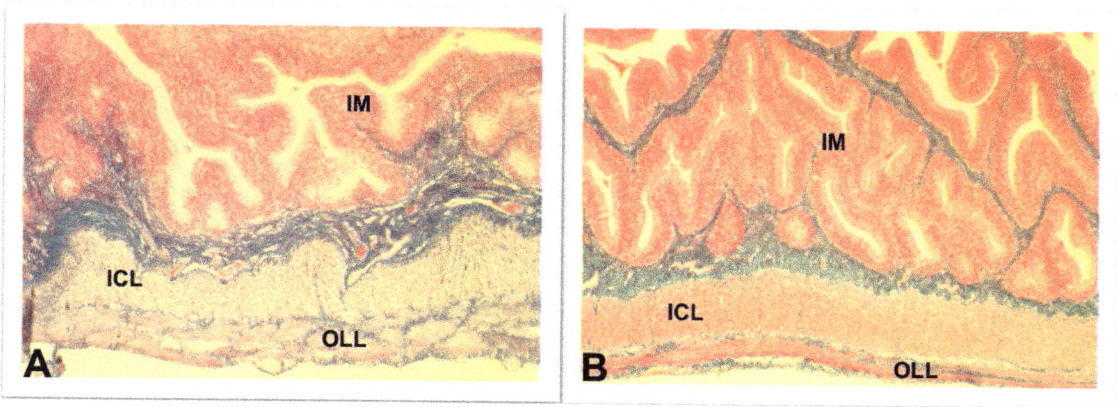

Figure 2. Light micrographs of the intestine of *Anguilla anguilla*. Mallory staining. (**A**) Control and (**B**) exposed specimens. In control specimens, the intestinal mucosa (IM) showed signs of histolysis, whereas the muscular layers, inner circular layer (ICL) and outer longitudinal layer (OLL), showed sign of degeneration. In cocaine exposed specimens, the intestinal mucosa (IM) was well organized and had many folds, and the muscular layers appeared thick and well stained. Magnification: 100×.

The chronic exposure to cocaine induced similar changes in the epithelial tissue of skin, intestine, and gills [69,80]. The epidermis appeared thickened, with fewer small mucous cells, and club cells devoid of their vacuole, to indicate a massive release phenomenon; moreover, the underlying basal lamina appeared thickened and folded (Figure 1). Ten days after the interruption of cocaine exposure, mucous and club cells recovered normal morphology, while the basal lamina and the thickness of the epidermis were still different from the normal situation.

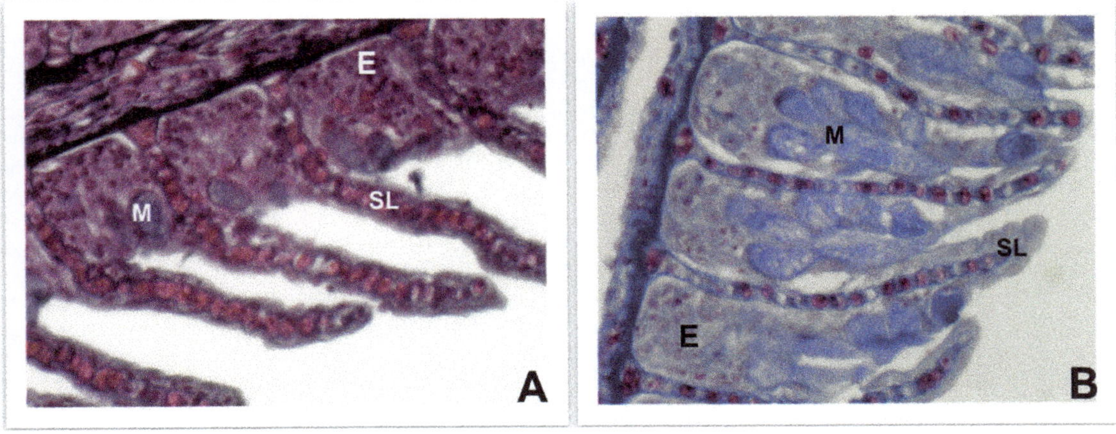

Figure 3. Light micrographs of the gills of *Anguilla anguilla*. Mallory staining. (**A**) Control and (**B**) exposed specimens. In exposed specimens, the epithelium (E) appeared thickened, compared to control, and contained many mucous (M) cells, also present in the secondary lamellae (SL), which appeared partially fused. Magnification: 400×.

The intestinal epithelium, that at silver stage was degenerating, appeared thickened and convoluted, with many mucous cells. Also, the muscular layers were well organized, with fibers increased in size (Figure 2), that, however, began to regress ten days after the end of exposure to cocaine, when instead the epithelium still appeared well organized.

Finally, the gills showed a hyperplasic epithelium in which the mucous cells had increased a lot and appeared not only along the branchial filament and the interlamellar epithelium, but also in the secondary lamellae (Figure 3). Moreover, also partial, and total fusion of secondary lamellae was found; however, ten days after the end of exposure to cocaine, the appearance of the epithelium returned to normal [80]. Overall, the changes observed in skin, intestine, and gills, such as hyperplasia of the epithelium, the increase in the number of mucous cells, and the fusion of the secondary lamellae in the gills, are considered progressive changes, that suggest a defensive reaction against cocaine, and an adaptation of the eels to it [81]. These results differ from those obtained in humans and other species; for example, in humans and rats a systemic cocaine exposure causes skin oxidation, vasculitis, infectious complications and many other dermatologic conditions [82], and, in humans, gastrointestinal complications were seen following cocaine administration [83].

Finally, as regards the gills, data are available only in invertebrates where cocaine, at environmental concentrations, did not alter the oxidative status in the gills of *Mytilus galloprovincialis* [84] whereas, in *Perna perna*, mitochondrial DNA damage was seen in the gills, despite their detoxifying and antioxidant activity [85]. These differences may reflect differences in dosage and/or type of treatment, and species-specific sensitivity to cocaine; although the exact mechanism of action of cocaine on the skin, intestines and gills of eel needs to be clarified, it is likely that the effects of cocaine involve hormonal changes in cortisol and prolactin. Indeed, in fish the epithelial cell proliferation is controlled by cortisol and PRL; the latter also controls mucus secretion by skin, intestine and gills, differentiation, and proliferation of mucous cells [68] and proliferation of vascular smooth muscle cells [86]. Therefore, both hormones could be involved in the observed changes, although a direct action of cocaine cannot be ruled out.

4.3.2. Skeletal Muscle

The skeletal muscle of European eel is composed by red and white fibers, the first located along the lateral line, the second forming the greatest volume of the body tissue. The diameter is greater in white fibers than in red ones, both fibers are surrounded by a reticular connective tissue and have myofibrils regularly aligned and parallel each other in the sarcoplasm [23]. Following a chronic exposure to cocaine [27], the skeletal muscle had a significant cocaine content [34]. In addition, cocaine significantly damaged muscle fibers, causing lacerations and transverse ruptures (Figure 4), that were still present ten days after the cessation of cocaine exposure. Moreover, an increase in the mean diameter of fibers was observed, mainly in the red fibers, that appeared more damaged, showing signs of swelling, rarefaction of myofibrils and disorganization of the contractile apparatus The morphological changes were accompanied by increases (1) in the activity of cytochrome oxidase (COX), marker of oxidative metabolism and caspase 3, marker of apoptosis activation; (2) in the serum levels of creatine kinase (CK), lactate dehydrogenase (LDH) and aspartate aminotransferase (AST), biomarkers of skeletal muscle damage [87]. All these parameters were still altered ten days after the cessation of cocaine exposure. Cocaine is known to induce rhabdomyolysis, a syndrome characterized by the breakdown of muscle fibers and the release into the blood of cytosolic components, such as enzymes, electrolytes, and myoglobin. It is supposed that cocaine either acts directly on the muscle or through repeated ischemia caused by the vasoconstrictor activity of this drug. Subsequent reperfusion and generation of free radicals would damage sarcolemma, leading to the increase of cytosolic enzyme release [88]. It's possible that a similar mechanism also works in eels, since the characteristics of the muscle observed, and the high circulating values of the enzymes measured suggest the presence of rhabdomyolysis, induced by cocaine exposure.

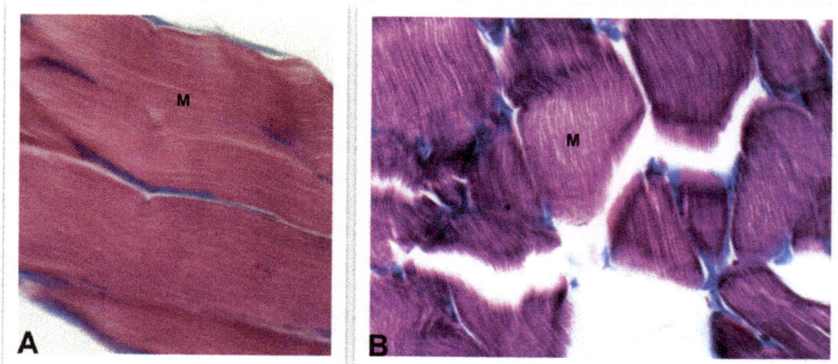

Figure 4. Light micrographs of the skeletal muscle of *Anguilla anguilla*. Mallory staining. (**A**) Control and (**B**) exposed specimens. In exposed specimens, the muscle fibres (M) showed less compactness and numerous breaks. Magnification: 400×.

4.3.3. Liver and Kidney

Liver and kidney are organs that play an important role in metabolization and excretion. In the eel, the liver parenchyma has anastomosing cords of polygonal hepatocytes, among which are present many sinusoids. At silver stage, the cytoplasm of hepatocytes contains large quantities of lipids (Figure 5).

Eel kidney has a pronephric portion with lymphoid and hematopoietic functions, and a posterior mesonephric kidney, with excretory and endocrine functions, where most nephrons are present. In the nephrons, a glomerulus within the Bowmans' capsule, proximal, distal, and collecting tubules are seen; moreover, kidney shows some melanomacrophage centers, containing melanin acting as scavenger for free radicals (Figure 6).

In both organs, cocaine induced numerous changes [89]. Liver and kidney showed signs of nuclear alterations, as kariolysis and pycnotic nuclei, and structural alterations, as necrotic areas, loss of cytoplasmic lipids and parenchymal cells in the liver; increase in the number of melanomacrophage centers (common in conditions of environmental stress), dilated renal tubules and reduced Bowman's space in the kidney. After the interruption to cocaine exposure, the liver showed signs of a gradual return to normal, while the kidney was still very altered. The morphological alterations of liver and kidney were accompanied by increases (1) in the activity of cytochrome oxidase (COX), marker of oxidative metabolism and caspase 3, marker of apoptosis activation, in both organs; (2) in liver glucose-regulated protein (GRP)78 expression, a well-known regulator of apoptosis and a marker of endoplasmic reticulum (ER)-stress [90]; (3) in the level of blood glucose, a classical marker of stress response, whose increase is typical of the response of fish to pollutants; (4) in the serum levels of alanine aminotransferase (ALT), biomarker of liver injury, and C-reactive protein (CRP), marker of inflammatory process. Many of these parameters were still altered after the interruption of cocaine exposure.

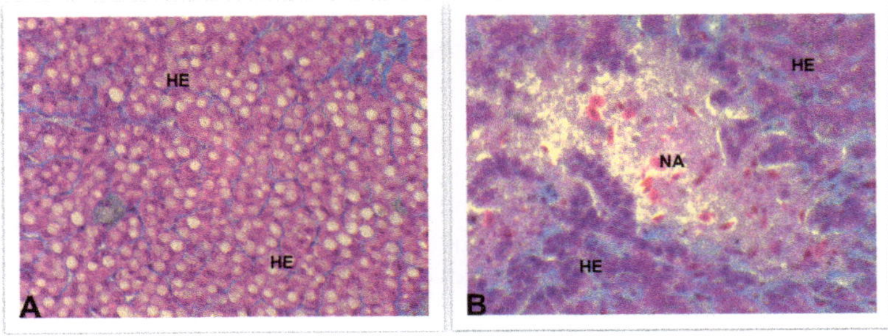

Figure 5. Light micrographs of the liver of *Anguilla anguilla*. Mallory staining. (**A**) Control and (**B**) exposed specimens. In exposed specimens, the hepatocytes (HE) were devoid of lipids; necrotic areas (NA) and loss of parenchymal cells were evident. Magnification: 400×.

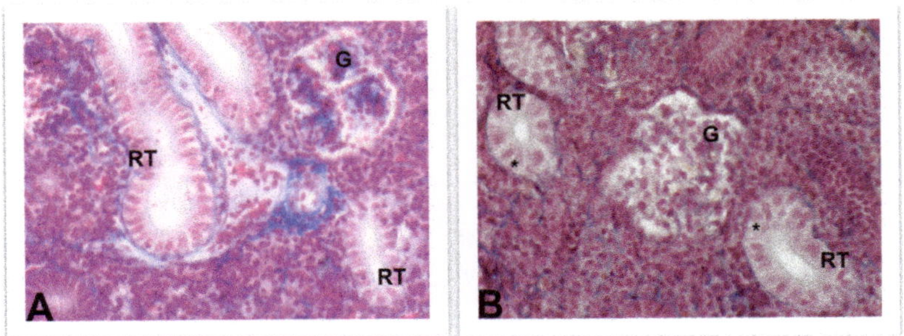

Figure 6. Light micrographs of the kidney of *Anguilla anguilla*. Mallory staining. (**A**) Control and (**B**) exposed specimens. In exposed specimens, the renal tubules (RT) showed nuclear alterations (*) and the structure of glomerulus (G) was disorganized. Magnification: 400×.

The damages observed in eel liver and kidney agree with the well-known hepatotoxicity and nephrotoxicity of cocaine when administered to humans, mammals, and fish [91]. Hepatotoxicity is believed to result from the production of oxygen reactive species (ROS) resulting from the metabolization of cocaine by hepatic esterases, while it is believed that nephrotoxicity may be partly associated with rhabdomyolysis, the release of large amounts of circulating myoglobin and its precipitation in the glomerular filtrate, resulting in damage to renal tubules. However, direct kidney damage caused by cocaine metabolization cannot be ruled out [92]. The increase in COX activity, in turn, might reflect the increase in specific mitochondrial enzyme activity and/or in mitochondrial protein mass, and could be related to an increased energy demand resulting from the metabolization activity of the liver and kidney. It has been observed that the hyperactivity of COX could activate the intrinsic pathway of apoptosis [93], and in fact, this hypothesis agrees with the increase of caspase 3 found in eels exposed to cocaine. Finally, the variations in all other parameters studied confirm that cocaine induced liver and kidney distress in eels.

4.3.4. Ovaries

In the eel, gonads lie along the entire length of the body cavity and the eggs come out of the abdominal pore since female gonads have no separate outlet. European eel is an asynchronous, determinate, batch spawner [94]; at the silver stage, in the ovaries, it was possible to observe previtellogenic (pvOos), early vitellogenic (evOos) and fully vitellogenic (fvOos) oocytes, the latter having yolk vesicles throughout the cytoplasm. The oocytes were surrounded by follicular cells and connective cells such as theca endocrine cells. The exposure to cocaine [74] caused several changes, as the prevalence in the ovary of pvOos and connective tissue, and the presence of small size follicles, compared to control animals, although the number of follicles was nearly identical in both control and exposed animals. Moreover, cocaine modified the localization and reduced the intensity of antibody-labeled signal of three key enzymes playing a key role in oogenesis: P450 aromatase, catalyzing the conversion of androgens into estrogens and regulating sexual differentiation [95]; 3-β-hydroxysteroid dehydrogenase (3 β-HSD), regulating the synthesis of progesterone [96]; 17 β-hydroxysteroid dehydrogenase (17 β-HSD), involved in estrogen synthesis and gametogenesis [97]. These results agree with the adverse effects of cocaine on reproduction, observed in rabbits [98], rhesus monkeys [99] and *Drosophila melanogaster* [100]. Cocaine may act directly on the gonads and/or its action may be mediated through decreased FSH and LH levels; however, the exact mechanism of cocaine action requires further study.

5. Conclusions

European eel is an endangered species whose survival is hampered by many different factors, such as overfishing, habitat loss, pest attack and water contamination. Our studies have shown that the presence of illicit drugs in the water, as cocaine, may also be a problem for this species; indeed, the alterations in nervous and endocrine systems, and in peripheral tissues, induced by cocaine, could decrease its ability to survive and its reproductive fitness. Moreover, the presence of cocaine in the muscle, which is the edible part of the animal, can be a problem not only for the eel, which needs a healthy muscle to complete reproductive migration, but also for human consumption of this fish. In addition, in water contaminated by drugs, these are present not individually, but as mixtures, often accompanied by other types of pollutants; all this could make the effects on aquatic fauna, of which the European eel is a representative, unpredictable. Our study was performed on a single metamorphic stage, the silver stage, and at a single temperature, 15 °C, suitable for adequate consumption of energy resources and proper development of the gonads, and it is possible that the results may be different under different conditions. Moreover, although the study has been conducted on eels, it is possible to imagine that similar damage can be induced by cocaine, and probably by other drugs present in waters, also on other species of fish. Therefore, it is of great importance that governments adopt a strategy of upgrading wastewater treatment plants on the one hand, and that they undertake an appropriate environmental remediation policy on the other hand, to preserve aquatic fauna; indeed, as our results showed, after the interruption of cocaine exposure, many vital parameters of eels improved, or otherwise showed a tendency to improve.

Author Contributions: Writing—original draft preparation, L.R. and M.K.F.; writing—review and editing, A.C.; visualization, L.R. and A.C.; supervision, I.C., L.L. and C.D.S.P.; funding acquisition, M.K.F. and C.D.S.P. All authors have read and agreed to the published version of the manuscript.

Funding: The experimental study was supported by departmental basic research funds of University of Naples Federico II (Italian Ministry of University and Research, MIUR). The study on the eels' gonads was funded by São Paulo Research Foundation (FAPESP #grants 2016/24033-3; 2019/20187-4); São Paulo Research Foundation (FAPESP #grant 2015/17329-0); Conselho Nacional de Desenvolvimento Científico e Tecnológico (CNPq #grant 309361/2019-2).

Institutional Review Board Statement: The experiments were carried out in accordance with EU Directive 2010/63/EU for animal experimentation and institutional guidelines for the care and use of laboratory animals and were authorized by General Direction of Animal Health and Veterinary Drugs of Italian Ministry of Health (Authorization No. 221/2015-PR and Authorization No. 22/2015-PR).

Data Availability Statement: The data presented in these studies are available on request from the corresponding author.

Conflicts of Interest: The authors declare that they have no known competing financial interests or personal relationships that could have appeared to influence the work reported in this paper.

References

1. UNODC. Book 2: Drug use and health consequences. In *World Drug Report*; (United Nations Publication, sales, No. E.22.XI.8); UNODC: Vienna, Austria, 2022.
2. UNODC. Book 5: Drug and the environment. In *World Drug Report*; (United Nations Publication, sales, No. E.22.XI.8); UNODC: Vienna, Austria, 2022.
3. Pal, R.; Megharaj, M.; Kirkbride, K.P.; Naidu, R. Illicit drugs and the environment—A review. *Sci. Total Environ.* **2013**, *463–464*, 1079–1092. [CrossRef] [PubMed]
4. Fontes, M.K.; Maranho, L.; Pereira, C.D.S. Review on the occurrence and biological effects of illicit drugs in aquatic ecosystems. *Environ. Sci. Pollut. Res.* **2020**, *27*, 30998–31034. [CrossRef]
5. Fontes, M.K.; Dourado, P.L.R.; de Campos, B.G.; Maranho, L.A.; de Almeida, E.A.; de Souza Abessa, D.M.; Pereira, C.D.S. Environmentally realistic concentrations of cocaine in seawater disturbed neuroendocrine parameters and energy status in the marine mussel *Perna perna*. *Comp. Biochem. Physiol. C* **2022**, *251*, 109198. [CrossRef]
6. Chen, L.; Guo, C.; Sun, Z.; Xu, J. Occurrence, bioaccumulation and toxicological effect of drugs of abuse in aquatic ecosystem: A review. *Environ. Res.* **2021**, *200*, 111362. [CrossRef]
7. Yedhu Krishnan, R.; Manikandan, S.; Subbaiya, R.; Biruntha, M.; Balachandar, R.; Karmegam, N. Origin, transport and ecological risk assessment of illicit drugs in the environment A review. *Chemosphere* **2023**, *311*, 137091. [CrossRef]
8. Grilly, M.G.; Salamone, J.D. *Drugs, Brain and Behavior*, 6th ed.; Pearson: Boston, MA, USA, 2012; pp. 132–178.
9. Kasprzyk-Hordern, B.; Dinsdale, R.M.; Guwy, A.J. The removal of pharmaceuticals, personal care products, endocrine disruptors and illicit drugs during wastewater treatment and its impact on the quality of receiving waters. *Water Res.* **2009**, *43*, 363–380. [CrossRef]
10. Seabra-Pereira, C.D.; Maranho, L.A.; Cortez, F.S.; Pusceddu, F.H.; Santos, A.R.; Ribeiro, D.A.; Cesar, A.; Guimarães, L.L. Occurrence of pharmaceuticals and cocaine in a Brazilian coastal zone. *Sci. Total Environ.* **2016**, *548*, 148–154. [CrossRef] [PubMed]
11. Davoli, E.; Zuccato, E.; Castiglioni, S. Illicit drugs in drinking water. *Curr. Opin. Environ. Sci. Health* **2019**, *7*, 92–97. [CrossRef]
12. Deng, Y.; Guo, C.; Zhang, H.; Yin, X.; Chen, L.; Wu, D.; Xu, J. Occurrence and removal of illicit drugs in different wastewater treatment plants with different treatment techniques. *Environ. Sci. Eur.* **2020**, *32*, 28. [CrossRef]
13. Davey, C.J.E.; Kraak, M.H.S.; Pratorius, A.; ter Laak, T.L.; van Wezel, A.P. Occurrence, hazard, and risk of psychopharmaceuticals and illicit drugs in European surface waters. *Water Res.* **2022**, *222*, 118878. [CrossRef]
14. Pisetta, A.M.; Roveri, V.; Lopes Guimarães, L.; de Oliveira, T.M.N.; Correia, A.T. First report on the occurrence of pharmaceuticals and cocaine in the coastal waters of Santa Catarina, Brazil, and its related ecological risk assessment. *Environ. Sci. Pollut. Res.* **2022**, *29*, 63099–63111. [CrossRef]
15. Heard, K.; Palmer, R.; Zahniser, N.R. Mechanisms of acute cocaine toxicity. *Open Pharmacol. J.* **2008**, *2*, 70–78. [CrossRef] [PubMed]
16. Volkov, N.D.; Wang, G.J.; Fischman, M.W.; Foltin, R.; Fowler, J.S.; Franceschi, D.; Franceschi, M.; Logan, J.; Gatley, S.J.; Wong, C.; et al. Effects of route administration on cocaine induced dopamine transporter blockade in the human brain. *Life Sci.* **2000**, *67*, 1507–1515. [CrossRef]
17. Cone, E.J.; Tsadik, A.; Oyler, J.; Darwin, W.D. Cocaine metabolism and urinary excretion after different routes of administration. *Ther. Drug Monit.* **1998**, *20*, 556–560. [CrossRef]
18. Zhang, Y.; Zhang, T.; Guo, C.; Lv, J.; Hua, Z.; Hou, S.; Zhang, Y.; Meng, W.; Xu, J. Drugs of abuse and their metabolites in the urban rivers of Beijing, China: Occurrence, distribution, and potential environmental risk. *Sci. Total Environ.* **2017**, *579*, 305–313. [CrossRef] [PubMed]
19. Zuccato, E.; Castiglioni, S.; Bagnati, R.; Chiabrando, C.; Grassi, P.; Fanelli, R. Illicit drugs, a novel group of environmental contaminants. *Water Res.* **2008**, *42*, 961–968. [CrossRef] [PubMed]
20. Thomas, K.V.; Silva, F.M.A.; Langford, K.H.; Souza, A.D.F.; Nizzeto, L.; Waichman, A.V. Screening for selected human pharmaceuticals and cocaine in the urban streams of Manaus, Amazonas, Brazil. *J. Am. Water Resour. Assoc.* **2014**, *50*, 302–308. [CrossRef]
21. Klosterhaus, S.L.; Grace, R.; Hamilton, M.C.; Yee, D. Method validation and reconnaissance of pharmaceuticals, personal care products, and alkylphenols in surface water, sediments, and mussels in an urban estuary. *Environ. Int.* **2013**, *54*, 92–99. [CrossRef] [PubMed]
22. Bevacqua, D.; Melià, P.; Gatto, M.; De Leo, G. A global viability assessment of the European eel. *Glob. Change Biol.* **2015**, *21*, 3323–3335. [CrossRef]

23. Tesch, S.W. *The Eel*, 3rd ed.; Blackwell Science, Ltd.: Oxford, UK, 2003; pp. 1–117.
24. Belpaire, C.; Goemans, G. The European eel *Anguilla anguilla*, a rapporteur of the chemical status for the water framework directive? *Vie et Milieu/Life Environ.* **2007**, *57*, 235–252.
25. Van Ginneken, V.; Palstra, A.; Leonards, P.; Nieveen, M.; van den Berg, H.; Flik, G.; Spanings, T.; Niemantsverdriet, P.; van der Thillart, G.; Murk, A. PCBs and the energy cost of migration in the European eel (*Anguilla anguilla* L.). *Aquat. Toxicol.* **2009**, *92*, 213–220. [CrossRef] [PubMed]
26. Bettinetti, R.; Galassi, S.; Quadroni, S.; Volta, P.; Capoccioni, F.; Ciccotti, E.; De Leo, G.A. Use of *Anguilla anguilla* for biomonitoring persisten organic pollutants (POPs) in brackish and riverine waters in central and southern Italy. *Water Air Soil Pollut.* **2011**, *217*, 321–331. [CrossRef]
27. Capaldo, A.; Gay, F.; Lepretti, M.; Paolella, G.; Martucciello, S.; Lionetti, L.; Caputo, I.; Laforgia, V. Effects of environmental cocaine concentrations on the skeletal muscle of the European eel (*Anguilla anguilla*). *Sci. Total Environ.* **2018**, *640–641*, 862–873. [CrossRef] [PubMed]
28. Poklis, A.; Maginn, D.; Barr, J.L. Tissue disposition of cocaine in man: A report of five fatal poisonings. *Forensic Sci. Int.* **1987**, *33*, 83–88. [CrossRef]
29. Nayak, P.K.; Misra, A.L.; Mulè, S.J. Physiological disposition and biotransformation of (3H) cocaine in acutely and chronically treated rats. *J. Pharmacol. Exp. Ther.* **1976**, *196*, 556–569. [PubMed]
30. Fontes, M.K.; de Campos, B.G.; Cortez, F.S.; Pusceddu, F.H.; Nobre, C.R.; Moreno, B.B.; Lebre, D.T.; Maranho, L.A.; Pereira, C.D.S. Mussels get higher: A study on the occurrence of cocaine and benzoylecgonine in seawater, sediment and mussels from a subtropical ecosystem (Santos Bay, Brazil). *Sci. Total Environ.* **2021**, *757*, 143808. [CrossRef] [PubMed]
31. Niklaus, S.; Kirla, K.T.; Kraemer, T.; Groh, K.; Schirmer, K.; Neuhauss, S. Cocaine accumulation in zebrafish eyes leads to augmented amplitudes in the electroretinogram. *Matters* **2017**, *3*, e201703000003. [CrossRef]
32. Miller, T.H.; Ng, K.T.; Bury, S.T.; Bury, S.E.; Bury, N.R.; Barron, L.P. Biomonitoring of pesticides, pharmaceuticals and illicit drugs in a freshwater invertebrate to estimate toxic or effect pressure. *Environ. Int.* **2019**, *129*, 595–606. [CrossRef]
33. Ondarza, P.M.; Haddad, S.P.; Avigliano, E.; Miglioranza, K.S.B.; Brooks, B.W. Pharmaceuticals, illicit drugs and their metabolites in fish from Argentina: Implications for protected areas influenced by urbanization. *Sci. Total Environ.* **2019**, *649*, 1029–1037. [CrossRef]
34. Capaldo, A.; Gay, F.; Maddaloni, M.; Valiante, S.; De Falco, M.; Lenzi, M.; Laforgia, V. Presence of cocaine in the tissues of the European eel, *Anguilla anguilla*, exposed to environmental cocaine concentrations. *Water Air Soil Pollut.* **2012**, *223*, 2137–2143. [CrossRef]
35. Yin, X.; Guo, C.; Deng, Y.; Jin, X.; Teng, Y.; Xu, J.; Wu, F. Tissue-specific accumulation, elimination, and toxicokinetics of illicit drugs in adult zebrafish (*Danio rerio*). *Sci. Total Environ.* **2021**, *792*, 148153. [CrossRef]
36. Arnok, P.; Singh, R.R.; Burakham, R.; Pérez-Fuentetaja, A.; Aga, D.S. Selective uptake and bioaccumulation of antidepressants in fish from effluent-impacted Niagara River. *Environ. Sci. Technol.* **2017**, *51*, 10652–10662. [CrossRef] [PubMed]
37. Schultz, M.M.; Furlong, E.T.; Kolpin, D.W.; Werner, S.L.; Schoenfuss, H.L.; Barber, L.B.; Blazer, V.S.; Norris, D.O.; Vajda, A.M. Antidepressant pharmaceuticals in two U.S. effluent-impacted streams: Occurrence and fate in water and sediment, and selective uptake in fish neural tissue. *Environ. Sci. Technol.* **2010**, *44*, 1918–1925. [CrossRef] [PubMed]
38. Schultz, M.M.; Painter, M.M.; Bartell, S.E.; Logue, A.; Furlong, E.T.; Wemer, S.L.; Schoenfuss, H.L. Selective uptake and biological consequences of environmentally relevant antidepressant pharmaceutical exposures on male fathead minnows. *Aquat. Toxicol.* **2011**, *104*, 38–47. [CrossRef] [PubMed]
39. Pyz-Lukasisk, R.; Chalabis-Mazurek, A.; Gondek, M. Basic and functional nutrients in the muscle of fish: A review. *Int. J. Food Prop.* **2020**, *23*, 1941–1950. [CrossRef]
40. Potter, G.; Smith, A.S.T.; Vo, N.T.K.; Muster, J.; Weston, W.; Bertero, A.; Maves, L.; Mack, D.L.; Rostain, A. A more open approach is needed to develop cell-based fish technology: It starts with zebrafish. *One Earth* **2020**, *3*, 54–64. [CrossRef]
41. Sébert, M.E.; Weltzien, F.A.; Moisan, C.; Pasqualini, C.; Dufour, S. Dopaminergic systems in the European eel: Characterization, brain distribution, and potential role in migration and reproduction. *Hydrobiologia* **2008**, *602*, 27–46. [CrossRef]
42. Dufour, S.; Sebert, M.E.; Weltzien, F.A.; Rousseau, K.; Pasqualini, C. Neuroendocrine control by dopamine of teleost reproduction. *J. Fish Biol.* **2010**, *76*, 129–160. [CrossRef]
43. Gay, F.; Maddaloni, M.; Valiante, S.; Laforgia, V.; Capaldo, A. Endocrine disruption in the European eel, *Anguilla anguilla*, exposed to an environmental cocaine concentration. *Water Air Soil Pollut.* **2013**, *224*, 1579. [CrossRef]
44. Lòpez-Patiño, M.A.; Yu, L.; Cabral, H.; Zhdanova, I.V. Anxiogenic effects of cocaine withdrawal in zebrafish. *Physiol. Behav.* **2008**, *93*, 160–171. [CrossRef]
45. Sofuoglu, M.; Nelson, D.; Babb, D.A.; Hatsukami, D.K. Intravenous cocaine increases plasma epinephrine and norepinephrine in humans. *Pharmacol. Biochem. Behav.* **2001**, *68*, 455–459. [CrossRef]
46. Perry, S.F.; Capaldo, A. The autonomic nervous system and chromaffin tissue: Neuroendocrine regulation of catecholamine secretion in non-mammalian vertebrates. *Auton. Neurosci.* **2011**, *165*, 54–66. [CrossRef]
47. Böhm, M.; Grässel, S. Role of proopiomelanocortin-derived peptides and their receptors in the osteoarticular system: From basic to translational research. *Endocr. Rev.* **2012**, *33*, 623–651. [CrossRef] [PubMed]
48. Norris, D.O. *Vertebrate Endocrinology*, 4th ed.; Academic Press: Amsterdam, The Netherlands, 2007.

49. Godoy, L.; Rossignoli, M.T.; Delfino-Pereira, P.; Garcia-Cairasco, N.; de Lima Umeoka, E.H. A comprehensive overview on stress neurobiology: Basic concepts and clinical implications. *Front. Behav. Neurosci.* **2018**, *12*, 127. [CrossRef]
50. Amano, M.; Mizusawa, N.; Okubo, K.; Amiya, N.; Mizusawa, K.; Chiba, H.; Yamamoto, N.; Takahashi, A. Cloning of corticotropin-releasing hormone (CRH) precursor cDNA and immunohistochemical detection of CRH peptide in the brain of the Japanese eel, paying special attention to gonadotropin-releasing hormone. *Cell Tissue Res.* **2014**, *356*, 243–251. [CrossRef] [PubMed]
51. Maugars, G.; Mauvois, X.; Martin, P.; Aroua, S.; Rousseau, K.; Dufour, S. New insights into the evolution of corticotropin-releasing hormone family with a special focus on teleosts. *Front. Endocrinol.* **2022**, *13*, 937218. [CrossRef]
52. Mello, N.K. Hormones, nicotine and cocaine: Clinical studies. *Horm. Behav.* **2010**, *58*, 57–71. [CrossRef] [PubMed]
53. Manetti, L.; Cavagnini, F.; Martino, E.; Ambrogio, A. Effects of cocaine on the hypothalamic-pituitary-adrenal axis. *J. Endocrinol. Investig.* **2014**, *37*, 701–708. [CrossRef] [PubMed]
54. Wendelar Bonga, S.E. The stress response in fish. *Physiol. Rev.* **1997**, *77*, 591–625. [CrossRef]
55. Dufour, S.; Burzawa-Gérard, E.; Le Belle, N.; Sbaihi, M.; Vidal, B. Reproductive endocrinology of the European eel, *Anguilla anguilla*. In *Eel Biology*; Aida, K., Tsukamoto, K., Yamauchi, K., Eds.; Springer: Tokyo, Japan, 2003; pp. 373–383.
56. Robinet, T.; Feunteun, E. Sublethal effects of exposure to chemical compounds: A cause for the decline in Atlantic eels? *Ecotoxicology* **2002**, *11*, 265–277. [CrossRef]
57. Degani, G.; Dosoretz, C. The effect of 3,3′,5–triiodo–L–thyronine and 17–a–methyltestosterone on growth and body composition of the glass stage of the eel (*Anguilla anguilla* L). *Fish Physiol. Biochem.* **1986**, *1*, 145–151. [CrossRef]
58. Deal, C.K.; Volkoff, H. The role of the thyroid axis in fish. *Front. Endocrinol.* **2020**, *11*, 596585. [CrossRef]
59. Peter, M.C.S. The role of thyroid hormones in stress response of fish. *Gen. Comp. Endocrinol.* **2011**, *172*, 198–210. [CrossRef] [PubMed]
60. Rosseau, K.; Le Belle, N.; Sbaihi, M.; Marchelidon, J.; Schmitz, M.; Dufour, S. Evidence for negative feedback in the control of eel growth hormone by thyroid hormones. *J. Endocrinol.* **2002**, *175*, 605–613. [CrossRef] [PubMed]
61. Budziszewska, B.; Jaworska-Feil, L.; Lasoń, W. The effect of repeated amphetamine and cocaine administration on adrenal, gonadal and thyroid hormone levels in the rat plasma. *Exp. Clin. Endocrinol.* **1996**, *104*, 334–338. [CrossRef] [PubMed]
62. Dhopesh, V.P.; Burke, W.M.; Maany, I.; Ravi, N.V. Effect of cocaine on thyroid functions. *Am. J. Drug Alcohol. Abuse* **1991**, *17*, 423–427. [CrossRef] [PubMed]
63. Lacy, M.E.; Utzschneider, K.M. Cocaine intoxication and thyroid storm: Similarity in presentation and implications for treatment. *J. Investig. Med. High Impact Case Rep.* **2014**, *2*, 2324709614554836. [CrossRef] [PubMed]
64. Redding, J.M.; De Luze, A.; Leloup-Hatey, J.; Leloup, J. Suppression of plasma thyroid hormone concentrations by cortisol in the European eel *Anguilla anguilla*. *Comp. Biochem. Physiol. A* **1986**, *83*, 409–413. [CrossRef]
65. Vijayan, M.M.; Flett, P.A.; Leatherland, J.F. Effects of cortisol on the in vitro hepatic conversion of thyroxine to triiodothyronine in brook charr (*Salvelinus fontinalis* Mitchill). *Gen. Comp. Endocrinol.* **1988**, *70*, 312–318. [CrossRef]
66. Saleem, M.; Martin, H.; Coates, P. Prolactin biology and laboratory measurement: An update on physiology and current analytical issues. *Clin. Biochem. Rev.* **2018**, *39*, 3–16.
67. Breves, J.P.; McCormick, S.D.; Karlstrom, R.O. Prolactin and teleost ionocytes: New insights into cellular and molecular targets of prolactin in vertebrate epithelia. *Gen. Comp. Endocrinol.* **2014**, *203C*, 21–28. [CrossRef]
68. Sakamoto, T.; McCormick, S.D. Prolactin and growth hormone in fish osmoregulation. *Gen. Comp. Endocrinol.* **2006**, *147*, 24–30. [CrossRef] [PubMed]
69. Gay, F.; Ferrandino, I.; Monaco, A.; Cerulo, M.; Capasso, G.; Capaldo, A. Histological and hormonal changes in the European eel (*Anguilla anguilla*) after exposure to environmental cocaine concentration. *J. Fish Dis.* **2016**, *39*, 295–308. [CrossRef]
70. Mello, N.K.; Mendelson, J.H.; Drieze, J.M.; Teoh, S.K.; Kelly, M.L.; Sholar, J.W. Effects of dopamine on prolactin: Interactions with cocaine self-administration by female rhesus monkeys. *J. Pharmacol. Exp. Ther.* **1994**, *270*, 1110–1120. [PubMed]
71. Olivereau, M. Dopamine, prolactin control, and osmoregulation in eels. *Gen. Comp. Endocrinol.* **1975**, *26*, 550–561. [CrossRef]
72. Fontaine, R.; Royan, M.R.; von Krog, K.; Weltzien, F.A.; Baker, D.M. Direct and indirect effect of sex steroids on gonadotrope cell plasticity in the teleost fish pituitary. *Front Endocrinol* **2020**, *11*, 605068. [CrossRef] [PubMed]
73. Tenegu, S.; Pranoty, A.; Mamta, S.K.; Senthilkumaran, B. Development and organization of gonadal steroidogenesis in bony fishes. A review. *Aquac. Fish* **2021**, *6*, 223–246. [CrossRef]
74. Fontes, M.K.; Rosati, L.; Di Lorenzo, M.; Pereira, C.D.S.; Maranho, L.A.; Laforgia, V.; Capaldo, A. Aquatic pollution and risks to biodiversity: The example of cocaine effects on the ovaries of *Anguilla anguilla*. *Animals* **2022**, *12*, 1766. [CrossRef]
75. Heesch, C.M.; Negus, B.H.; Bost, J.E.; Kaffer, J.H.; Snyder, N.W.; Eichhorn, E.J. Effects of cocaine on anterior pituitary and gonadal hormones. *J. Pharmacol. Exp. Ther.* **1996**, *278*, 1195–1200.
76. Mello, N.K.; Mendelson, J.H.; Negus, S.S.; Kelly, M.; Knudson, I.; Roth, M.E. The effects of cocaine on gonadal steroid hormones and LH in male and female rhesus monkeys. *Neuropsychopharmacology* **2004**, *29*, 2024–2034. [CrossRef] [PubMed]
77. Geeraerts, C.; Belpaire, C. The effects of contaminants in European eel: A review. *Ecotoxicology* **2010**, *19*, 239–266. [CrossRef]
78. Ghattas, S.M.; Yanai, T. Light microscopical study on the skin of European eel (*Anguilla anguilla*). *World J. Fish Mar. Sci.* **2010**, *2*, 152–161.
79. Bragadeswaran, S.; Thangaraj, S. Hemolytic and antibacterial studies on skin mucus of eel fish, *Anguilla anguilla* Linnaeus, 1758. *Asian J. Biol. Sci.* **2011**, *4*, 272–276. [CrossRef]

80. Capaldo, A.; Gay, F.; Laforgia, V. Changes in the gills of the European eel (*Anguilla anguilla*) after chronic exposure to environmental cocaine concentration. *Ecotoxicol. Environ. Saf.* **2019**, *169*, 112–119. [CrossRef] [PubMed]
81. Bernet, D.; Schmidt, H.; Meier, W.; Burkhardt-Holm, P.; Wahli, T. Histopathology in fish: Proposal for a protocol to assess aquatic pollution. *J. Fish Dis.* **1999**, *22*, 25–34. [CrossRef]
82. Brewer, J.D.; Meyes, A.; Bostwick, J.M.; Hamacher, K.L.; Pittelkow, M.R. Cocaine abuse: Dermatologic manifestations and therapeutic approaches. *J. Am. Acad. Dermatol.* **2008**, *59*, 483–487. [CrossRef]
83. Chander, B.; Aslanian, H.R. Gastric perforations associated with the use of crack cocaine. *Gastroenterol. Hepatol.* **2010**, *6*, 733–735.
84. De Felice, B.; Parolini, M. Effects of single and combined exposure to cocaine and benzoylecgonine on the oxidative status of *Mytilus galloprovincialis*. *Environ. Toxicol. Pharmacol.* **2020**, *80*, 103475. [CrossRef]
85. dos Santos Barbosa Ortega, A.; Maranho, L.A.; Nobre, C.R.; Moreno, B.B.; Guimarães, R.S.; Temponi Lebre, D.; de Souza Abessa, D.M.; Ribeiro, D.A.; Pereira, C.D.S. Detoxification, oxidative stress, and cytogenotoxicity of crack cocaine in the brown mussel *Perna perna*. *Environ. Sci. Pollut. Res.* **2019**, *26*, 27569–27578. [CrossRef]
86. Sauro, N.D.; Zorn, M.E. Prolactin induces proliferation of vascular smooth muscle cells through a protein kinase C-dependent mechanism. *J. Cell Physiol.* **1991**, *148*, 133–138. [CrossRef] [PubMed]
87. Brancaccio, P.; Lippi, G.; Maffulli, N. Biochemical markers of muscular damage. *Clin. Chem. Lab. Med.* **2010**, *48*, 757–767. [CrossRef] [PubMed]
88. Brazeau, G.A.; McArdle, A.; Jackson, M.J. Effects of cocaine on leakage of creatine kinase from skeletal muscle: In vitro and in vivo studies in mice. *Life Sci.* **1995**, *57*, 1569–1578. [CrossRef] [PubMed]
89. Capaldo, A.; Gay, F.; Caputo, I.; Lionetti, L.; Paolella, G.; Di Gregorio, I.; Martucciello, S.; Di Lorenzo, M.; Rosati, L.; Laforgia, V. Effects of environmental cocaine concentrations on COX and caspase-3 activity, GRP-78, ALT, CRP and blood glucose levels in the liver and kidney of the European eel (*Anguilla anguilla*). *Ecotoxicol. Environ. Saf.* **2021**, *208*, 111475. [CrossRef] [PubMed]
90. Hotamisligil, G.S. Endoplasmic reticulum stress and the inflammatory basis of metabolic disease. *Cell* **2010**, *140*, 900–917. [CrossRef]
91. Arinc, E.; Bozcaarmutlu, A. Catalyzation of cocaine N-demethylation by cytochromes P4502B, P4503A, and P4502D in fish liver. *J. Biochem. Mol. Toxicol.* **2003**, *17*, 169–176. [CrossRef]
92. Valente, M.J.; Henrique, R.; Vilas-Boas, V.; Silva, R.; de lourdes Bastos, M.; Carvalho, F.; Guedes de Pinho, P.; Carvalho, M. Cocaine-induced kidney toxicity: An in vitro study using primary cultured human proximal tubular epithelial cells. *Arch. Toxicol.* **2012**, *86*, 249–261. [CrossRef]
93. Hüttemann, M.; Helling, S.; Sanderson, T.H.; Sinkler, C.; Samavati, L.; Mahapatra, G.; Varughese, A.; Lu, G.; Liu, J.; Ramzan, R.; et al. Regulation of mitochondrial respiration and apoptosis through cell signaling: Cytochrome c oxidase and cytochrome c in ischemia/reperfusion injury and thermoregulation. *Biochim. Biophys. Acta* **2012**, *1817*, 598–609. [CrossRef] [PubMed]
94. Tyler, C.R.; Sumpter, J.P. Oocyte growth and development in teleosts. *Rev. Fish Biol. Fish* **1996**, *6*, 287–318. [CrossRef]
95. Blakemore, J.; Naftolin, F. Aromatase: Contributions to physiology and disease in women and men. *Physiology* **2016**, *31*, 258–269. [CrossRef] [PubMed]
96. Kostic, T.S.; Stojknov, N.J.; Bjelic, M.M.; Mihajlovic, A.I.; Janjic, M.M.; Andric, S.A. Pharmacological doses of testosterone upregulated androgen receptor and 3-beta-hydroxysteroid dehydrogenase/delta-5-delta-4 isomerase and impaired Leydig cells steroidogenesis in adult rats. *Toxicol. Sci.* **2011**, *121*, 397–407. [CrossRef]
97. Aranyakanout, C.; Ijiri, S.; Hasegawa, Y.; Adachi, S. 17 β-Hydroxysteroid dehydrogenase type 12 is responsible for maturation inducing steroid synthesis during oocyte maturation in *Nile tilapia*. *Gen. Comp. Endocrinol.* **2020**, *290*, 113399. [CrossRef] [PubMed]
98. Kaufmann, R.A.; Savoy-Moore, R.; Sacco, A.G.; Subramanian, M.G. The effect of cocaine on oocyte development and the follicular microenvironment in the rabbit. *Fertil. Steril.* **1990**, *54*, 921–926. [CrossRef] [PubMed]
99. Potter, D.A.; Moreno, A.; Luther, M.F.; Eddy, C.A.; Siler-Khodr, T.M.; King, T.S.; Schenken, R.S. Effects of follicular-phase cocaine administration on menstrual and ovarian cyclicity in rhesus monkeys. *Am. J. Obstet. Ginecol.* **1998**, *178*, 118–125. [CrossRef]
100. Willard, S.S.; Koss, C.M.; Cronmiller, C. Chronic cocaine exposure in *Drosophila*: Life, cell death and oogenesis. *Dev. Biol.* **2006**, *296*, 150–163. [CrossRef] [PubMed]

fishes

Article

Recombinant Gonadotropins to Induce Oocyte Development In Vitro and In Vivo in the European Eel *Anguilla anguilla*

Pauline Jéhannet [1], Arjan P. Palstra [1,*], Ignacio Giménez Nebot [2], Henk Schipper [3], William Swinkels [4], Leon T. N. Heinsbroek [5] and Hans Komen [1]

1 Animal Breeding & Genomics Department, Wageningen University & Research, 3700 AH Wageningen, The Netherlands
2 Rara Avis Biotec S.L., Calle Moratín, 17—4°, 46002 Valencia, Spain
3 Experimental Zoology Group, Wageningen University & Research, 6700 AH Wageningen, The Netherlands
4 Palingkwekerij Koolen BV, Hongarijesedijk 12, 5571 XC Bergeijk, The Netherlands
5 Wageningen Eel Reproduction Experts B.V., Mennonietenweg 13, 6702 AB Wageningen, The Netherlands
* Correspondence: arjan.palstra@wur.nl

Abstract: Commonly, female European eels are injected weekly with pituitary extract (PE) from carp (CPE) or salmon (SPE) to induce sexual maturation. However, a PE is a mixture of gonadotropins and other hormones that are not specific for eel and rapidly cleared from circulation. The aim of this study was therefore to test the effects of highly stable eel-specific recombinant gonadotropins (rGTHs) on oocyte development in vitro and in vivo in European eels. For the in vitro trial, the dose–effect responses of maturing eel oocytes on CPE and recombinant luteinizing hormone (rLH) were studied before and after 12 and 18 h of incubation. For the in vivo experiment, sexual maturation was stimulated by treatment with (i) CPE, (ii) recombinant follicle-stimulating hormone (rFSH) followed by CPE and (iii) rFSH followed by rLH. For the in vitro experiment, the expression of the nuclear progestin receptor 2 (*pgr2*) was induced by rLH, implying that rLH was preparing the oocyte for ovulation. For the in vivo experiment, the females treated with rGTHs had high gonadosomatic index (GSI) values (rFSH-CPE: 75, 77; rFSH-rLH: 80) in comparison with the females injected with CPE (50–60), suggesting that rFSH strongly induced vitellogenic growth. Larvae were produced for all treatment groups and for the first time by rGTH treatment alone but dose and timing still need optimization.

Keywords: fish reproduction; recombinant gonadotropin; vitellogenesis; oocyte maturation; ovulation

Citation: Jéhannet, P.; Palstra, A.P.; Giménez Nebot, I.; Schipper, H.; Swinkels, W.; Heinsbroek, L.T.N.; Komen, H. Recombinant Gonadotropins to Induce Oocyte Development In Vitro and In Vivo in the European Eel *Anguilla anguilla*. *Fishes* **2023**, *8*, 123. https://doi.org/10.3390/fishes8030123

Academic Editor: Jose Martin Pujolar

Received: 24 January 2023
Revised: 17 February 2023
Accepted: 20 February 2023
Published: 22 February 2023

1. Introduction

Gametogenesis is controlled by two pituitary gonadotropins (GTHs): follicle-stimulating hormone (FSH) and luteinizing hormone (LH), as reviewed in [1,2]. While FSH is involved in vitellogenesis and the stimulation of oocyte growth, LH promotes oocyte maturation and ovulation. During vitellogenesis, the oocyte grows considerably by accumulating yolk globules within its ooplasm, as reviewed in [1,2]. Briefly, FSH binds to its ovarian receptor (Fshr) and promotes the production of 17β-estradiol (E2). Upon the binding of E2 to the hepatic nuclear estrogen receptors, the production of vitellogenin is stimulated which is then incorporated and cleaved into small units of yolk proteins within the oocyte [3,4]. Following the vitellogenic growth phase, the oocyte matures, as reviewed in [1,5], which is characterized by the (1) migration of the germinal vesicle (GV) from the center of the oocyte towards its surface where the nuclear envelope disintegrates (i.e., germinal vesicle breakdown (GVBD)), (2) hydration of the oocyte which results from the cleavage of the yolk into small amino acids that increase the osmotic pressure thereby allowing water influx in the oocyte [6], and (3) fusion of the lipid droplets that coalesce to a few large globules [7]. For inducing maturation, LH binds to its ovarian receptors (Lhrcg1 and Lhrcg2) to promote the production of the maturing-inducing hormone (MIH). In turn, MIH binds its membrane progestin receptors (mPRs) to induce GVBD. Following the LH surge-induced maturation, the MIH mediates ovulation, as reviewed in [8], via its nuclear progestin receptors (Pgr1 and Pgr2). LH stimulates the production of prostaglandins, which act through their receptors, such as the Ptger4b, to trigger the release of the oocyte from its surrounding follicles, as reviewed by [9].

Although recent advances in eel reproduction protocols has led to the regular production of larvae [10–13], the life cycle of the European eel is not closed yet in captivity. Generally, eels are weekly injected with SPE or CPE to induce oocyte growth [7,14–16]. Subsequently, females are injected with DHP, as the main MIH in eels [17], to induce oocyte maturation and ovulation. With the current protocol, egg and larvae quality is often poor, as exemplified by the low hatching rates, low larval survival and abundant occurrence of malformations [13,18,19]. Therefore, there is an urgent need to improve the artificial reproduction protocols to obtain high-quality eggs that will develop into larvae that are able to grow, survive and metamorphose into juvenile glass eels.

PE is a mixture of FSH, LH, growth hormone (GH), prolactin (PRL), somatolactin (SL) and thyroid-stimulating hormone (TSH) [20–24]. While the gonadotropins directly control gametogenesis, the secondary hormones, such as GH and PRL, regulate other physiological processes but also play a role in reproductive physiology [25–29]. In fish, GTHs have been reported to control the release of secondary pituitary hormones, such as GH [30], SL [31] and PRL [32]. Contrary to using the PE of species other than eel, rGTHs are species specific, and the dosages of the gonadotropins are controllable. Like other synthetic hormones, rGTHs do not pose a disease transmission risk as may PE [33]. rGTHs have been developed and tested in several other fish species. In Japanese eels, rGTHs were initially produced by *Drosophila* S2 cells, but these only moderately induced the development of ovarian tissues and testis in vivo, despite their great effectiveness in vitro [34]. This low biological activity in vivo is probably due to the high clearance rate from the circulation of recombinants produced by insect cells in vertebrates [35]. For increasing recombinant production, rGTHs have been produced by infecting silkworm larvae with baculovirus that contained cDNAs encoding eel gonadotropin subunits [36,37]. With these rGTHs, spermatogenesis was induced but the males did not produce sperm [36]. rGTHs produced in the Chinese hamster ovary (CHO) have been found to induce spermatogenesis but also spermiation in European eels [38]. Recently, Ramos-Júdez et al. [39] reported that the rGTHs produced in the CHO were successful in inducing oocyte development from previtellogenesis to oocyte maturation in the flathead grey mullet *Mugil cephalus*, and they were able to produce eggs and larvae that developed into juveniles. These results suggest that rGTHs produced in the CHO have a higher bioactivity and half-life than the ones produced in insect cells and silkworm pupae.

The main objective of this study was to establish the stimulatory effects of eel-specific rGTHs (rFSH and rLH) that were produced in CHO cells on oocyte development in vitro and in vivo in European eels. For the in vitro trial, the dose and effect response of the oocytes to rLH in boosting maturation and ovulation were assessed and compared with CPE as the control. For the in vivo experiment, sexual maturation was stimulated by treatment with (i) CPE as the control, (ii) rFSH followed by CPE and (iii) rFSH followed by rLH.

2. Materials and Methods

2.1. Ethics

The animal study protocol was approved by the Central Committee for Animal Experiments (project number: AVD401002017817), the DEC (Animal Experiments Committee) and the IvD (Authority for Animal Welfare) (experiment numbers: 2017.D-0007.001–5).

2.2. Broodstock Conditioning

2.2.1. Males

Both wild and farmed males (80–120 g) were used for the experiment. Wild males were caught by fyke net in the Harinxma canal (The Netherlands). The farmed males were obtained from Palingkwekerij Koolen B.V. (Bergeijk, The Netherlands). The males were transferred to the animal experimental facilities of Wageningen University and Research. The males were anaesthetized with 2-phenoxyethanol (2 mL in 10 L water), PIT-tagged (Trovan, DorsetID, Aalten, The Netherlands) and housed with the females.

2.2.2. Females

Young elvers of 10 g were obtained from Palingkwekerij Koolen B.V. (Bergeijk, The Netherlands) and transferred to the CARUS aquaculture facilities of Wageningen University and Research. The elvers were placed in 400 L tanks that were kept at 24 °C under dimmed light conditions. The elvers were feminized [40] by feeding them with 2 mm pellets (Alltech-Coppens, Helmond, The Netherlands) that were coated with E2 (Sigma Aldrich, Saint Louis, MI, USA) over a 6 month period. Feminized eels appear to be more sensitive to hormonal treatment [41], which shortens the generation time to two years. After that period, they were fed with a custom-made broodstock diet (protein 525 g kg^{-1}, fat 98 g kg^{-1} and ash 76 g kg^{-1}) for an additional 6 months. Then, premature females of 330 $\pm$ 43 g were selected, transferred to seawater (Tropic Marine, 36 ppt) and no longer fed. They were subjected to simulated migration in a swim gutter for 2 months, with a slightly adjusted protocol from [42]: constant swimming at a speed of 0.51 m s^{-1} for ~2700 km in the dark at daily alternating temperatures between 10 and 15 °C. After the simulated migration, the eels were randomly selected, anaesthetized with 2-phenoxyethanol (2 mL in 10 L water), PIT-tagged (Trovan, DorsetID, Aalten, The Netherlands) and housed in seawater at a salinity of 36 ppt, a water temperature of 16 °C and under dark conditions. Each eel was injected with an implant containing 17α-methyltestosterone (5 mg) and E2 (2 mg) in the peritoneal cavity for two months to induce the start of vitellogenesis [43].

2.3. Production of rGTHs

Homologous single chain recombinant European eel rFSH and rLH were produced by Rara Avis Biotec S.L. (Valencia, Spain) using in-house technology, as previously reported [38]. Briefly, the CHO cells were transfected with the expression constructs encoding fusion proteins containing the entire coding sequence of *A. anguilla* FSHβ (GenBank accession no. AAN73407.1) or LHβ (GenBank accession no. CAA43374.1) subunits, the 28 carboxyl-terminal amino acids of the human chorionic gonadotropin (hCG) β subunit as a linker and the mature sequence of the *A. anguilla* glycoprotein hormone α subunit (GenBank accession no. CAA43373.1). The secreted recombinant hormones were purified from the culture medium by ion exchange chromatography, concentrated and stored at −80 °C until use.

2.4. In Vitro Experiment of Ovarian Tissues

2.4.1. Biometrics

Before starting the weekly CPE injections, ten females were anaesthetized and measured for body length (BL), body weight (BW), body girth (BG) and eye diameters horizontal (Edh) and vertical (Edv) to calculate the Fulton's condition factor (K), body girth index (BGI) and the eye index (EI):

Condition factor (K) = $100 * (BW/BL^3)$;

BW: body weight (g) (BL): body length (cm);

Body girth index (BGI) = BG/BL;

BG: body girth (cm);

EI = $100 * (((Edh + Edv) * 0.25)^2 \pi * (10 * BL)^{-1})$, from [44];

Edh: eye horizontal diameter (mm); Edv: eye vertical diameter (mm).

2.4.2. Artificial Maturation

The females were weekly anesthetized and injected with 20 mg kg^{-1} CPE (Catfish, Den Bosch, The Netherlands) that was dissolved in a physiological salt solution (0.9% NaCl; 20 mg mL^{-1}). From week 7 onwards, two days after each injection, the eels were weighed to determine whether oocyte hydration had commenced, as indicated by an increased body weight index:

BWI = 100 ∗ (BW/BW at the moment of first injection);

BW: body weight (g).

When the BWI increased to above 110, the females were anaesthetized to take an ovarian biopsy by inserting a cannula through the cloaca. The oocyte development was graded on a scale from 1 to 7, according to [7]. When most of the oocytes were in stage 3, ovarian tissue was sampled for the in vitro experiment.

2.4.3. In Vitro Dose–Response Effects of Ovarian Tissue

The ovarian tissue (~5 g) was obtained by perforating the body wall cavity (5 cm anterior from the cloaca) with a needle that had an inner diameter of 2.3 mm. The ovarian biopsy was placed in a Petri dish containing ice-cold Leibovitz's L15 culture medium (Thermo Fisher Scientific, Waltham, MA, USA) supplemented with 2.5 g L^{-1} HEPES, 0.1 g L^{-1} streptomycin and 100,000 IU L^{-1} penicillin (Pen Strep, ThermoFisher, Waltham, MA, USA). The oocytes were dispersed in the culture medium by pipetting the ovarian pieces. For one of the females, ~20 oocytes were fixed in cold 4% paraformaldehyde overnight, washed in PBS and then placed in 70% ethanol to check for the presence of follicle cells. For each female, ~60 oocytes were stocked in each well of two 24-well culture plates in 1 mL medium supplemented with the treatment solution (CPE: 1.25, 12.5 and 125 µg mL^{-1}; rLH: 10, 100, 1000 ng mL^{-1}; concentrations were based on the results of a pilot study) or with 1 mL of hormone-free media with the plates as duplicates. Just before incubation, another oocyte sample was taken as the control for microscopy and gene expression analysis. One culture plate was incubated for 12 h and one plate was incubated for 18 h, as based on [45]. Both plates were incubated at 16 °C to be consistent with the water temperature of the in vivo experiment. After the incubation periods, the oocytes were sampled for microscopy and gene expression analysis.

2.4.4. Microscopy Analysis

For each female, the oocytes were stained with Serra's fixative (ethanol:formalin:acetic at 6:4:1, diluted 20 times in PBS) for 3 min to stain the germinal vesicle for assessment of the percentage of oocytes displaying GVBD. The stained oocytes were then photographed using a digital camera connected to a microscope at 2x magnification. The lipid and oocyte diameters of ~20 oocytes were measured with the free ImageJ software version 1.52 [46]. The lipid droplet diameter was measured according to [47]. Upon visual assessment, for each oocyte, the ten largest lipid droplets were measured and the five maximum values averaged. At the final stage of lipid coalescence, the lipid droplets fused together to form a few large oil globules [7]. When the lipid droplet diameter exceeded 190 µm, the diameter was based solely on the largest lipid droplet diameter. For each droplet, the diameter was measured only once, since their shapes were mostly spherical. For the oocyte diameter, the maximum diameter was measured, as the oocyte could sometimes be elliptical.

2.4.5. Histology

From the ~20 oocytes that were fixed in 4% paraformaldehyde to check for the presence of the follicle cells, 10 oocytes were randomly selected, placed in 0.9% agarose (type VII, low gelling; Sigma Aldrich, Saint Louis, MO, USA), dehydrated via an ethanol series and embedded in paraffin. The embedded oocytes were then sectioned using a microtome (Microm, HM 350, ThermoFisher Scientific, Waltham, MA, USA) into 5 µm thick sections. The oocytes were stained with Mayers Haematoxylin–Eosin and photographed with a DFC450c color camera attached to a Leica DM6b microscope.

2.4.6. Gene Expression Analysis

The primers used for the gene expression analysis are listed in Table 1. The mPRα, mPRAL2 and Ptger4b protein sequences from zebrafish (*Danio rerio*) were aligned with *A. anguilla* using the NCBI tBLASTx tool to predict the gene sequence. The primers that were not previously described were designed using Primer3 v0.4.0 [48,49].

Table 1. Primers used for each target gene. T^0: annealing temperature; bp: base pair. The primer efficiency of *mPRα, mPRAL1, mPRAL2, mPRδ* and *lhcgr2* could not be determined, since the expression of these genes were too low (ct > 30).

Gene	Accession Number	Primer Sequence	T^0	Length	Efficiency	Source
elf1	EU407825	FW: CCCCTGCAGGATGTCTACAA RV: AGGGACTCATGGTGCATTTC	64	152 bp	96%	[50]
pgr1	AFV13730.1	FW: AGTTTGCCAATCTCCAGGTG RV: ATCAAACTGTGGCTGGCTCT	60	107 bp	101%	[51]
pgr2	AFV13731.1	FW: GCCTCTGGATGTCACTACGG RV: CCGGCACAAAGGTAGTTCTG	60	95 bp	94%	[51]
mPRα	XM_035410974.1	FW: GCCAGTACAGCCGAGTTCTATTT RV: GTCCTATCCAAGCCGTGATTT	62	141 bp		This study
mPRAL1	XM_035428761.1	FW: CTGGCCTACATGAGCTTCAG RV: CCCACGTAGTCCAGGAAGAA	62	92 bp		[51]
mPRAL2	XM_035409137.1	FW: AATCGTCATGGAGAGGCTTG RV: GTACCCGCTGTGGATGTAGG	62	157 bp		This study
mPRγ	XM_035394888.1	FW: AAACAGCACCTTCCACCTGT RV: TGCAGAAACGGTAAGCCAAG	60	102 bp	92%	[51]
mPRδ	XM_035430606.1	FW: GCAGCTTCCAGATGACCAAT RV: GCAGCATGTAGACCAGCAGA	60	147 bp		[51]
fshr	LN831181	FW: CCTGGTCGAGATAACAATCACC RV: CCTGAAGGTCAAACAGAAAGTCC	63	173 bp	109%	[52]
lhcgr1	LN831182	FW: GCGGAAACACAGGGAGAAC RV: GGTTGAGGTACTGGAAATCGAAG	60	155 bp	101%	[53]
lhcgr2	LN831183	FW: TCAACAACCTCACCAATCTCTCT RV: GCAGTGAAGAAATAGCCGACA	62	162 bp	106%	This study
ptger4b	XM_035392436.1	FW: ATTGAGAAGGTGAAGTGCCTGT RV: AGAATGTTTGAGAGGTGCTGGT	62	169 bp	105%	This study
arα	FR668031	FW: AGGAAGAACTGCCCCTCTTG RV: ATTTGCCCGATCTTCTTCAG	62	90 bp	93%	[50]

For each female, ~20 oocytes were kept in RNAlater (Ambion Inc., Huntingdon, UK), refrigerated at 5 °C overnight and stored at −80 °C until RNA extraction. The total tissue was homogenized with a tissue lyzer (Qiagen, Tissuelizer II) and extracted with 1 mL Trizol (Invitrogen, CA, USA) according to the manufacturer's protocol. Traces of DNA were digested with the ISOLATE II RNA Mini Kit (Bioline, London, UK) according to the manufacturer's protocol. The RNA purity was then assessed using a Nanodrop and Bioanalyzer. The 260/280 ratios and RIN values were, on average, 2.1 ± 0.07 and 9.3 ± 0.6, respectively. Complementary DNA (125 ng uL^{-1}) was generated with Superscript III Reverse transcriptase according to the manufacturer's protocol (ThermoFisher, Waltham, MA, USA).

Gene expression analysis was performed by quantitative real-time PCR using SYBR on a QuantStudio Real-Time PCR system. The reactions were heated at 95 °C for 2 min, followed by 40 cycles of denaturation (95 °C, 5 s), annealing (60–64 °C, 10 s) and extension (72 °C, 5 s). A melt curve analysis was performed, and the products were electrophoresed on agarose gel to check for product specificity. When the CT values were below 30, standard curves were generated by diluting cDNA at 1:5 for *elf1*, *pgr1*, and *pgr2* at 1:2 for *lhr1*, *lhr2*, *fshr*, *ara* and *mPRδ* to determine the primer efficiency. Since some mPRs (*mPRα*, *mPRAL1*, *mPRAL2* and *mPRδ*) had CT values higher than 30, the primer efficiency could not be determined for these genes. When the CT values of a sample could not be determined due to the fact of a low expression, the CT values were set at 35. The data are expressed as the fold change using the $2^{-\Delta\Delta CT}$ method [54]. The transcript levels of each target were normalized over the elongation factor 1 (*elf1*), since its expression was stable across the various treatments (ANOVA, $p = 0.536$).

2.5. In Vivo Experiment

2.5.1. Biometrics and Ultrasound

Before starting the weekly CPE injections, thirty females were anaesthetized and measured for BW, BG, BL, Edh and Edv to calculate the K, BGI and EI [44]. The females were then scanned by ultrasonography using a compact and portable system (MyLabFive™Vet with a LA435 probe, Esaote, Genoa, Italy) according to [55]. The females were scanned from head to tail on the ventral side to determine the gonadosomatic index (GSI) according to the formula developed in [56]:

$$\begin{aligned}\text{GSI} = (\log\left(GM_{E(LM)}\right) \\ &= 0.07062 + 0.55102 \times \log(GL_E) \\ &+1.11327 \times \log(\textit{mean gonad area}) + \varepsilon \\ G_L: \text{gonad length, } GM: \text{gonad mass, } \varepsilon \text{ as a variance } \sigma 2 = 0.05\end{aligned}$$

2.5.2. Artificial Induction of Sexual Maturation

The thirty female eels were randomly and equally divided over three groups and treated according to the schematic representation presented in Figure 1. The females of group 1 (N = 10) were weekly injected with CPE (Catvis, Den Bosch, The Netherlands) at a dose of 20 mg kg^{-1} and were considered the control group (Figure 1A). The females of group 2 (N = 10) were weekly injected with (1) 12 µg rFSH from week 1 to week 12, (2) 12 µg rFSH and 10 mg kg^{-1} CPE from week 13 to week 15, (3) 6 µg rFSH and 20 mg kg^{-1} CPE from week 16 to week 18, and (4) 20 mg kg^{-1} CPE from week 18 onwards (Figure 1B). The females of group 3 (N = 10) were weekly injected with (1) 12 µg rFSH from week 1 to week 12, (2) 12 µg rFSH and 10 µg rLH from week 13 to week 15, (3) 6 µg rFSH and 20 µg rLH from week 16 to week 18, and (4) 20 µg rLH from week 18 onwards (Figure 1C). As described before, from week 7 onwards, the females were anesthetized and weighed. When the BWI increased to above 110, ovarian tissue was sampled and photographed using a binocular. When the oocytes were approximately at stage 3, the oocyte maturation was boosted by injecting CPE at a dose of 20 mg kg^{-1} for groups 1 and 2 or by injecting rLH at a dose of 20 µg for group 3. The next day, a new biopsy was obtained and photographed under the binocular. When the oocyte development showed sufficient progression to stage 4, the eels were anesthetized and injected at eight locations in the ovarium with DHP at a dose of 2 mg kg^{-1}. The females were then placed in a 373 L spawning tank in which the water was gradually increased in temperature from 18 to 20 °C and kept at 36 ppt under dark conditions. In the timespan of 11–16 h after receiving the DHP injection, the females were regularly checked for egg release. When ovulation was noticed by egg release following slight abdominal pressure, females were slightly anesthetized and stripped by applying gentle pressure on the abdomen to collect the eggs in dry bowls. From these eggs, a small sample was taken and photographed under the binocular.

The males were anesthetized and injected with hCG (Sigma Aldrich, Saint Louis, MO, USA) at a dose of 1000 IU [57] and returned to the 373 L tanks kept at 36 ppt and 16 °C. Between 8 and 10 weeks later, these males spermiated. At the moment of DHP injection of the females (so 10–12 h before fertilization), 6–10 males from the stock of spermiating males received their follow-up dose of 250 IU hCG [58] and were put in the tank with the females. Just before fertilization, 2–3 mL sperm was freshly collected by stripping the males and then diluted in 45 mL artificial eel plasma [59].

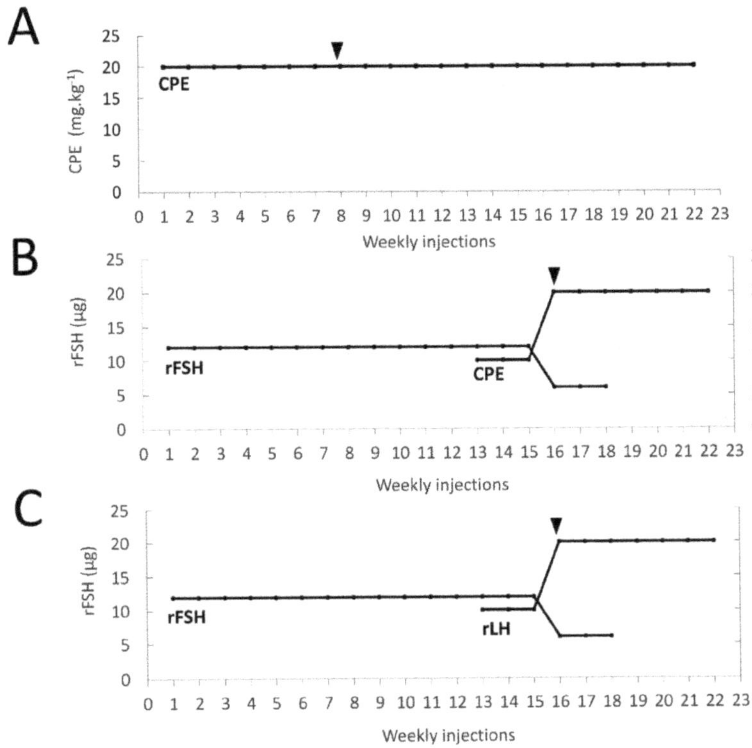

Figure 1. Schematic representations of the treatment administrated to each of the three groups of feminized eels, with (**A**) group 1: CPE (N = 10); (**B**) group 2: rFSH-CPE (N = 10); (**C**) group 3: rFSH-rLH (N = 10). For group 1, females were weekly injected with 20 mg kg^{-1} CPE. For group 2, females were weekly injected with 12 µg rFSH (week 1–12), 12 µg rFSH and 10 mg kg^{-1} CPE (week 13–15), 6 µg rFSH and 20 mg kg^{-1} CPE (week 16–18), and 20 mg kg^{-1} CPE from week 18 onwards. For group 3, females were weekly injected with 12 µg rFSH (week 1–12), 12 µg rFSH and 10 µg rLH (week 13–15), 6 µg rFSH and 20 µg rLH (week 16–18), and 20 µg rLH from week 18 onwards. Females started to release eggs (black arrows) at week 8 for group 1 and week 16 for groups 2–3.

2.5.3. Fertilization and Hatchery Practices

The diluted sperm and stripped eggs were gently mixed, and 300 mL of artificial sea salt water (Tropic Marine, 36 ppt, 18 °C) was added to the mixed gametes for activation. After 5 min, the eggs were placed in 3 L beakers filled with artificial seawater (36 ppt, 18 °C) under dark conditions. After 1 h, the percentage of floating eggs was estimated in relation to the eggs that were sinking. From the 3 L beakers, the floating layer was collected, rinsed gently over a sieve and transferred into beakers filled with fresh artificial seawater. Per beaker, 10–20 g of the eggs were kept in the suspension by gentle aeration and dead material was removed every 12 h. After hatching (48–60 h post fertilization), the water in the beakers was gently mixed to uniformly disperse the larvae and estimate their number. The larvae were stocked in two plankton nets (300 µm mesh) that were each hanging in a 100 L white conic tank connected to a 338 L recirculating system. The number of larvae that hatched was estimated, and the larval longevity (i.e., the number of days post-hatch that larvae survived) was monitored daily.

2.6. Statistical Analysis

All data were analyzed using R (version 3.2.4; R foundation for statistical computing, Vienna, Austria). The BGI, EI, K, GSI and normalized copy numbers of each target gene were compared between treatments using the Kruskal–Wallis test followed by a pairwise Dunn post hoc test with Benjamini–Hochberg correction for multiple comparisons among groups. Residuals from the BW, BL, BG, oocyte diameter, lipid diameter and the number of weekly injections that were needed to reach sexual maturation were tested for normality with the Shapiro–Wilk test. After testing the normality with the Shapiro–Wilk test, only BG and BL followed a normal distribution. BG and BL were therefore compared between the CPE, rFSH-CPE and rFSH-rLH treatments using one-way ANOVA. BW, the number of injections that were needed to reach sexual maturation and the oocyte and lipid diameters were compared using Kruskal–Wallis followed by a pairwise Dunn post hoc test with Benjamini–Hochberg correction for the multiple comparisons among groups. For the number of injections needed to reach maturation, the booster and DHP injections were not included in the statistical analysis. The differences were considered significant when $p < 0.05$.

3. Results

3.1. In Vitro Experiment

3.1.1. Biometrics and Histology

The females used for the in vitro experiment had a BL of 53 ± 3 cm, BW of 279 ± 44 g, BGI of 0.20 ± 0.01, EI of 8.9 ± 1.3 and K of 0.19 ± 0.02. The ovarian tissue sampled before the in vitro experiment was composed of folliculated and defolliculated oocytes (Figure 2). While some oocytes had theca and granulosa cells that were easily distinguishable by their respective flat and round shapes (Figure 2A), others were lacking follicle cells (Figure 2B).

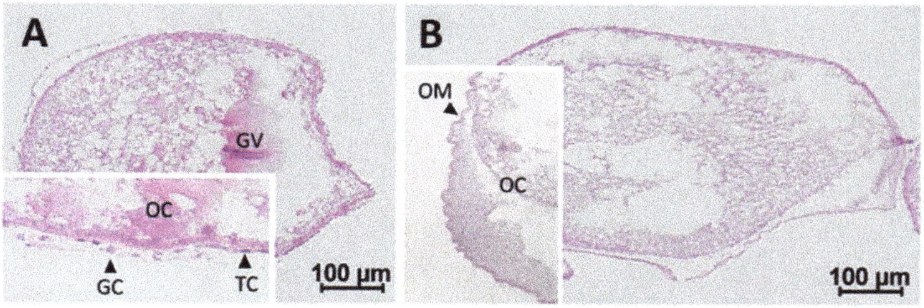

Figure 2. Maturing oocytes in European eel: (**A**) folliculated; (**B**) defolliculated oocytes. The insets zoom in on parts of the section. OC, oocyte; GV, germinal vesicle; OM, oocyte membrane; TC, theca cells; GC; granulosa cells.

3.1.2. GVBD, Hydration and Lipid Fusion

Both CPE and rLH did not induce GVBD, hydration ($p = 0.990$) and lipid fusion at 16 °C ($p = 0.980$) (for an overview, see Figure 3). Large variations in hydration and lipid fusion were observed between the females in the in vitro ovarian tissues (Figures S1–S2).

3.1.3. Expression of Nuclear and Membrane Progestin Receptors

Pgr1 expression remained stable with time and treatment ($p = 0.504$, Figure S3). *Pgr2* expression increased with the treatment dose of rLH ($p < 0.001$) (Figure 4). The treatment with 1000 ng rLH after 12 h of incubation significantly increased the expression of *pgr2* when compared to responses of the 0 h ($p = 0.032$), 12 h ($p = 0.012$) and 18 h controls ($p = 0.021$). Similar results were obtained after 18 h of incubation when comparing the dose–responses at 1000 ng rLH vs. the 0 h control ($p = 0.014$), at 1000 ng rLH vs. 12 h control ($p = 0.009$) and at 1000 ng rLH vs. the 18 h control ($p = 0.015$). The CPE treatment tended to cause an increase in *pgr2* expression in a dose-dependent manner, but the results were not significant.

When comparing the CPE and rLH treatments at various dosages, significant differences were detected but were ignored since they were not between comparable dosages.

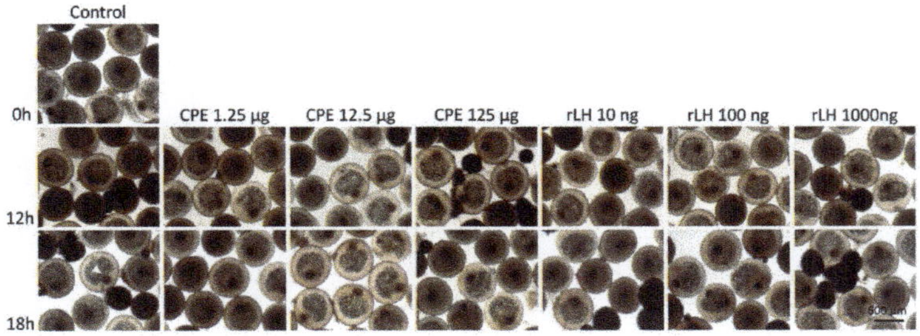

Figure 3. Maturing eel oocytes when treated with various dosages of CPE (0, 1.25, 12.5 and 125 µg mL^{-1}) and rLH (0, 10, 100 and 1000 ng mL^{-1}) in vitro at the start of incubation (0 h) and after 12 and 18 h of incubation. Oocytes were fixed with Serra's solution to stain the nucleus (white arrow). Oocytes incubated without hormone (control) had a visible nucleus after 0, 12 and 18 h of incubation. Similarly, oocytes incubated with CPE and rLH still had a visible nucleus showing that both CPE and rLH did not induce GVBD. In addition, the oocyte and lipid diameters did not change over time, neither with treatment. Still, variations in the oocyte and lipid diameters and variations in the position of the nucleus (center and periphery) were observed between oocyte batches of different females but independent of the treatment.

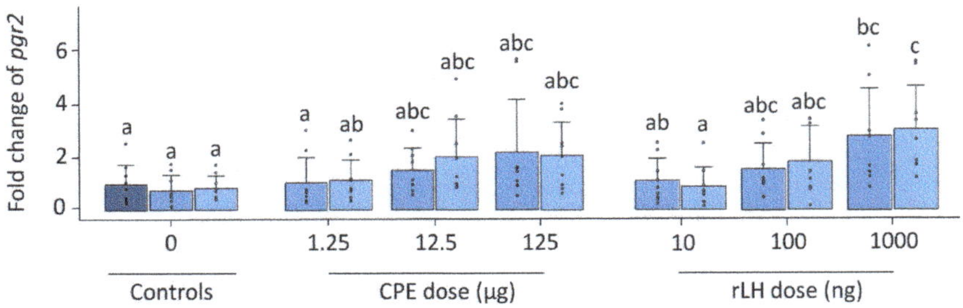

Figure 4. Expression of the nuclear progestin receptor *pgr2* when treated with various dosages of CPE (0, 1.25, 12.5 and 125 µg mL^{-1}) and rLH (0, 10, 100 and 1000 ng mL^{-1}) in vitro at the start of incubation (dark blue) and after 12 (mild blue) and 18 (light blue) hours of incubation. The receptor expression was normalized to the 0 h control and expressed as the fold change. The expression was compared between time points and dosages of CPE and rLH. The expression of *pgr2* was upregulated after incubation with 1000 ng rLH after 12 and 18 h incubation when compared to the controls. The bars with no overlap in letters are significantly different from each other ($p < 0.05$). Data are presented as bar plots with an average ± standard deviation and individual data points as circles. The experiments were performed on N = 9 eels, with the exception of 100 ng rLH (N = 8 eels).

For the expression of the membrane progestin receptors *mPRα* ($p = 0.108$), *mPRAL1* ($p = 0.477$), *mPRγ* ($p = 0.999$) and *mPRδ* ($p = 0.261$), no significant differences were found when comparing the treatments with CPE and rLH at various dosages at 0 h and after 12 and 18 h of incubation (Figures S4–S5, S7–S8). Significant differences were detected for *mPRAL2* ($p = 0.022$, Figure S6) but were ignored since all CT values were above 35 cycles.

3.1.4. Expression of Gonadotropin, Androgen and Prostaglandin Receptors

In vitro, the *fshr* expression remained stable with the treatment and time ($p = 0.624$, Figure S9). The expression of *lhcgr1* changed over time and with the treatment ($p < 0.001$). The expression of *lhcgr1* decreased over time when comparing the dose–response at 18 h vs. the 0 h controls ($p = 0.007$) (Figure 5). For the CPE treatment, the expression of *lhcgr1* decreased when comparing the dose–responses at 1.25 μg CPE after 18 h of incubation vs. the 0 h control (p = 0.014) and the dose–response at 125 μg CPE after 18 h of incubation with the 0 h control (p = 0.029). For the rLH treatment, the dose–response at 10 ng rLH after 18 h of incubation decreased *lhcgr1* expression compared with the 0 h control (p = 0.005). Similar results were found when comparing the dose–response at 100 ng rLH after 18 h incubation vs. the 0 h control (p = 0.023), the dose–response at 1000 ng rLH after 12 h of incubation vs. the 0 h control (p = 0.006), and the dose–response at 1000 ng rLH after 18 h of incubation vs. the 0 control (p = 0.006). Overall, the dose–response effects between the treatments were very similar, except for one slight difference between the dose–response at 1.25 μg CPE at 12 h vs. 10 ng rLH at 18 h (p = 0.005). The expression of *lhcgr2* (p = 0.815), *ara* (p = 0.910) and *ptger4b* (p = 0.978) remained stable with the treatment and time (Figures S10–S12).

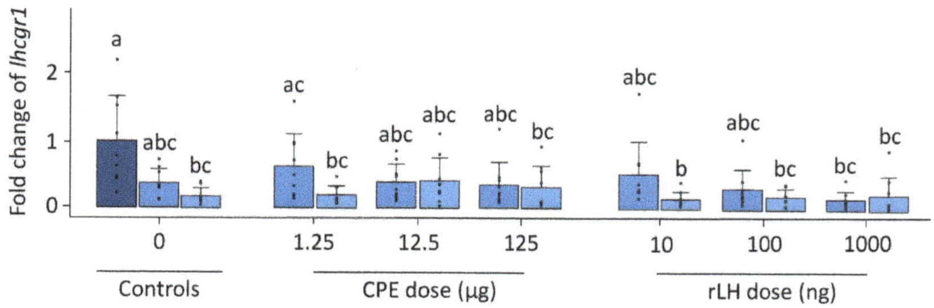

Figure 5. Expression of the luteinizing hormone receptor *lhcgr1* when treated with various dosages of CPE (0, 1.25, 12.5 and 125 μg mL^{-1}) and rLH (0, 10, 100 and 1000 ng mL^{-1}) in vitro at the start of incubation (dark blue) and after 12 (mild blue) and 18 (light blue) hours of incubation. The receptor expression was normalized to the 0 h control and expressed as the fold change. The expression was compared between the time points and dosages of CPE and rLH. The expression of *lhcgr1* decreased after 18 h of incubation when compared to the 0 h control. Bars with no overlap of letters are significantly different from each other (p < 0.05). The data are presented as bar plots with the average ± standard deviation and individual data points as circles. The experiments were performed on N = 9 eels.

3.2. *In Vivo Experiment*

3.2.1. Biometrics and GSI

An overview of the biometric measurements (BL, BW, BGI, EI, K and GSI) of the three treatment groups (CPE, rFSH-CPE and rFSH-rLH) prior to the start of the experiment is shown in Table 2. No significant differences between the treatment groups existed.

Table 2. Biometrics of the three treatment groups (CPE, rFSH-CPE and rFSH-rLH). BL, body length in cm; BW, body weight in g; BGI, body girth index; EI, eye index; K, condition factor; GSI, gonadosomatic index.

Biometric	CPE (N = 10)	rFSH-CPE (N = 10)	rFSH-rLH (N = 10)
BL (cm)	53 ± 3	55 ± 3	53 ± 1
BW (g)	284 ± 42	299 ± 32	286 ± 26
BGI	0.21 ± 0.01	0.20 ± 0.01	0.21 ± 0.01
EI	9.6 ± 1.2	9.4 ± 1.0	9.4 ± 1.2
K	0.19 ±0.02	0.18 ±0.02	0.19 ±0.02
GSI	5.6 ± 1.9	5.1 ± 2.3	4.2 ± 1.2

3.2.2. Sexual Maturation

The treatment affected the number of weekly injections to reach the oocyte maturation stage (p < 0.001) when crossing the threshold of a BWI of 110, indicating a significant hydration response. The females injected with CPE required significantly fewer weekly injections (9 ± 2 injections) to reach the oocyte maturation stage than the females injected with the rFSH-CPE (18 ± 1 injections, $p = 0.008$) and rFSH-rLH (20 ± 3 injections, $p < 0.001$) treatments (Figure 6). Increases in the oocyte diameter and lipid diameter were observed after the booster injection, after DHP injection and at stripping for each of the treatments (Figure 7).

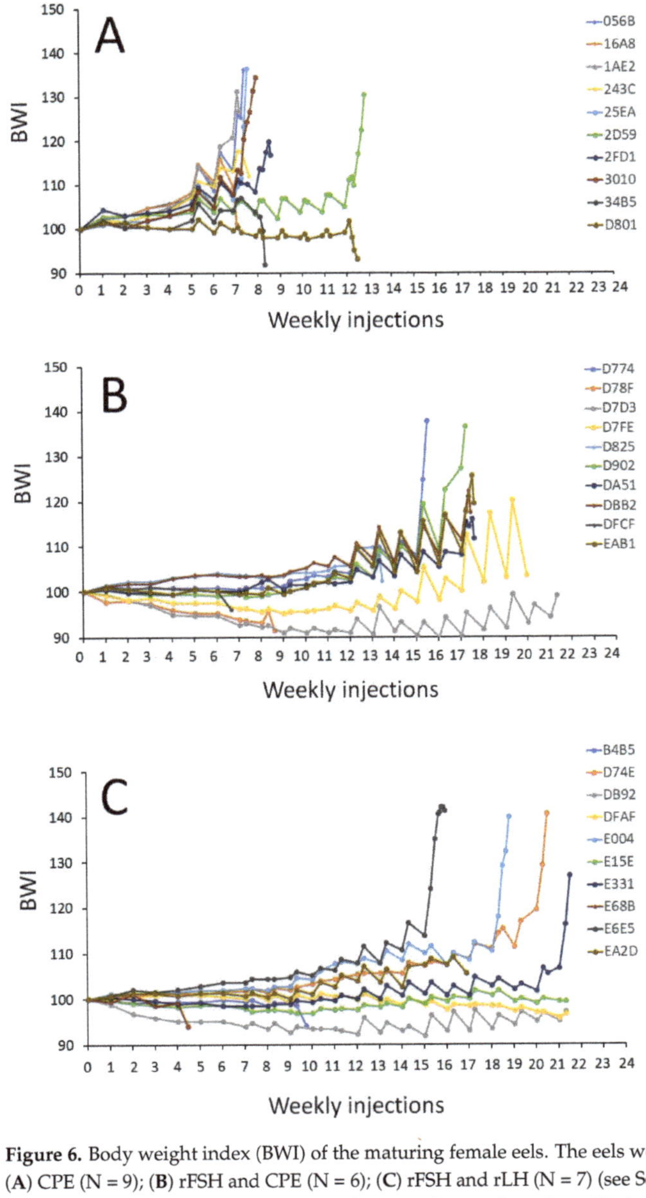

Figure 6. Body weight index (BWI) of the maturing female eels. The eels were weekly injected with (**A**) CPE (N = 9); (**B**) rFSH and CPE (N = 6); (**C**) rFSH and rLH (N = 7) (see Section 2 for a description of the treatment protocol). The BWI values vs. the number of weekly injections are shown. The number of weekly injections was different between the treatments ($p < 0.001$). Females treated with CPE needed fewer injections to reach maturation than females injected with rFSH-CPE ($p = 0.008$) and rFSH-rLH ($p < 0.001$). Note that the booster and DHP injections are shown but not included in the analyses.

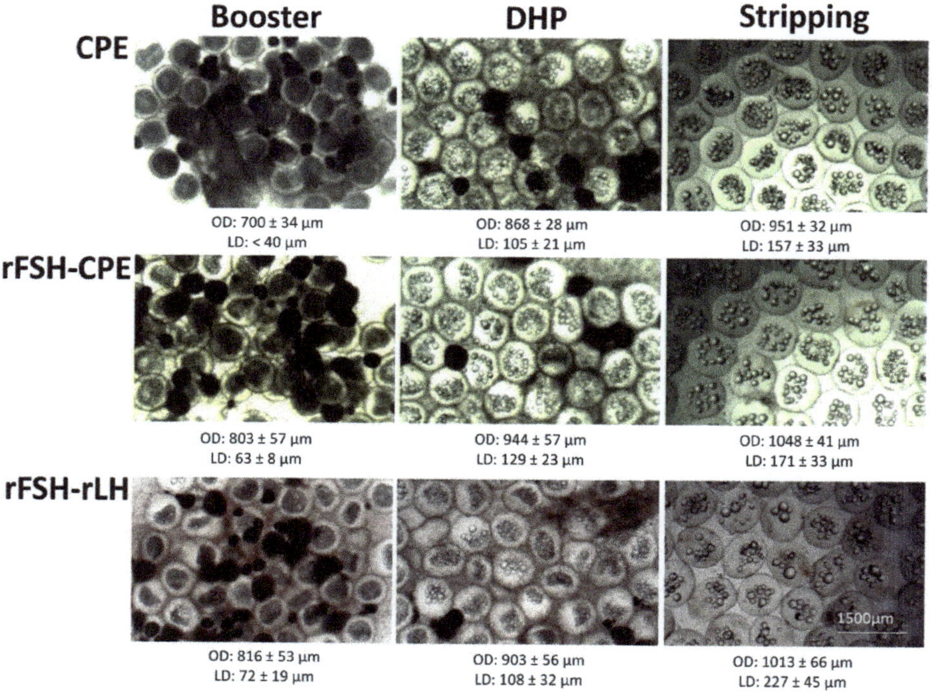

Figure 7. Oocytes of eels treated with CPE, rFSH-CPE and rFSH-rLH at the moments of booster injection, DHP injection and stripping (see Section 2 for a description of the treatment protocol). For each treatment, increases in oocyte diameter and lipid diameter were observed between the moments of DHP injection and booster injection and between stripping and DHP injection. At the bottom of each picture, the average ± standard deviation of the respective oocyte diameter (OD) and lipid diameter (LD) is given. No differences between treatments were observed.

3.2.3. Reproductive Success

An overview of the reproductive success of females treated with CPE, rFSH-CPE and rFSH-rLH is shown in Table 3. Of the ten eels that were treated with CPE, one eel died before reaching oocyte maturation after the 9th weekly injection (GSI: 22), seven eels matured after 7–9 weekly injections (GSI: 28–60) and two matured after the 13th weekly injection (GSI: 30, 38). Of the nine eels that matured, three died before DHP injection, two died after DHP injection (GSI: 60, 50) and four eels could be stripped 12 h after DHP injection. Two of these eels produced larvae that survived until 6 and 18 days post-hatch (dph). Of the ten eels that were treated with rFSH-CPE, four eels died before reaching maturation after 7–22 weekly injections (GSI: 11–27), and the remaining six eels matured after 16–20 weekly injections (GSI: 20–77). Of the six females that sexually matured, one died before DHP injection, three died after DHP injection (GSI: 51, 75, 77) and two eels could be stripped after DHP injection. One of these two eels produced larvae that survived for 2 dph. Of the ten eels that were treated with rFSH-rLH, three females died before reaching maturation after 5–22 weekly injections (GSI: 18–23) and seven matured after 16–22 weekly injections (GSI: 19–80). Of the seven eels that sexually matured, three died before DHP injection, one died after DHP injection (GSI: 80) and three could be stripped 14.5 h after DHP injection. One of these three females produced larvae that survived up to 2 dph but showed developmental abnormalities, such as notochord deformities, necrosis and pericardial oedemas.

Table 3. Reproductive success of the three treatment groups (CPE, rFSH-CPE and rFSH-rLH). Tag, PIT-tag code; Inj., number of weekly injections to reach maturation; BWI_1, BWI at booster moment; BWI_2, BWI at DHP injection; BGI_1, BGI at booster moment; BGI_2, BGI at DHP injection; t(h), hours after DHP injection; GSI, gonadosomatic index of females that died before/during maturation; Floaters, percentage of floaters after 1 h fertilization; Larvae, number of larvae after hatching; Fate, fate of the females.

Treatment	Tag	Inj.	BWI_1	BWI_2	BGI_1	BGI_2	t(h)	GSI	Floaters (%)	Larvae	Fate
CPE	34B5	9						22			†
CPE	D801	13						30			††
CPE	16A8	7						31			††
CPE	243C	8						28			††
CPE	1AE2	8		131		0.29		60			†††
CPE	2FD1	9	118	120	0.26	0.27		50			†††
CPE	3010	8	131	134	0.29	0.30	13	11 *	40		
CPE	2D59	13	122	130	0.26	0.28	12	38 *	10		
CPE	056B	8	125	136	0.25	0.26	12	20 *	60	>50	18 dph larvae
CPE	25EA	8	112	136	0.23	0.26	12	15 *	80	<50	6 dph larvae
rFSH-CPE	D78F	9						11			†
rFSH-CPE	D7D3	22						27			†
rFSH-CPE	D825	14						22			†
rFSH-CPE	DFCF	7						14			†
rFSH-CPE	D7FE	20						37			††
rFSH-CPE	DA51	18	114	116	0.24	0.26		51			†††
rFSH-CPE	DBB2	18		122		0.27		77			†††
rFSH-CPE	EAB1	18	121	126	0.30	0.31		75			†††
rFSH-CPE	D902	17		137		0.26		58 *			
rFSH-CPE	D774	16	125	138	0.25	0.29	13	20 *	70	<50	2 dph larvae
rFSH-rLH	B4B5	10						23			†
rFSH-rLH	DFAF	22						23			†
rFSH-rLH	E68B	5						18			†
rFSH-rLH	DB92	22						32			††
rFSH-rLH	E15E	22						31			††
rFSH-rLH	EA2D	17						41			††
rFSH-rLH	E331	22	116	127	0.26	0.29		80			†††
rFSH-rLH	E004	19	132	140	0.26	0.28	13	32 *	0		
rFSH-rLH	E6E5	16	141	141	0.28	0.29	14	19 *	10		
rFSH-rLH	D74E	21	129	141	0.27	0.28	16.5	27 *	3	<50	2 dph larvae

* GSI values reflecting the amount of ovarian tissue that was still left in the females after stripping; † died before sexual maturation; †† died during sexual maturation; ††† died after DHP injection.

4. Discussion

In the present study, we demonstrated that (1) eel-specific rGTHs (i.e., rFSH and rLH) produced in CHO cells drive sexual maturation in vivo; (2) rFSH strongly induces vitellogenesis in vivo; (3) rLH increased oocyte sensitivity to DHP by inducing the expression of *pgr2* in vitro; and (4) larvae were produced for the first time with rGTHs alone in European eels. These findings provide important insights into the mechanisms of oocyte development in European eels that will help in improving the reproduction protocol of this species and, ultimately, help close the life cycle of this species in captivity.

The rGTHs induced oocyte development from early vitellogenesis to oocyte maturation in female European eels. The females that reached maturation had GSI values ranging between 37–77 and 31–80 for the rFSH-CPE and rFSH-rLH treatments, respectively. The high GSI values in the females treated with rGTHs are in sharp contrast to a study by Kazeto et al. [34], who found that both rGTHs led to only very moderate changes in inducing sexual maturation in vivo in Japanese eels. These authors reported that the females treated with rFSH and rLH (100 $\mu g\ kg^{-1}$) for 8 weeks had mean GSI values of 1.51 ± 0.09 and 1.59 ± 0.08, respectively. The low GSI values in the study by Kazeto et al. [34] in comparison with the values of our study probably resulted from a difference in the systems used for rGTH production. While the study by Kazeto et al. [34] used rGTHs produced by insect cells that are known to be rapidly cleared from the circulation in vertebrates [35], we used rGTHs produced in mammalian cells. A comparison between the study by Kazeto et al. [34] and ours suggests a higher biological activity of rGTHs produced in mammalian cells to induce in vivo effects, as also recently reported in flathead grey mullet [39].

Three females treated with rGTHs had very high GSI values (rFSH-CPE: 75, 77; rFSH-rLH: 80) in comparison to the females injected with CPE (50–60). While the GSI values of eels obtained by applying the routine protocol of weekly CPE injections were within the range (30–60) of what has been observed in artificially reproduced European eels [7], high GSI values as observed in the females treated with rGTHs have not been reported yet. The high GSI values were the specific result of the rFSH treatment, because they were found in both the rFSH + CPE and rFSH + rLH groups. FSH increases the ovary weight by inducing oogonia proliferation and oocyte growth, as reviewed in [1,2]. In previous studies, histological analysis of ovarian tissues has shown that rFSH induced the formation of yolk globules in Japanese eel [34], flathead grey mullet [39] and Manchurian trout [60]. However, why were these GSI values in the rFSH-treated eels so much higher than what we observed for the CPE-treated eels? Should we inject CPE or rLH sooner and/or in higher doses since rFSH induces strong vitellogenic growth? Or does rFSH better synchronize oocyte development, and is the asynchronous oocyte development that we generally observed in the CPE-treated eels a flaw of the CPE treatment? In nature, the European eel may exhibit synchronous oocyte development and spawn eggs in a single batch. This hypothesis is supported by our earlier observations that CPE-treated eels that could be stripped almost empty were the most successful in their reproduction [7]; however, this is in contrast to our observations on a spontaneously matured female that had an estimated GSI of 47 and with only half of the oocytes hydrated and matured [61]. The mature females caught in the act of spawning in the Sargasso should provide a conclusive answer to this issue. Other explanations may be that the more stable rGTHs should be injected at longer than the weekly intervals of CPE or that the rGTHs lead to enlarged oocytes impairing the oocyte quality. Future studies should provide further insight into these issues.

Independent of the treatment, the mortality among the experimental eels was relatively high, as 50% of the females died during the course of the weekly injections and 40% of the remaining eels died after the DHP injection. This may originate from the fact that the experimental eels were not wild eels but feminized eels that were still young and perhaps not sufficiently conditioned as good broodstock. The wild silver eels that we used for experimenting were caught during their seaward migration at an age of at least 7 years old [62], and the feminized eels were used just one year after they arrived as young elvers. In our facilities, silvering is induced by subjecting them to a simulated migration, but at the end of this migration, they may not be at the stage of the wild silver eels yet. Their oocytes are still in previtellogenic stages with fewer lipid droplets [43] than in oocytes of the migrant wild eels, particularly when wild eels themselves are subjected to long-term swimming exercise, increasing the lipid deposition in the oocytes even further [63]. However, after the co-implantation with 17MT and E2, the feminized females had GSI values that reflected the initiation of vitellogenesis. Similar GSI values were reported by Palstra et al. [43] after steroid implantation in European eels. Therefore, the feminized eels in this study may have died because they were still too young for the induction of sexual maturation.

Both CPE and rLH did not induce hydration and lipid fusion in vitro after 18 h of incubation. The lack of histological changes during oocyte maturation in vitro is inconsistent with our in vivo results after the moment of boostering. Twenty-four hours after the females were injected with an extra CPE or rLH injection, the oocyte diameter increased, the cytoplasm became translucent and the lipid droplets fused together. In addition to the absence of effects on hydration and lipid coalescence, both CPE and rLH did not induce GVBD in vitro in European eels, which is in sharp contrast with previous studies in eels [36,37,64]. Kim et al. [64], who used the same rLH concentrations as our study, showed that rLH (1000 ng mL^{-1}) and SPE (1 mg mL^{-1}) induced approximately 30% GVBD in vitro at 20 °C after 24 h of incubation. The discrepancy between our study and the one by Kim et al. [64] might be explained by the difference in temperature (16 vs. 20 °C), which has been found to modulate the progression of oocyte development in eels [65,66]. In Japanese eels, oocytes matured faster at 20 °C than at 15 °C [66]. Similarly, female European eels reared at higher temperatures (18 and 21 °C) matured faster than those kept in water at 15 °C [65]. We performed the in vitro dose–response experiment at 16 °C in this study to be consistent with the water temperature of the in vivo experiment. A longer incubation time might have been necessary to induce histological changes in vitro at this temperature.

The expression of the membrane progestin receptors (*mPRα, mPRAL1, mPRAL2, mPRγ* and *mPRδ*) remained relatively constant in the in vitro trial of the maturing oocytes, which is in sharp contrast with the increased expression of the mPRs during spermatogenesis in European eels [51]. Our results indicate that the *mPRs* that are essential for the completion of oocyte maturation in fish [67,68] may already be present on the surface of the oocyte membrane. Similar results were found in vitro for *mPRa* in medaka, zebrafish and catfish [69–71] and for *mPRβ* in rainbow trout [72], zebrafish and catfish [73]. The upregulation of *mPRa* mRNA has been reported in seatrout after 8 h of incubation with hCG as an Lh analogue [67]. The lack of *mPR* overexpression in our study suggests that the upregulation of the mRNA levels is not required during oocyte maturation in eels. However, as shown in other fish species (spotted sea trout [67,74]; zebrafish [75]), it is possible that Lh increases mPR protein levels during oocyte maturation in female European eels. The protein level and transcript abundance of the various mPRs should receive more research attention in future studies to gain deeper insights into the mechanisms of oocyte maturation in European eels.

The expression of *lhcrg1* but not *lhcrg2* decreased over time in the in vitro trial of maturing oocytes. In Japanese eels, Kazeto et al. [76] showed that ovarian *lhr* transcripts increased during artificial maturation by the late vitellogenic stage. The expression of *fshr* remained stable in the in vitro trial of the maturing oocytes. In rainbow trout, a sharp increase in *fshr* expression just before oocyte maturation and ovulation was observed when compared with its expression at the end of vitellogenesis [77]. Like *fshr, ara* remained stable in the in vitro trial of the maturing oocytes. This result is in contrast with the study by Bobe and coauthors [78], who showed that *ara* expression decreased during oocyte maturation in rainbow trout. Androgen receptors have been shown to be important for oocyte maturation in zebrafish [79,80]. The decrease in *lhcrg1* and the lack of changes in the *fshr* and *ara* expressions in the in vitro maturing oocytes may be explained by the low incubation temperature.

Following the DHP injection, several females injected with rGTHs successfully ovulated. Larvae were produced, for the first time, from eels treated with rGTHs. Still, the egg quality must have been low, as many embryos had abnormal development, failed to hatch and continued their development inside the egg shell until their demise. It is unclear from our study whether this low reproductive success is an effect of using rGTHs, since low hatching rates [19] and deformities [18] have been routinely observed in European eels and may originate from the used broodstock eels, the method of maturation induction and the hatchery and nursery conditions. In vitro, one marker of ovulation, *pgr2*, was induced by rLH, which is consistent with previous studies on mammals, as reviewed in [81], and fish (medaka [70]; zebrafish [82,83]). The results suggest that this process is highly conserved in vertebrates. In zebrafish, pgr $^{-}/^{-}$ females showed normal oocyte development, but mature oocytes were trapped within the follicular layers and, therefore, these mutants failed to ovulate [82]. In fish, Pgr has been identified as essential for the upregulation of the *ptgr4b* gene [70,82]. In our study, LH-induced *pgr* expression was not associated with a higher *ptger4b* expression, which is not in line with the results obtained in medaka [70] and zebrafish [82]. Hagiwara et al. [70,71] proposed a mechanistic model for the expression of *ptger4b* in LH-primed preovulatory oocytes. Briefly, LH via its ovarian receptor induces *pgr* expression that translates into the Pgr protein within a few hours after the LH surge. Pgr binds to DHP which, in turn, increases *ptgr4b* expression. In this model, the authors assumed that DHP was present at high levels in the granulosa cells to induce *ptgr4b* in medaka, which is in contrast with eels that have low DHP levels within their ovarian follicles [84]. Therefore, the lack of DHP production by the follicular layers in European eels likely explains why the *ptgr4b* expression remained stable in the rLH-stimulated oocytes in our study. When considering the induced expression of *pgr2* by rLH, it can be concluded that LH is preparing the oocyte for ovulation by increasing oocyte sensitivity to DHP.

5. Conclusion

In European eels, the newly developed eel-specific rGTHs (i.e., rFSH and rLH) successfully induced maturation and led to the production of eggs, embryos and larvae, but the dose and timing still need optimization. In the in vivo experiment, rFSH strongly induced vitellogenic growth up to high GSI values of 70–80, which has never been observed before. In the in vitro experiment with maturing oocytes, the rLH induced the *pgr2* expression to prepare the oocyte for ovulation by increasing the oocyte sensitivity to DHP. Our study provides important insights into the mechanisms of oocyte maturation and ovulation that can help improve the artificial reproduction protocol in European eels.

Supplementary Materials: The following supporting information can be downloaded at: https://www.mdpi.com/article/10.3390/fishes8030123/s1, Figure S1: Oocyte diameters when treated with various dosages of CPE (0, 1.25, 12.5 and 125 μg mL^{-1}) and rLH (0, 10, 100 and 1000 ng mL^{-1}) in vitro at the start of incubation (dark blue) and after 12 (mild blue) and 18 (light blue) hours of incubation. Both CPE and rLH did not increase the oocyte diameter at any dose (p = 0.990), which indicates that hydration did not occur in vitro. The data are presented as bar plots with the average ± standard deviation of the independent experiments performed on N = 10 eels; Figure S2: Lipid diameters when treated with various dosages of CPE (0, 1.25, 12.5 and 125 μg mL^{-1}) and rLH (0, 10, 100 and 1000 ng mL^{-1}) in vitro at the start of incubation (dark blue) and after 12 (mild blue) and 18 (light blue) hours of incubation. Both CPE and rLH did not increase the lipid diameter at any dose (p = 0.980), which indicates that lipid fusion did not occur in vitro. The data are presented as bar plots with the average ± standard deviation of the independent experiments performed on N = 10 eels; Figure S3: Expression of the nuclear progestin receptor *pgr1* when treated with various dosages of CPE (0, 1.25, 12.5 and 125 μg mL^{-1}) and rLH (0, 10, 100 and 1000 ng mL^{-1}) in vitro at the start of incubation (dark blue) and after 12 (mild blue) and 18 (light blue) hours of incubation. The receptor expression was normalized to the 0 h control and expressed as the fold change. The expression was compared between the time points and dosages of CPE and rLH. Both CPE and rLH did not change the *pgr1* expression at any dose (p = 0.504). The data are presented as bar plots with the average ± standard deviation and the individual data points as circles. The experiments were performed on N = 9 eels; Figure S4: Expression of the membrane progestin receptor *mPRα* when treated with various dosages of CPE (0, 1.25, 12.5 and 125 μg mL^{-1}) and rLH (0, 10, 100 and 1000 ng mL^{-1}) in vitro at the start of incubation (dark blue) and after 12 (mild blue) and 18 (light blue) hours of incubation. The receptor expression was normalized to the 0 h control and expressed as the fold change. The expression was compared between the time points and dosages of CPE and rLH. Both CPE and rLH did not change the *mPRα* expression at any dose (p = 0.108). The data are presented as bar plots with the average ± standard deviation and the individual data points as circles. The experiments were performed on N = 9 eels; Figure S5: Expression of the membrane progestin receptor *mPRAL1* when treated with various dosages of CPE (0, 1.25, 12.5 and 125 μg mL^{-1}) and rLH (0, 10, 100 and 1000 ng mL^{-1}) in vitro at the start of incubation (dark blue) and after 12 (mild blue) and 18 (light blue) hours of incubation. The receptor expression was normalized to the 0 h control and expressed as the fold change. The expression was compared between the time points and dosages of CPE and rLH. Both CPE and rLH did not change the *mPRAL1* expression at any dose (p = 0.477). The data are presented as bar plots with the average ± standard deviation and the individual data points as circles. The experiments were performed on N = 9 eels; Figure S6: Expression of the membrane progestin receptor *mPRAL2* when treated with various dosages of CPE (0, 1.25, 12.5 and 125 μg mL^{-1}) and rLH (0, 10, 100 and 1000 ng mL^{-1}) in vitro at the start of incubation (dark blue) and after 12 (mild blue) and 18 (light blue) hours of incubation. The receptor expression was normalized to the 0 h control and expressed as the fold change. The expression was compared between the time points and dosages of CPE and rLH. Significant differences were detected (p = 0.022) but ignored since all CT values were above 35 cycles. The data are presented as bar plots with the average ± standard deviation and the individual data points as circles. The experiments were performed on N = 9 eels; Figure S7: Expression of the membrane progestin receptor *mPRδ* when treated with various dosages of CPE (0, 1.25, 12.5 and 125 μg mL^{-1}) and rLH (0, 10, 100 and 1000 ng mL^{-1}) in vitro at the start of incubation (dark blue) and after 12 (mild blue) and 18 (light blue) hours of incubation. The receptor expression was normalized to the 0 h control and expressed as the fold change. The expression was compared between the time points and dosages of CPE and rLH. Both CPE and rLH did not change the *mPRδ* expression at any dose (p = 0.261). The data are presented as bar plots with the average ± standard deviation and the individual datapoints as circles. The experiments were performed on N = 9 eels; Figure S8: Expression of the membrane progestin receptor *mPRγ* when treated with various dosages of CPE (0, 1.25, 12.5 and 125 μg mL^{-1}) and rLH (0, 10, 100 and 1000 ng mL^{-1}) in vitro at the start of incubation (dark blue) and after 12 (mild blue) and 18 (light blue) hours of incubation. The receptor expression was normalized to the 0 h control and expressed as the fold change. The expression was compared between the time points and dosages of CPE and rLH. Both CPE and rLH did not change *mPRγ* expression at any dose (p = 0.999). The data are presented as bar plots with the average ± standard deviation and the individual data points as circles. The experiments were performed on N = 9 eels; Figure S9: Expression of the follicle-stimulating hormone receptor *fshr* when treated with various dosages of CPE (0, 1.25, 12.5 and 125 μg mL^{-1}) and rLH (0, 10, 100 and 1000 ng mL^{-1}) in vitro at the start of incubation (dark blue) and after 12 (mild blue) and 18 (light blue) hours of incubation. The receptor expression was normalized to the 0 h control and expressed as the fold change. The expression was compared between the time points and dosages of CPE and rLH. Both CPE and rLH did not change the *fshr* expression at any dose (p = 0.624). The data are presented as bar plots with the average ± standard deviation and the individual data points as circles. The experiments were performed on N = 9 eels; Figure S10: Expression of the luteinizing hormone receptor *lhcgr2* when treated with various dosages of CPE (0, 1.25, 12.5 and 125 μg mL^{-1}) and rLH (0, 10, 100 and 1000 ng mL^{-1}) in vitro at the start of incubation (dark blue) and after 12 (mild blue) and 18 (light blue) hours of incubation. The receptor expression was normalized to the 0 h control and expressed as the fold change. The expression was compared between the time points and dosages of CPE and rLH. Both CPE and rLH did not change the *lhcgr2* expression at any dose (p = 0.815). The data are presented as bar plots with the average ± standard deviation and the individual data points as circles. The experiments were performed on N = 9 eels; Figure S11: Expression of the androgen receptor *ara* when treated with various dosages of CPE (0, 1.25, 12.5 and 125 μg mL^{-1}) and rLH (0, 10, 100 and 1000 ng mL^{-1}) in vitro at the start of incubation (dark blue) and after 12 (mild blue) and 18 (light blue) hours of incubation. The receptor expression was normalized to the 0 h control and expressed as the fold change. The expression was compared between the time points and dosages of CPE and rLH. Both CPE and rLH did not change the *ara* expression at any dose (p = 0.910). The data are presented as bar plots with the average ± standard deviation and the individual data points as circles. The experiments were performed on N = 9 eels; Figure S12: Expression of the prostaglandin receptor *ptger4b* when treated with various dosages of CPE (0, 1.25, 12.5 and 125 μg mL^{-1}) and rLH (0, 10, 100 and 1000 ng mL^{-1}) in vitro at the start of incubation (dark blue) and after 12 (mild blue) and 18 (light blue) hours of incubation. The receptor expression was normalized to the 0 h control and expressed as the fold change. The expression was compared between the time points and dosages of CPE and rLH. Both CPE and rLH did not change the *ptger4b* expression at any dose (p = 0.978). The data are presented as bar plots with the average ± standard deviation and individual data points as the circles. The experiments were performed on N = 9 eels.

Author Contributions: Conceptualization, A.P.P., P.J., I.G.N., L.T.N.H. and H.K.; methodology, A.P.P., P.J., I.G.N., L.T.N.H. and H.K.; experimental work, P.J. and L.T.N.H.; gene expression analysis, P.J; histology: P.J. and H.S.; formal analysis, P.J.; investigation, P.J., A.P.P. and H.K.; data curation, P.J. and L.T.N.H.; writing—original draft preparation, P.J.; writing—review and editing, A.P.P. and H.K.; supervision, A.P.P. and H.K.; project administration, A.P.P.; funding acquisition, A.P.P. and W.S. All authors have read and agreed to the published version of the manuscript.

Funding: This study was funded as the project OptimAAL (grant number: 18971000009) by the DUPAN foundation, as well as the Dutch Ministry of Economic Affairs and the European Union, European Maritime and Fisheries Fund.

Institutional Review Board Statement: The animal study protocol was approved by the Central Committee for Animal Experiments (project number: AVD401002017817), the DEC (Animal Experiments Committee) and the IvD (Authority for Animal Welfare) (experiment numbers: 2017.D-0007.001–5).

Data Availability Statement: Raw data supporting the conclusions of this article will be made available by the authors, without undue reservation, to any qualified researcher.

Acknowledgments: The authors wish to thank A. De Wit, L. Kruijt, K. Laport and B. Dibbits for the technical support related to the gene expression analysis; animal care takers and zootechnicians W. Nusselder, T. Spanings, M. ter Veld, S. Visser, M. van Loon and T. Wiegers of the animal experimental facilities at CARUS for monitoring fish health and water quality. The results of this study have been presented at the Aquaculture Europe congress of 2021 and reported in an abstract (https://aquaeas.org/Program/PaperDetail/38729, accessed on 21 February 2022), as well as pitched in an overview presentation at the WCGALP congress of 2022 and reported in the proceedings (https://www.wageningenacademic.com/pb-assets/wagen/WCGALP2022/21_005.pdf, accessed on 21 February 2022).

Conflicts of Interest: The coauthors I.G., W.S. and L.H. are employees of a company. The other authors have no competing interest.

References

1. Planas, J.V.; Swanson, P. Physiological function of gonadotropins in fish. In *Fish Reproduction*; Rocha, M.J., Arukwe, A., Kapoor, B.G., Eds.; Sciences Publisher: Enfield, CT, USA, 2008; pp. 37–66.
2. Lubzens, E.; Young, G.; Bobe, J.; Cerdà, J. Oogenesis in teleosts: How fish eggs are formed. *Gen. Comp. Endocrinol.* **2010**, *165*, 367–389. [CrossRef]
3. Sire, M.F.; Babin, P.J.; Vernier, J.M. Involvement of the lysosomal system in yolk protein deposit and degradation during vitellogenesis and embryonic development in trout. *J. Exp. Zool.* **1994**, *269*, 69–83. [CrossRef]
4. Montserrat, N.; González, A.; Méndez, E.; Piferrer, F.; Planas, J.V. Effects of follicle stimulating hormone on estradiol-17b production and P-450 aromatase (CYP19) activity and mRNA expression in brown trout vitellogenic ovarian follicles in vitro. *Gen. Comp. Endocrinol.* **2004**, *137*, 123–131. [CrossRef] [PubMed]
5. Nagahama, Y.; Yamashita, M. Regulation of oocyte maturation in fish. *Develop. Growth Differ.* **2008**, *50*, S195–S219. [CrossRef] [PubMed]
6. Greeley, M.S.; Calder, D.R.; Wallace, R.A. Changes un teleost yolk proteins during oocyte maturation: Correlation of yolk proteolysis with oocyte hydration. *Comp. Biochem. Physiol.* **1986**, *84*, 1–9. [CrossRef]
7. Palstra, A.P.; Cohen, E.G.H.; Niemantsverdriet, P.R.W.; van Ginneken, V.J.T.; van den Thillart, G.E.E.J.M. Artificial maturation and reproduction of European silver eel: Development of oocytes during final maturation. *Aquaculture* **2005**, *249*, 533–547. [CrossRef]
8. Takahashi, T.; Hagiwara, A.; Ogiwara, K. Follicle rupture during ovulation with an emphasis on recent progress in fish models. *Reproduction* **2019**, *157*, R1–R13. [CrossRef]
9. Takahashi, T.; Hagiwara, A.; Ogiwara, K. Prostaglandins in teleost ovulation: A review of the roles with a view to comparison with prostaglandins in mammalian ovulation. *Mol. Cell Endocrinol.* **2018**, *461*, 236–247. [CrossRef]
10. Di Biase, A.; Lokman, P.M.; Govoni, N.; Casalini, A.; Emmanuele, P.; Parmeggiani, A.; Mordenti, O. Co-treatment with androgens during artificial induction of maturation in female eel, *Anguilla anguilla*: Effect on egg production and early development. *Aquaculture* **2017**, *479*, 508–515. [CrossRef]
11. Asturiano, J.F. Improvement on the reproductive control of the European eel. In *Reproduction in Aquatic Animals: From Basic Biology to Aquaculture Technology*, 1st ed.; Yoshida, M., Asturiano, F., Eds.; Springer: Singapore, 2020. [CrossRef]
12. Politis, S.N.; Syropoulou, E.; Benini, E.; Bertolini, F.; Sørensen, S.R.; Miest, J.J.; Butts, I.A.E.; Tomkiewicz, J. Performance thresholds of hatchery produced European eel larvae reared at different salinity regimes. *Aquaculture* **2021**, *539*, 736651. [CrossRef]
13. Jéhannet, P.; Palstra, A.P.; Heinsbroek, L.T.N.; Kruijt, L.; Dirks, R.P.; Swinkels, W.; Komen, H. What goes wrong during early development of artificially reproduced European eel *Anguilla anguilla*? Clues from the larval transcriptome and gene expression patterns. *Animals* **2021**, *11*, 1710. [CrossRef] [PubMed]
14. Yamamoto, K.; Yamauchi, K. Sexual maturation of Japanese eel and production of eel larvae in the aquarium. *Nature* **1974**, *251*, 220–222. [CrossRef] [PubMed]
15. Lokman, P.M.; Young, G. Induced spawning and early ontogeny of New Zealand freshwater eels (*Anguilla dieffenbachii* and *A. australis*). *N. Z. J. Mar. Freshw. Res.* **2000**, *34*, 135–145. [CrossRef]
16. Kottmann, J.S.; Jørgensen, M.G.P.; Bertolini, F.; Loh, A.; Tomkiewicz, J. Differential impacts of carp and salmon pituitary extracts on induced oogenesis, egg quality, molecular ontogeny and embryonic developmental competence in European eel. *PLoS ONE* **2020**, *15*, e0235617. [CrossRef] [PubMed]
17. Adachi, S.; Ijiri, S.; Kazeto, Y.; Yamauchi, K. Oogenesis in the Japanese eel, *Anguilla japonica*. In *Eel Biology*; Aida, K., Tsukamoto, K., Yamauchi, K., Eds.; Springer: Tokyo, Japan, 2003; pp. 301–317. [CrossRef]

18. Politis, S.N.; Mazurais, D.; Servili, A.; Zambonino-Infante, J.L.; Miest, J.J.; Tomkiewicz, J.; Butts, I.A.E. Salinity reduction benefits European eel larvae: Insights at the morphological and molecular level. *PLoS ONE* **2018**, *13*, e0198294. [CrossRef] [PubMed]
19. Da Silva, F.F.G.; Jacobsen, C.; Kjørsvik, E.; Støttrup, J.G.; Tomkiewicz, J. Oocyte and egg quality indicators in European eel: Lipid droplet coalescence and fatty acid composition. *Aquaculture* **2018**, *496*, 30–38. [CrossRef]
20. Kawauchi, H.; Moriyama, S.; Yasuda, A.; Yamaguchi, K.; Shirahata, K.; Kubota, J.; Hirano, T. Isolation and characterization of chum salmon growth hormone. *Arch. Biochem. Biophys.* **1986**, *244*, 542–552. [CrossRef] [PubMed]
21. Degani, G.; Boker, R.; Jackson, K. Growth hormone, gonad development, and steroid levels in female carp. *Comp. Biochem. Physiol. Part C Toxicol. Pharmacol.* **1996**, *115*, 133–140. [CrossRef]
22. Moriyama, S.; Swanson, P.; Larsen, D.A.; Miwa, S.; Kawauchi, H.; Dickhoff, W.W. Salmon thyroid-stimulating hormone: Isolation, characterization, and development of a radioimmunoassay. *Gen. Comp. Endocrinol.* **1997**, *108*, 457–471. [CrossRef]
23. Onuma, T.; Ando, H.; Koide, N.; Okada, H.; Urano, A. AEffects of salmon GnRH and sex steroid hormones on expression of genes encoding growth hormone/prolactin/somatolactin family hormones and a pituitary-specific transcription factor in masu salmon pituitary cells in vitro. *Gen. Comp. Endocrinol.* **2005**, *143*, 129–141. [CrossRef]
24. Minegeshi, Y.; Dirks, R.P.; de Wijze, D.L.; Brittijn, S.A.; Burgerhout, E.; Spaink, H.P.; van den Thillart, G.E.E.J.M. Quantitative bioassays for measuring biologically functional gonadotropins based on eel gonadotropic receptors. *Gen. Comp. Endocrinol.* **2012**, *178*, 145–152. [CrossRef]
25. Le Gac, F.; Blaise, O.; Fostier, A.; Le Bail, P.Y.; Loir, M.; Mourot, B.; Weil, C. Growth hormone (GH) and reproduction: A review. *Fish Physiol. Biochem.* **1993**, *11*, 219–293. [CrossRef] [PubMed]
26. Planas, J.V.; Swanson, P.; Rand-Weaver, M.; Dickhoff, W.W. Somatolactin stimulates *in vitro* gonadal steroidogenesis in coho salmon, *Oncorhynchus kisutch*. *Gen. Comp. Endocrinol.* **1992**, *87*, 1–5. [CrossRef]
27. Benedet, S.; Björnsson, B.T.; Taranger, G.L.; Andersson, E. Cloning of somatolactin alpha, beta forms and the somatolactin receptor in Atlantic salmon: Seasonal expression profile in pituitary and ovary of maturing female broodstock. *Reprod. Biol. Endocrinol.* **2008**, *6*, 42. [CrossRef] [PubMed]
28. Whittington, C.M.; Wilson, A.B. The role of prolactin in fish reproduction. *Gen. Comp. Endocrinol.* **2013**, *191*, 123–136. [CrossRef] [PubMed]
29. Kumar, R.S.; Trant, J.M. Piscine glycoprotein hormone (gonadotropin and thyrotropin) receptors: A review of recent development. *Comp. Biochem. Physiol. B Biochem. Mol. Biol.* **2001**, *129*, 347–355. [CrossRef]
30. Wong, A.O.L.; Zhou, H.; Jiang, Y.; Ko, W.K.W. Feedback regulation of growth hormone synthesis and secretion in fish and the emerging concept of intrapituitary feedback loop. *Comp. Biochem. Physiol. Part A Mol. Integr. Physiol.* **2006**, *144*, 284–305. [CrossRef] [PubMed]
31. Pandofli, M.; Pozzi, A.G.; Canepa, M.; Vissio, P.G.; Shimizu, A.; Maggese, M.C.; Lobo, G. Presence of β-Follicle-Stimulating Hormone and β Luteinizing Hormone transcripts in the brain of *Cichlasoma dimerus* (Perciformes: Cichlidae). *Neuroendocrinology* **2009**, *89*, 27–37. [CrossRef]
32. Lin, C.; Jiang, X.; Hu, G.; Ko, W.K.; Wong, A.O.L. Grass carp prolactin: Molecular cloning, tissue expression intrapituitary autoregulation by prolactin and paracrine regulation by growth hormone and luteinizing hormone. *Mol. Cell Endocrinol.* **2015**, *399*, 267–283. [CrossRef]
33. Mylonas, C.C.; Fostier, A.; Zanuy, S. Broodstock management and hormonal manipulations of fish reproduction. *Gen. Comp. Endocrinol.* **2010**, *165*, 516–534. [CrossRef]
34. Kazeto, Y.; Kohara, M.; Miura, T.; Miura, C.; Yamaguchi, S.; Trant, J.M.; Adachi, S.; Yamauchi, K. Japanese eel follicle-stimulating hormone (Fsh) and luteinizing hormone (Lh): Production of biologically active recombinant Fsh and Lh by Drosophila S2 cells and their differential actions on the reproductive biology. *Biol. Reprod.* **2008**, *79*, 938–946. [CrossRef] [PubMed]
35. Brooks, S.A. Appropriate glycosylation of recombinant proteins for human use. Implication of choice of expression system. *Mol. Biotech.* **2004**, *28*, 241–255. [CrossRef] [PubMed]
36. Kobayashi, M.; Hayakawa, Y.; Park, W.; Banba, A.; Yoshizaki, G.; Kumamaru, K.; Kagawa, H.; Kaki, H.; Nagaya, H.; Sohn, Y.C. Production of recombinant Japanese eel gonadotropins by baculovirus in silkworm larvae. *Gen. Comp. Endocrinol.* **2010**, *16*, 379–386. [CrossRef] [PubMed]
37. Choi, J.H.; Kim, D.J.; Hong, S.M.; Jo, S.J.; Min, K.S.; Sohn, Y.C.; Lee, J.M.; Kusakabe, T. Molecular analysis and bioactivity of luteinizing hormone from Japanese eel, *Anguilla japonica*, produced in silkworm pupae. *Biotechnol. Bioprocess. Eng.* **2016**, *21*, 381–388. [CrossRef]
38. Peñaranda, D.S.; Gallego, V.; Rozenfeld, C.; Herranz-Jusdado, J.G.; Pérez, L.; Gómez, A.; Giménez, I.; Asturiano, J.F. Using specific recombinant gonadotropins to induce spermatogenesis and spermiation in the European eel (*Anguilla anguilla*). *Theriogenology* **2018**, *107*, 6–20. [CrossRef]
39. Ramos-Júdez, S.; Giménez, J.; Gumbau-Pous, J.; Arnold- Cruañes, L.S.; Estévez, A.; Duncan, N. Recombinant Fsh and Lh therapy for spawning induction of previtellogenic and early spermatogenic arrested teleost, the flathead grey mullet (*Mugil cephalus*). *Sci. Rep.* **2022**, *12*, 6563. [CrossRef]
40. Chai, Y.; Tosaka, R.; Abe, T.; Sago, K.; Sago, Y.; Hatanaka, E.; Ijiri, S.; Adachi, S. The relationship between the developmental stage of oocytes in various seasons and the quality of the egg obtained by artificial maturation in the feminized Japanese eel *Anguilla japonica*. *Aquac. Sci.* **2010**, *58*, 269–278. [CrossRef]
41. Ijiri, S.; Tsukamoto, K.; Chow, S.; Kurogi, H.; Adachi, S.; Tanaka, H. Controlled reproduction in the Japanese eel (*Anguilla japonica*), past and present. *Aquac. Europe.* **2011**, *36*, 13–17.

42. Mes, D.; Dirks, R.P.; Palstra, A.P. Simulated migration under mimicked photothermal conditions enhances sexual maturation of farmed European eel (*Anguilla anguilla*). *Aquaculture* **2016**, *452*, 367–372. [CrossRef]
43. Palstra, A.P.; Bouwan, L.J.; Jéhannet, P.; Kruijt, L.; Schipper, H.; Blokland, M.H.; Swinkels, W.; Heinsbroek, L.T.N.; Lokman, P.M. 17α-methyltestosterone and 17β-estradiol implants for the induction of vitellogenesis in feminized European silver eels (*Anguilla anguilla* L.). *Front. Genet.* **2022**, *13*, 969202. [CrossRef]
44. Pankhurst, N.W. Relation of visual changes to the onset of sexual maturation in the European eel *Anguilla anguilla* (L.). *J. Fish Biol.* **1982**, *21*, 127–140. [CrossRef]
45. Kagawa, H.; Horiuchi, Y.; Kasuga, Y.; Kishi, T. Oocyte hydration in the Japanese eel (*Anguilla japonica*) during meiosis resumption and ovulation. *J. Exp. Zool.* **2011**, *311*, 752–762. [CrossRef]
46. Schindelin, J.; Arganda-Carreras, I.; Frise, E.; Kaynig, V.; Longair, M.; Pietzsch, T.; Preibisch, S.; Rueden, C.; Saalfeld, S.; Schmid, B.; et al. Fiji: An open-source platform for biological-image analysis. *Nat. Methods* **2012**, *9*, 676–682. [CrossRef] [PubMed]
47. Unuma, T.; Hasegawa, N.; Sawaguchi, S.; Tanaka, T.; Matsubara, T.; Nomura, K.; Tanaka, H. Fusion of lipid droplets in Japanese eel oocytes: Stage classification and its use as a biomarker for induction of final oocyte maturation and ovulation. *Aquaculture* **2011**, *322–323*, 142–148. [CrossRef]
48. Koressaar, T.; Remm, M. Enhancements and modifications of primer design program Primer3. *Bioinformatics* **2007**, *23*, 1289–1291. [CrossRef]
49. Untergasser, A.; Cutcutache, I.; Koressaar, T.; Ye, J.; Faircloth, B.C.; Remm, M.; Rozen, S.G. Primer3–new capabilities and interfaces. *Nucl. Acids Res.* **2012**, *40*, e115. [CrossRef]
50. Setiawan, A.N.; Lokman, P.M. The use of reference gene selection programs to study the silvering transformation in a freshwater eel *Anguilla australis*: A cautionary tale. *BMC Mol. Biol.* **2010**, *11*, 75. [CrossRef]
51. Morini, M.; Peñaranda, D.S.; Vílchez, M.; Nourizadeh-Lillabadi, R.; Lafont, A.G.; Dufour, S.; Asturiano, J.F.; Weltzien, F.A.; Pérez, L. Nuclear and membrane progestin receptors in the European eel: Characterization and expression in vivo through spermatogenesis. *Comp. Biochem. Physiol. Part A Mol. Integr. Physiol.* **2017**, *207*, 79–92. [CrossRef]
52. Zadmajid, V.; Falahatimarvast, A.; Damsteegt, E.L.; Setiawan, A.N.; Ozaki, Y.; Shoae, A.; Lokman, P.M. Effects of 11-ketotestosterone and temperature on inhibin subunit mRNA levels in the ovary of the shortfinned eel, *Anguilla australis*. *Comp. Biochem. Phys. B.* **2015**, *187*, 14–21. [CrossRef]
53. Maugars, G.; Dufour, S. Demonstration of the coexistence of duplicated LH receptors in teleosts, and their origin in ancestral Actinopterygians. *PLoS ONE* **2015**, *10*, e0135184. [CrossRef]
54. Livak, K.J.; Schmittgen, T.D. Analysis of relative gene expression data using real- time quantitative PCR and the $2^{-\Delta\Delta CT}$ method. *Methods* **2001**, *25*, 402–408. [CrossRef]
55. Jéhannet, P.; Heinsbroek, L.T.N.; Palstra, A.P. Ultrasonography to assist with timing of spawning in European eel. *Theriogenology* **2017**, *101*, 73–80. [CrossRef]
56. Bureau du Colombier, S.; Jacobs, L.; Gesset, C.; Elie, P.; Lambert, P. Ultrasonography as a non-invasive tool for sex determination and maturation monitoring in silver eels. *Fish. Res.* **2015**, *164*, 50–58. [CrossRef]
57. Kahn, I.A.; Lopez, E.; Leloup-Hâtey, J. Induction of spermatogenesis and spermiation by a single injection of human chorionic gonadotropin in intact and hypophysectomised immature European eel (*Anguilla anguilla* L.). *Gen. Comp. Endocrinol.* **1987**, *68*, 91–103. [CrossRef]
58. Pérez, L.; Asturiano, J.F.; Tomás, A.; Zegrari, S.; Barrera, R.; Espinós, F.J.; Navarro, J.C.; Jover, M. Induction of maturation and spermiation in the male European eel: Assessment of sperm quality throughout treatment. *J. Fish Biol.* **2005**, *57*, 1488–1504. [CrossRef]
59. Peñaranda, D.S.; Pérez, L.; Gallego, V.; Barrera, R.; Jover, M.; Asturiano, J.F. European eel sperm diluent for short-term storage. *Reprod. Domest. Anim.* **2010**, *45*, 407–415. [CrossRef] [PubMed]
60. Ko, H.; Park, W.; Kim, D.J.; Kobayashi, M.; Sohn, Y.C. Biological activities of recombinant Manchurian trout FSH and LH: Their receptor specificity, steroidogenic and vitellogenic potencies. *J. Mol. Endocrinol.* **2007**, *38*, 99–111. [CrossRef]
61. Palstra, A.P.; Jéhannet, P.; Heinsbroek, L.T.N.; Lokman, P.M.; Vesala, S.; Tulonen, J.; Lakka, T.; Saukkonen, S. First observation of a spontaneously matured female European eel (*Anguilla anguilla*). *Sci. Rep.* **2020**, *10*, 2339. [CrossRef]
62. Palstra, A.P. Energetic Requirements and Environmental Constraints of Reproductive Migration and Maturation of European Silver eel (Anguilla anguilla L.). 2006. Available online: https://hdl.handle.net/1887/4926 (accessed on 23 January 2023).
63. Palstra, A.P.; van den Thillart, G.E.E.J.M. Swimming physiology of European silver eels (*Anguilla anguilla* L.): Energetic costs and effects on sexual maturation and reproduction. *Fish Physiol. Biochem.* **2010**, *36*, 297–322. [CrossRef]
64. Kim, N.S.L.; Park, C.W.; Kim, D.W.; Park, H.K.; Byambaragchaa, M.; Lee, N.S.; Hong, S.M.; Seo, M.Y.; Kang, M.H.; Min, K.S. Production and characterization of monoclonal antibodies against recombinant tethered follicle-stimulating hormone from Japanese eel *Anguilla japonica*. *Gen. Comp. Endocrinol.* **2016**, *233*, 8–15. [CrossRef]
65. Kucharczyk, D.; Czarkowski, T.; Nowosad, J.; Targonska, K.; Kupren, K.; Wyszomirska, E.; Kujawa, R.; Horváth, L.; Müller, T. Influence of temperature on successful European eel female maturation under controlled conditions. *Turkish J. Fish. Aquat. Sci.* **2016**, *16*, 477–482. [CrossRef]
66. Tanaka, T.; Adachi, S.; Nomura, K.; Tanaka, H.; Unuma, T. Effects of rearing temperature manipulation on oocyte maturation progress in Japanese eel. *Fish. Sci.* **2021**, *87*, 681–691. [CrossRef]

67. Zhu, Y.; Rice, C.D.; Pang, Y.; Pace, M.; Thomas, P. Cloning, expression, and characterization of a membrane progestin receptor and evidence it is an intermediary in meiotic maturation of fish oocytes. *Proc. Natl. Acad. Sci. USA* **2003**, *100*, 2231–2236. [CrossRef]
68. Thomas, P.; Pang, Y.; Zhu, Y.; Detweiler, C.; Doughty, K. Multiple rapid progestin actions and progestin membrane receptor subtypes in fish. *Steroids* **2004**, *69*, 567–573. [CrossRef]
69. Kazeto, Y.; Goto-Kazeto, R.; Trant, J.M. Membrane-bound progestin receptors in channel catfish and zebrafish ovary: Changes in gene expression associated with the reproductive cycles and hormonal reagents. *Gen. Comp. Endocrinol.* **2005**, *142*, 204–211. [CrossRef] [PubMed]
70. Hagiwara, A.; Ogiwara, K.; Katsu, Y.; Takahashi, T. Luteinizing hormone-induced expression of *Ptger4b*, a prostaglandin E2 receptor indispensable for ovulation of the medaka *Oryzias latipes*, is regulated by a genomic mechanism involving nuclear progestin receptor. *Biol. Reprod.* **2014**, *90*, 1–14. [CrossRef] [PubMed]
71. Hagiwara, A.; Ogiwara, K.; Takahashi, T. Expression of membrane progestin receptors (mPRs) in granulosa cells of medaka preovulatory follicles. *Zool. Sci.* **2016**, *33*, 98–105. [CrossRef]
72. Mourot, B.; Nguyen, T.; Fostier, A.; Bobe, J. Two unrelated putative membrane-bound progestin receptors, progesterone membrane receptor component I (PGMRCI) and membrane progestin receptor (mPR) beta, are expressed in the rainbow trout oocyte and exhibit similar ovarian expression patterns. *Reprod. Biol. Endocrinol.* **2006**, *4*, 6. [CrossRef] [PubMed]
73. Thomas, P.; Pinter, J.; Das, S. Upregulation of the maturation-inducing steroid membrane receptor in spotted seatrout ovaries by gonadotropin during oocyte maturation and its physiological significance. *Biol. Reprod.* **2001**, *64*, 21–29. [CrossRef]
74. Thomas, P.; Dressing, G.; Pang, Y.; Berg, H.; Tubbs, C.; Benninghoff, A.; Doughty, K. Progestin, estrogen and androgen G-protein coupled receptors in fish gonads. *Steroids* **2006**, *71*, 310–316. [CrossRef]
75. Aizen, J.; Pang, Y.; Harris, C.; Converse, A.; Zhu, Y.; Aguirre, M.A.; Thomas, P. Roles of progesterone receptor membrane component 1 and membrane progestin receptor alpha in regulation of zebrafish oocyte maturation. *Gen. Com. Endocrinol.* **2018**, *263*, 51–61. [CrossRef]
76. Kazeto, Y.; Mayuko, K.; Ryoata, T.; Koichiro, G.; Yokoyama, M.; Miura, C.; Miura, T.; Adachi, S.; Yamauchi, K. Molecular characterization and gene expression of Japanese eel (*Anguilla anguilla*) gonadotropin receptors. *Zool. Sci.* **2012**, *29*, 204–211. [CrossRef]
77. Sambroni, E.; Le Gac, F.; Breton, B.; Lareyre, J.J. Functional specificity of the rainbow trout (*Oncorhynchus mykiss*) gonadotropin receptors as assayed in mammalian cell line. *J. Endocrinol.* **2007**, *195*, 213–228. [CrossRef]
78. Bobe, J.; Nguyen, T.; Jalabert, B. Targeted gene expression profiling in rainbow trout (*Oncorhynchus mykiss*) Ovary during maturational competence acquisition and oocyte maturation. *Biol. Reprod.* **2004**, *71*, 73–82. [CrossRef] [PubMed]
79. Crowder, C.M.; Lassiter, C.S.; Gorelick, D.A. Nuclear androgen receptor regulates testes organization and oocyte maturation in zebrafish. *Endocrinology* **2018**, *159*, 980–993. [CrossRef]
80. Li, J.; Huang, D.; Sun, X.; Li, X.; Cheng, C.H.K. Zinc mediates the action of androgen in acting as a downstream effector of luteinizing hormone on oocyte maturation in zebrafish. *Biol. Reprod.* **2019**, *100*, 468–478. [CrossRef] [PubMed]
81. Robker, R.; Akison, L.K.; Russell, D.L. Control of oocyte release by progesterone-regulated gene expression. *Nucl. Recept. Signal.* **2009**, *7*, nrs-07012. [CrossRef]
82. Tang, H.; Liu, Y.; Li, J.; Yin, Y.; Li, G.; Chen, Y.; Li, S.; Zhang, Y.; Lin, H.; Liu, X.; et al. Gene knockout of nuclear progesterone receptor provides insights into the regulation of ovulation by Lh signalling in zebrafish. *Sci. Rep.* **2016**, *6*, 28545. [CrossRef]
83. Liu, D.T.; Carter, N.J.; Wu, X.J.; Hong, W.S.; Chen, S.X.; Zhu, Y. Progestin and nuclear progestin receptor are essential for upregulation of metalloproteinase in zebrafish preovulatory follicles. *Front. Endocrinol.* **2018**, *9*, 517. [CrossRef] [PubMed]
84. Yamauchi, K. Studies of gonadal steroids involved in final gonadal maturation in the Japanese eel, *Anguilla japonica*, a review. *Int. Revue Ges. Hydrobiol.* **1990**, *75*, 859–860. [CrossRef]

fishes

MDPI

Article

Temporal Pattern of the Occurrence of Japanese Glass Eels (*Anguilla japonica*) in the Pearl River Estuary

Fangmin Shuai [1,*], Jie Li [1], Shunchao Yu [2] and Jian Yang [3]

1 Pearl River Fisheries Research Institute, Chinese Academy of Fishery Sciences, Guangzhou 510380, China
2 Key Laboratory of the Pearl River Estuary Regulation and Protection of Ministry of Water Resources, Pearl River Hydraulic Scientific Research Institute, Guangzhou 510611, China
3 Key Laboratory of Ecological Environment and Resources of Inland Fisheries, Freshwater Fisheries Research Center, Chinese Academy of Fishery Sciences, Wuxi 214081, China
* Correspondence: shuai6662000@aliyun.com

Abstract: Japanese eels (*Anguilla japonica*) are a typical migratory fish species with high commercial importance. The Pearl River estuary in southern China is an important natural growing ground for Japanese glass eels, but limited information on Japanese glass eel population characteristics is available, despite their ecological importance. In this paper, we examined the annual patterns of the occurrence of Japanese glass eels in the Pearl River estuary from 2011 to 2022. The most frequently occurring Japanese glass eel's total length is 5.3 cm. The collecting period extended from December to February, and the collection catch-per-unit-effort (CPUE) decreased significantly from 2011 to 2022. The generalized linear model (GLM) indicated that daily changes in Japanese glass eel collection were significantly affected by tidal range, water temperature, and lunar distance. The catch peak appeared when the tidal range rose to 1.7 m, and the water temperature dropped below 8 °C on the full moon days. Overall CPUE analysis showed no significant periodic and inter-annual variability in the period 2011–2022, with the *ARIMA* model suggesting that the CPUE is expected to remain stable but low in the coming years (2023–2026), although recruitment ultimately depends on the overall spawning stock.

Keywords: *Anguilla japonica*; glass eels; Pearl River estuary; temporal pattern; CPUE; *ARIMA* model

Key Contribution: A low but stable abundance of glass eels was recorded in the Pearl River estuary over the period 2001–2022. The collection catch-per-unit-effort (CPUE) decreased significantly. Daily changes in Japanese glass eel collection were significantly affected by tidal range, water temperature, and lunar distance.

Citation: Shuai, F.; Li, J.; Yu, S.; Yang, J. Temporal Pattern of the Occurrence of Japanese Glass Eels (*Anguilla japonica*) in the Pearl River Estuary. *Fishes* **2023**, *8*, 256. https://doi.org/10.3390/fishes8050256

Academic Editor: Jose Martin Pujolar

Received: 10 April 2023
Revised: 4 May 2023
Accepted: 8 May 2023
Published: 11 May 2023

1. Introduction

Global fishery resources are being increasingly threatened by environmental stressors and habitat deterioration, which result from various human activities [1–3]. Migratory fishes, in particular, are facing increased human threats worldwide (such as loss of habitat and the obstruction of migration pathways) that affect their migration and population characteristics. Thus, understanding migration dependency, including the number of offspring ultimately contributing to population regeneration, is essential to clarify population replenishment mechanisms and is crucial for refining the stock protection of *Anguilla* species [4–6].

The Japanese eel is widely distributed throughout East Asia, from south-eastern China, Korea, and North to the Pacific coast of the Hokkaido Island of Japan [7]. Japanese eels migrate to their fixed spawning grounds (at the southern West Mariana Ridge) [8,9] after reaching sexual maturity. After hatching, the Japanese glass eels (called leptocephali) are transported by the westward North Equatorial Current (NEC) and the northward Kuroshio Current toward the continental shelf, where they metamorphose into glass eels, becoming pigmented elvers in estuaries [10,11]. However, the Japanese eels have sharply dropped in many different regions due to overfishing (including growth overfishing and recruitment overfishing) and habitat loss [12–14]. The Japanese eel (*Anguilla japonica*) is an occasional species in the Pearl River water system, from the estuary to the upper reaches of the Hongshuihe River.

The Japanese eel is considered to be a luxury food in China. Thus, it is an important aquaculture species with high commercial importance due to its palatability and resulting popularity with consumers. China is one of the largest aquaculture and export countries of Japanese eel in the world. The annual output of Japanese eel in China is about 220,000 tons, accounting for about 70% of the world's total output, and thus, this species supports an important fishery industry in China [15].

However, artificial breeding technologies for this species have not yet been established due to their complicated life cycle, although some success has been achieved recently [16,17]. The glass eels required for aquaculture are all dependent on the exploitation of wild resources, which has led to unsustainable glass eel fishing patterns in different Asian estuaries and coastal waters [18]. As a result, continuous high-pressure seedling fishing has led to further depletion of the natural Japanese glass eel population. Wild populations have drastically decreased, or disappeared completely, from most estuaries since the 1960s [12,19,20]. Since approximately 2010, annual recruitment has decreased by as much as 90% compared to catches in the 1960s. Japanese eel has also been listed as endangered on the International Union for Conservation of Nature (IUCN) Red List of Threatened Species [21].

As Japanese eel resources continue to diminish in aquatic ecosystems worldwide, it is necessary to understand the recruitment dynamics and the environmental factors that affect the annual abundance of Japanese eels. Obtaining such knowledge is a crucial step towards effectively managing, conserving, and restoring Japanese eel resources. Local monitoring of annual glass eels arrivals is an essential step to understanding the recruitment trend.

The Pearl River estuary is characterized by an average annual temperature of 23 °C, with diverse and abundant aquatic resources, and is an important pathway for glass eels to enter the Pearl River for their continental life phase [22]. However, the Pearl River estuary has experienced numerous anthropogenic disturbances due to the massive increase in the human population and overexploitation of the coastline. Such disturbances are known to negatively impact the upstream migration of glass eels [18]. However, research on glass eels in the Pearl River estuary is scarce, with few studies reported on their populations in local journals [18].

Therefore, based on an 11-year survey of the Pearl River estuary from 2011 to 2022, the purpose of this study was to (1) identify the temporal pattern of the occurrence of glass eels in the Pearl River estuary; (2) detect the relationships between abiotic factors and the CPUE of glass eels; and (3) describe future migration trends of Japanese glass eel in the Pearl River Estuary for this species to achieve early management knowledge using advanced mathematical methods. The main objective of this study was to improve the understanding of the recruitment patterns of glass eels in the Pearl River estuary and put forward targeted conservation strategies to ultimately contribute to the sustainable use of this species.

2. Materials and Methods

2.1. Study Site

Our study site was located in a traditional Japanese glass eel collecting zone of the Pearl River estuary in Jiwan District, Zhuhai (21°59′52″ N, 113°28′31″ E) (Figure 1). This area is an important pathway for glass eels to migrate upstream to the Pearl River for their continental life phase. The study site is also characterized by an irregular semidiurnal tidal cycle, experiencing two ebb and two flood tides each day, with a tidal range of 0.95–3.41 m.

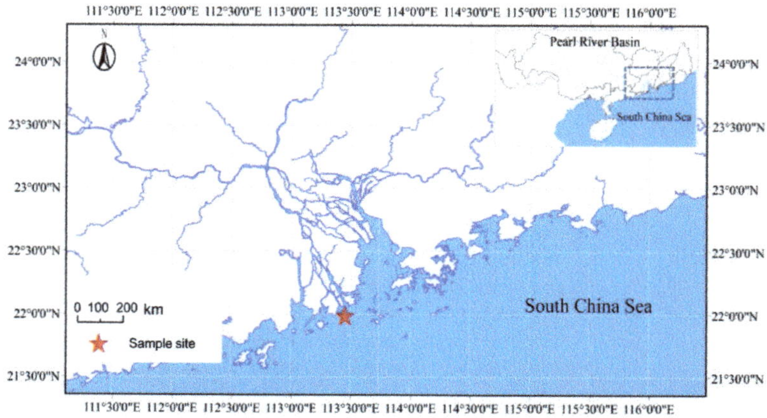

Figure 1. Location of the glass eel collecting site in the Pearl River estuary.

2.2. Data Collection

Japanese glass eel samples were collected from the study site during the migration season (from 1 December to 28 February) from 2011 to 2022 using a custom-made set net. The fixed set net was made of polyethylene thread and was in the shape of a tapered, gradually shrinking from the net mouth to the rear end. The net was 9 m long and consisted of three sections with different mesh sizes (from the entrance to the Japanese glass eel-collecting box, mesh 0.8 mm, 0.5 mm, and 0.355 mm were used, respectively). The tail end of the net mouth was connected with a floating seedling collecting box (Figure 2). The net mesh of the seedling collecting box was made of 0.25 mm mesh. This type of net is widely used in southern China to catch glass eels. Two expanded polystyrene floats were installed at the net mouth so that the net mouth was always oriented to face toward the current, and the fishing efficiency was the same during the ebb and flood tide. The fishing depth varied according to the state of the tide but was usually midwater, ranging from 0.5–1.2 m below the surface, as tidal currents flowed past due to the upward buoyancy of the expanded polystyrene floats.

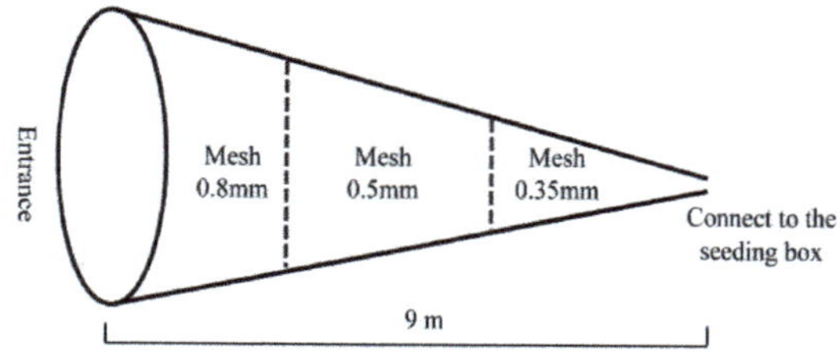

Figure 2. Schematic of the investigation set net.

Because Japanese glass eel migration occurs nearly exclusively at night with the rising tide [7], surveys were conducted on a rising tide at night. For each year of the study, the net was set 2 h before high water in the early evening, with the nets fishing until the tide turned to ebb every day.

The collected individuals were immediately measured for their total length (TL, to the nearest 1 mm) and wet body weight (BW, to the nearest 1 mg) on the day they were harvested. The catch depends on how much water passes through the net. At a constant abundance with high current speed, the catch will be large. With the same abundance and lower current speed, the catch will be lower. Thus, the CPUE of Japanese glass eel was quantified as the average number of individuals collected per net each day over a full moon cycle in this study. This will lead to a relatively constant and comparable effort and can be a useful measurement of changes in the abundance of Japanese glass eel over longer time periods, thus stably reflecting the long-term time series data for glass Japanese recruitment in this location, allowing us to compare and assess the time series of recruitment in south China.

We used a method for the staging of the glass eels according to Tesch and White (2008) [23] and Fukuda et al. (2013) [24]. That is, the glass eel stage has the following characteristics: pigmentation on top of the brain, the surface behind the eyes, the anterior surface reaches the eyes, rostral pigment develops to post brain, clear preanal development of mediolateral pigment, postanally over almost entire dorsum, mediolateral pigment reaches middle of the tail, clear development of preanal ventrolateral pigmentation. The elver stage has the following characteristics: Pigment rows along the myosepta are becoming indistinct, the lateral line is still recognizable, the individual melanophores on the head, behind and below the eyes, and the lower jaw are becoming indistinct.

Tide level (m) data were collected from http://global-tide.nmdis.org.cn// (accessed on 14 May 2022). In this analysis, we used the tidal level range data, which were the differences between the highest and the lowest tidal level values on each day. Dissolved oxygen (DO, $mg{\cdot}L^{-1}$), pH, chemical oxygen demand (COD, $mg{\cdot}L^{-1}$), inorganic nitrogen (IN, $mg{\cdot}L^{-1}$), active phosphate (AP, $mg{\cdot}L^{-1}$) and petroleum oil (Oil, $mg{\cdot}L^{-1}$) were selected as water quality parameters of the local habitat, and the data were collected from https://www.mee.gov.cn/ (accessed on 20 May 2022). River discharge (m^3/s) data were obtained through automatic measurement at a gauging station from the Information Center of the Ministry of Water Resources (available at http://xxfb.mwr.cn/index.html) (accessed on 22 May 2022). Salinity levels ($mg{\cdot}L^{-1}$) were converted from the water chlorine content, and the conversion formula was: Salinity = 1.81 $\times$ Chlorine content. Water chlorine content ($mg{\cdot}L^{-1}$) and water temperature (T, °C) were determined by a real-time online chlorine ion detector and a real-time online temperature detector (MADSUR -B6710Cl).

We also analyzed the effect of the lunar distance on the CPUE of glass eels. The lunar distance is the distance between the moon and the earth, represented by the units of earth radii, referring to the moon's perigee (at about 56 earth radii) and apogee (at about 63.8 earth radii). Lunar distance data were collected from the lunar package of the R language. All environment data were organized on a 1-day basis to match the Japanese glass eel collecting time.

The glass eel migration period was defined as the interval in days between the first day that a Japanese glass eel was collected and the last day that a Japanese glass eel was collected in a given year. The date of maximum yield was defined as the date on which the most glass eels were collected each year. We used Spearman rank correlation to evaluate the changes in the CPUE and arrival season each year.

2.3. Data Analysis

A generalized linear model (GLM) was used to assess how environmental factors impact the CPUE of glass eels by selecting the CPUE of glass eels as an independent variable and selecting the tidal range, salinity, DO, pH, COD, IN, AP, Oil, temperature, and lunar distance as response variables. As glass eels are only active at night, GLM analyses were only performed for the night during the main periods of Japanese glass eel entrance in the Pearl River estuary each studied year. Multi-collinearity among environmental factors was tested based on variance-inflation (VIF) < 5, and the model was optimized to: glass eels CPUE ~ log10Salinity + tide range + lunar distance + pH + DO + COD + IN + AP + Oil + Temperature.

2.4. Time Series Analyses

The Auto Regressive Integrated Moving Average (ARIMA) model is the most general class of model for forecasting a time series [25]. ARIMA can be viewed as a "filter" that tries to separate the signal from the noise, and the signal is then extrapolated into the future to obtain forecasts. ARIMA is based on the assumption that the response series is stationary;

that is, the mean and variances of the series are independent of time. It can provide more accurate projections than the ones obtained using weights. The ARIMA model is particularly suitable for predicting nonlinear data and can provide accurate operational forecasts of annual commercial catches. Thus, this model has been an important tool for forecasting fish populations for decades [26].

In this study, we employed a seasonal ARIMA model to analyze the complex Japanese glass eel data that were characterized by an obvious seasonality and non- linear regression. The seasonal ARIMA models were ARIMA(p,d,q) $(P,D,Q)_m$, where p indicates the autoregressive (AR) order, d is the value of differencing orders, q is the moving average (MA) order, and P, D, and Q indicate the seasonal order of AR, differencing, and MA, respectively [26,27].

We differentiated the series until it was stationary. The seasonality period was 52. The Japanese glass eel data were not stationary and presented a high degree of seasonality. Therefore, the series was decomposed separately into the trend effect, seasonal effects, and random variability. We selected the appropriate AR and MA order based on the Autocorrelation function (ACF) and Partial Autocorrelation Function (PACF) [26], and selected the most appropriate prediction model based on the smallest Akaike's Information Criterion (AIC) value.

Autocorrelation is the correlation between a time series and the same time series lag. ACF measures the amount of linear dependence between observations in a time series. While partial autocorrelations are the correlation coefficients between the basic time series and the same time series lag, PACF gives an unrestricted parametrization and represents the plot of partial coefficients of correlation of time series and lags of itself. Both ACF and PACF are commonly used in evaluating a time series variable's dependency on its past. The number of AR and MA terms of the stationary time series is determined by examining the patterns of the graphs of ACFs and PACFs. The confidence limits are provided when ACF or PACF are significantly different from zero [26].

Akaike information criterion is defined as AIC = nlog(MSE) + 2k, where n is the sample size, MSE is the mean square error, and k is the total number of estimable parameters. It provides guidelines for choosing the best possible model from a set of competing models. The established model should be parsimonious and use as few model parameters as possible so that the model fulfills all the diagnostic checks [28]. Hence, the model with minimum AIC is closer to the best possible choice, and the minimum AIC is selected as the optimal model fit to a given data. The sample autocorrelation function (ACF) and partial autocorrelation function (PACF) are useful qualitative tools to assess the presence of autocorrelation at individual lags. The Ljung–Box test is a more quantitative way to test for autocorrelation at multiple lags jointly. In this study, the Ljung Box test was mainly used to evaluate the assumptions of ARIMA models to ensure that the residuals are independent of each other.

Because the samples were collected over two years (December of the previous year, January and February of the following year), we classify the samples collected in the previous year into the next year and take them as the number of samples in the next year when making ARIMA predictions. For example, the samples collected in December in the year 2011 are classified as 2012 samples. All analyses were performed using R 4.1.13 Software [28]. Variables were considered statistically significant at $p < 0.05$.

3. Results

3.1. Population Characteristics of Glass Eels

A total of 42,974 individual glass eels were sampled during the present study. The glass eels collected in our sample ranged in size from 4.8 to 6.0 TL cm with a mean of 5.4 $\pm$ 0.3 cm (SE; Figure 3a). The most frequently occurring Japanese glass eel size was 5.3 cm. Weight ranged from 0.092 to 0.157 g, with a mean of 0.119 $\pm$ 0.02 g (SE; Figure 3b). The most frequently occurring Japanese glass eel weight is 0.11 g. The body color was transparent and not easy to observe with the naked eye, although the pigment began to appear (Figure 4). It can be inferred that the collected glass eels are between glass eels and elver eels.

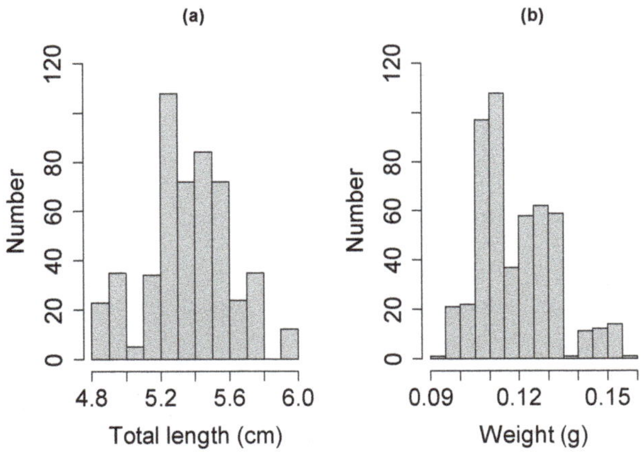

Figure 3. Length frequency distribution (**a**) and weight frequency distribution (**b**) of glass eels collected from the Pearl River Estuary. The measured number of glass eels is 503.

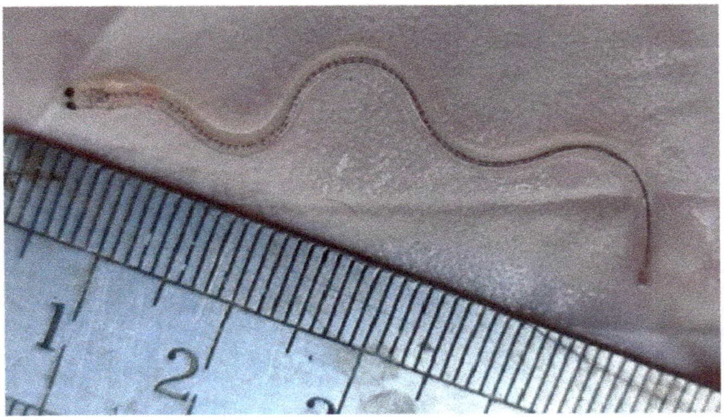

Figure 4. Photo of Japanese glass eel collected in the Pearl River estuary.

Based on the data collected between 2011 and 2022, there was a recurring pattern and steady seasonality within each year. The collection period of glass eels in the Pearl River estuary extended from December of the previous year to February of the subsequent year. The earliest occurrence of glass eels was on 20 December 2021, and the latest was on 5 March 2014. The annual catch of glass eels in each net fluctuated substantially, ranging from 8014 individuals in 2013 to 2256 in 2019. The peak catches lasted for about 15 days, primarily from 10 January to 30 January, and the daily catch remained high (>200 individuals per net). The CPUE of glass eels collected among these peak days accounting for more than 50% of the total amount collected in the whole year. In addition, low catches always appeared toward the start or end of the fishing season, and it appeared that the tidal range had less influence on Japanese glass eel catches during those times. Japanese glass eel CPUE varied significantly from year to year (Figure 5). The CPUE decreased significantly from 2011 to 2022 (Spearman's rho, one-tailed, $p = 0.03$).

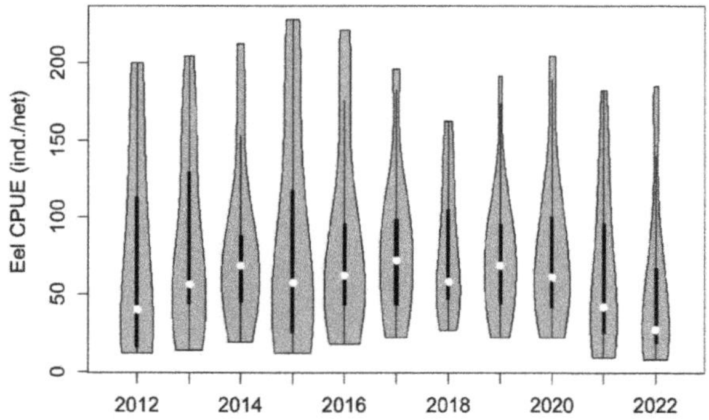

Figure 5. Violin plot of the occurrence of glass eels. The vertical length of each violin graph represents the duration of the Japanese glass eel collection time. The horizontal width of the graph represents the probability of the occurrence of a certain CPUE value. White dots represent the median of CPUE, and the small black rectangular boxes represent the quantiles of CPUE.

3.2. Relationships between Japanese Glass Eel CPUE and Environmental Factors

The GLM model showed that the CPUE was significantly affected by the tidal range ($p = 0.044$), water temperature ($p < 0.001$), and lunar distance ($p = 0.037$), while the impact of other environmental factors (including local water quality; i.e., salinity, DO, pH, COD, inorganic nitrogen, active phosphate, and Oil) was not significant (Figure 6). Thus, tidal range, water temperature and lunar distance contributed to the CPUE of glass eels in the Pearl River estuary.

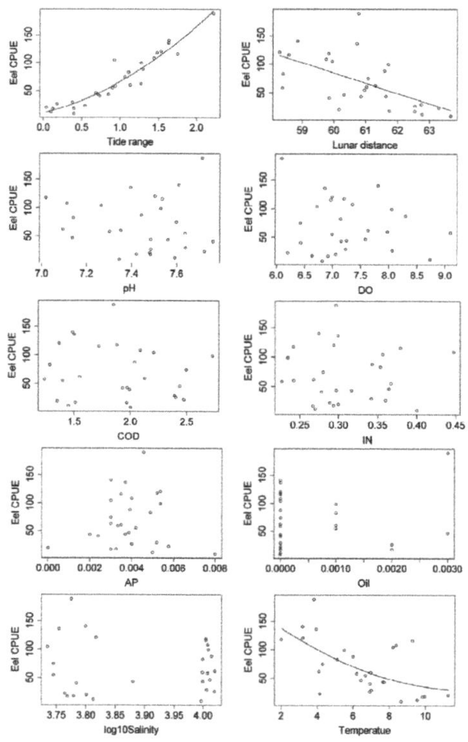

Figure 6. The effects of environmental factors on the catches of glass eels during 2011–2022.

The study revealed that the catch peak often occurred with a sharp rise in the tidal range when it was close to 1.7 m. The lunar distance was also a significant variable, showing a negative relationship with CPUE. Thus, there will be a large number of glass eels on the full moon days with smaller lunar distances in the Pearl River estuary. Water temperature also exhibited a negative relationship with Japanese eel CPUE. When the water temperature was below 8 °C, glass eels were collected in large numbers (more than 50 individuals per net per day) (Figure 6).

3.3. *Recruitment Dynamics Forecasting*

The histogram of forecast errors is stochastic, and their distributions fluctuate around zero (Figure 7). The two horizontal lines in the ACF and PACF plots designate the 95% confidence intervals for the estimated autocorrelation and partial autocorrelation coefficients. The significant spike at lag 0 in the ACF is indicative of a regression AR(0) component. A significant spike at lag 1 in the PACF indicates that an additional non-seasonal term, the MA(1) component, needs to be included in the model (Figure 7). Thus, we select an $\text{ARIMA}(0,0,1)(0,1,0)_{208}$ model as our prediction model based on the smallest AIC value. The forecast errors are shown in Figure 7. Nearly all of the spikes were within the significance limits, and thus, the residuals appear to be white noise. A Ljung–Box test also showed that the residuals have no remaining autocorrelations, and the forecast errors exhibited a normal distribution. All such data indicated that the model worked well.

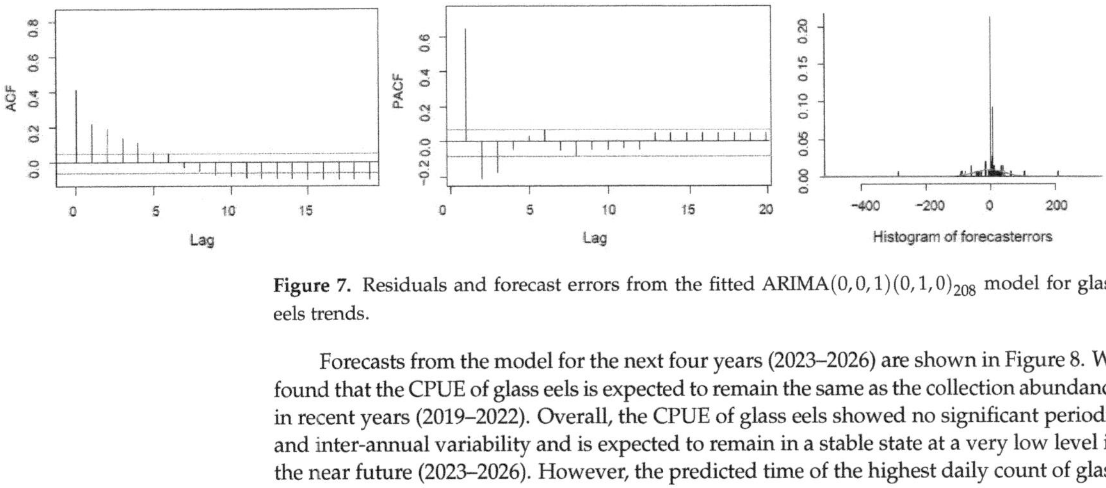

Figure 7. Residuals and forecast errors from the fitted $\text{ARIMA}(0,0,1)(0,1,0)_{208}$ model for glass eels trends.

Forecasts from the model for the next four years (2023–2026) are shown in Figure 8. We found that the CPUE of glass eels is expected to remain the same as the collection abundance in recent years (2019–2022). Overall, the CPUE of glass eels showed no significant periodic and inter-annual variability and is expected to remain in a stable state at a very low level in the near future (2023–2026). However, the predicted time of the highest daily count of glass eels will occur later each year (Spearman's rho, one-tailed, $p = 0.04$) (Figure 9).

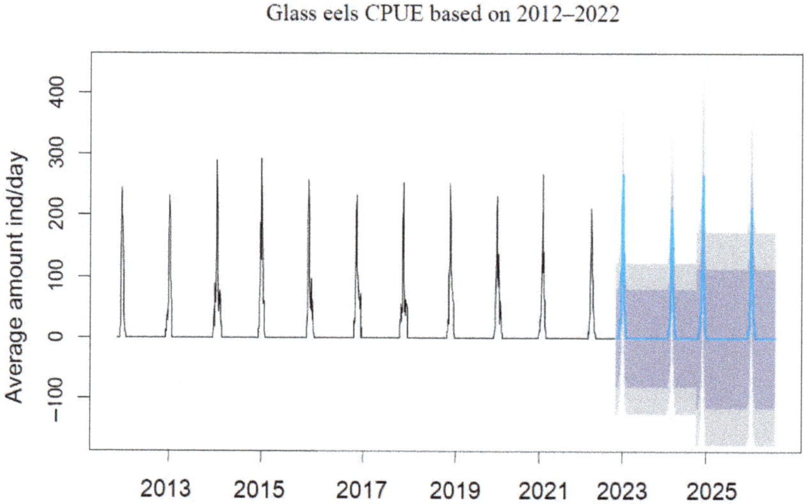

Figure 8. Forecasts of the CPUE of glass eels from 2012 to 2026. The bright blue line is the forecast, and the dark gray area and the light gray area are the 80% and 95% confidence levels, respectively.

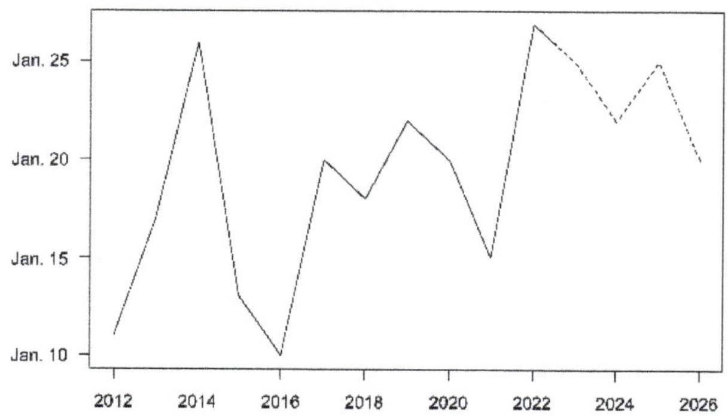

Figure 9. The days on which the maximum number of glass eels were collected. The solid line represents the actual collection data, while the dotted line represents the predicted data.

4. Discussion

4.1. Synchronization of Spawning with the New Moon

It has been widely reported that lunar phase, tidal cycle, day–night rhythm, moonlight, water temperature, salinity, turbidity, water odor, and rainfall can affect the recruitment dynamic of glass eels [29,30], although contradictory results exist between studies [31]. Our results indicated that daily changes in glass eel CPUE were significantly affected by lunar distance, tide range, and water temperature. Under the "New Moon Hypothesis", glass eels do not arrive continuously but in pulses or batches that follow the new moon [32], as has been observed in Taiwan [33], with peaks occurring between the last quarter and first quarter moon periods, with a near one-month periodicity.

Our results indicated that Japanese glass eels exhibited a negative preference for the lunar distance during the migrating season. In our study, CPUE was high during small lunar distances (full moon) and low during large lunar distances (quarter moon). Early studies on *A. japonica* showed that early developmental glass eels stages (VA and VB) were light-sensitive and did not migrate under a full moon [30]. However, a recent study by Cresci [34] on European eels showed that glass eels use lunar cues for orientation and that the arrival of glass eels is lunar dependent on new and full moons. Therefore, the photopathic behavior of glass eels may play an important role in catching abundance. However, moon and half-moon catches seem to differ across studies. In Grey River, New Zealand, the study of Laffaille et al. [31] showed higher catches of *A. japonica* and *A. marmorata* in most months during the new moon relative to the full moon. However, in some months, glass eels collected during the full moon were much higher than during the new moon at Yakushima Island, Japan [35]. This might be due to glass eels at early development stages being non-pigmented and lucifugous, highly influenced by light levels, and hence, showing a tendency to avoid the strong moonlight during the full moon period [36]. On the other hand, glass eels at late development stages are pigmented and with some phototaxis [37]. Accordingly, horizontal distribution experiments found a clear negative phototaxis in leptocephali and metamorphosing glass eels, but no phototaxis was detected after metamorphosis into glass eels [38]. Besides, the effect of the lunar cycle and hence moonlight intensity is modulated by cloud cover and turbidity; the lunar effect is not observed in highly turbid estuaries [20,39], such as in the Yangtze Estuary, China [40].

In regards to tidal range, CPUE peaks often appeared at times corresponding to the full moon. A large tidal range means that there is a large hydrodynamic force, which can easily send glass eels (with weak swimming ability) from tidal flat areas into the freshwater environment [35]. This is a critical step during which glass eels move to freshwater and exchange orientation and depth as the tide changes [34]. However, the daily catch was not always proportional to the increased tidal range at the beginning or end of the fishing season. At the end of the fishing season, in particular, older and more-pigmented glass eels with swimming ability are able to migrate into freshwater without the aid of tides.

Many studies have also shown that water temperature is an important factor that affects eel migration [30,31,40–42]. For instance, glass eels migrating upstream in a New Zealand river showed a clear preference for water temperatures between 12 and 20 °C, with an optimum of 16.5 °C [41]. The results of our study show that larger migrations occur in winter when the water temperature drops below 8 °C, which triggers the migration of glass eels from the sea to the Pearl River estuary. Our finding is quite similar to the results of previous studies with glass eel catches peaking in winter when the temperature was 6–7 °C in the Geum River Estuary, South Korea [30], and 6.2–8.5 °C in the Yangtze Estuary, China [43]. On the other hand, sea surface temperatures in the South China Sea have increased up to 0.44 °C/decade in recent decades [44,45], which might cause glass eels to migrate later in the season. Interestingly, our modeling approach suggested a shift to a later recruitment season, which could be explained by this temperature increase. This is because cold seawater is essential to induce early sexual development in catadromous migratory eels, such as for the increased proliferation and differentiation of specific spermatogonial cells [46]. Eels require cold seawater to promote the activity of the brain–pituitary–gonad (BPG) axis, which induces silvering (eels transform from yellow to silver) [47]. Hence the delay in Yellow eel migration due to increased water temperature might be detrimental as it could indirectly affect reproduction. In addition, the increased temperature has the potential to decrease larval numbers. Some studies have reported that the survival of glass eels declined with the progression of the migratory season [48,49]. Late-season glass eels did not have sufficient energy reserves to transition effectively to the new environment [50].

4.2. Stable but Low Recruitment of Japanese Eel in the Pearl River Estuary

The present study analyzed the daily variation in the catch of Japanese glass eels over 11 successive recruitment seasons (2001–2022) in the Pearl River estuary in China. CPUE analysis suggested a low but stable abundance of glass eels in the Pearl River estuary in the period 2001–2022, with no significant variability within years and across years. Moreover, ARIMA modeling suggested that in the next four years (2023–2026), abundance will also remain low, although recruitment ultimately depends on the overall spawning stock, and future recruitment cannot be predicted on the basis of local recruitment alone. While catches were stable in the period studied, in comparison with historical data, the observed drop in CPUE is apparent.

What are the causes of the low recruitment of Japanese eels in the area? The decline of eel stocks is a global phenomenon affecting not only Japanese eels but also European and American eels [51–54]. Many causes have been put forward to explain the declining trend in eel populations, mostly anthropogenic, including habitat destruction, construction of dams, fishing, and human-introduced parasites and viruses [55–58]. One of the plausible causes of the decline of Japanese eels might be the construction of dams along the Pearl River. At present, there are 32 hydropower stations, each with an installed capacity greater than 100 MW, located in the Pearl River system. The issue with dams is that they might block and impede the spawning migration of adults. The yellow eels live in the rivers for approximately 5–8 years [59]. Later in life, they have a second metamorphosis into silver eels and then migrate back to the spawning site in the waters west of the Mariana Islands, 15° N 140° E [8]. If dams impede the spawning migration, they might contribute that fewer individuals reach the spawning site, resulting in lower recruitment. The fact that dams can stop the spawning migration of silver eels has been observed in European eels, for instance, the dams in the Sardinian rivers' network, Sardinia (Italy) [60], in the River Fremur of northern Brittany (France) [61] and in the Rance estuary (Brittany, France) [62]. Those dams disrupt river connectivity, consequently impeding fish movements to reach feeding and spawning habitats.

In addition, while usually regarded as a tough species that can survive in poor-quality waters, as a fatty fish, eels are particularly sensitive to contamination. Because of its high-fat content and local benthic feeding behavior, the feeding stage is considered extremely prone to the bioaccumulation of pollutants. Most contaminants are highly concentrated in lipid stores [63] and affect lipid metabolism [64–66]. Apart from acute fish toxicity, i.e., the effects of short-term exposure to a chemical, often associated with sudden fish casualties, bioaccumulation of chemicals inside the fish may result in chronic fish toxicity, with sublethal effects being apparent only at specific periods of the fish life (e.g., during maturation of the gonads (endocrine disruption), starvation, reproduction and offspring development) [67]. In particular, heavy metals seriously impact the fitness of eels. In European eels, a significant negative correlation between heavy metal pollution load and condition was observed, suggesting an impact of pollution on the health of sub-adult eels [68].

In our study area, the Pearl River estuary is enriched in nutrients during the high inflows. The concentrations of dissolved inorganic nitrogen and dissolved inorganic phosphorus have gradually increased over the past two decades [69]. High levels of Cu, As, Pb, Cr, and Hg are found, and Cu and Cd pose ecological risks in the Pearl River estuaries [70]. These heavy metals are genotoxic for European eels [71]. Using transcriptomics, the study of Pujolar et al. (2012) [72] on European eels comparing animals from polluted and non-polluted areas suggested that bioaccumulation of pollutants, including heavy metals, can seriously impair the spawning migration of adults. Individuals from polluted environments showed an up-regulation of genes related to detoxification but a down-regulation of metabolic genes involved in the mitochondrial respiratory chain and oxidative phosphorylation. Although the authors did not measure metabolism, the study suggests that pollutants might have a significant effect, possibly resulting in a low energetic status of the fish, pointing to a poor quality of spawners that could jeopardize spawning migration and reproduction. A high contaminant burden could also impair normal reproduction or affect larval development since lipids and lipophilic contaminants are mainly mobilized toward the gonads during the spawning migration [73].

Moreover, the main anthropogenic factor put forward to explain the declining global trend in stocks of American, European and Japanese eels is over-fishing [52,56,74]. In the case of Japanese eels, the average annual eel fishing catch in Japan dropped from 130 tons each year in the 1960s to 7 tons each year in the 1990s [19]. Wild populations have drastically decreased or disappeared completely from most estuaries since the 1960s [12]. Japanese eel has been listed as endangered on the International Union for Conservation of Nature (IUCN) Red List of Threatened Species [21]. In the case of the Pearl River, fisheries have also experienced long-term overfishing, as well as many other numerous anthropogenic disturbances since the 1950s. The current status of the Japanese eel stock is worrying [75].

Currently, in the Pearl River, there are two fishing ban periods, one for freshwater (from 1 March to 30 June) and one for marine (from 1 May to 15 August). Those do not coincide with the time at which glass eels migrate into the Pearl River (December–February). What could be done to alleviate the situation? One alternative would be to ban fishing completely. However, this is not realistic because Japanese eels are important aquaculture species in China. The annual output of eel in China is about 220,000 tons, accounting for about 70% of the world's total output. Japanese eels support an important fishery industry in China [15]. It is also difficult to alter the banning period so that it covers the migration period (December–February), as the current bans are 1 March to 30 June in freshwater and 1 May to 15 August in oceans. These times are quite remote from the glass eels' migrating time. Another solution could be the introduction of quotas, limiting the number of fish that can be caught. In the case of New Zealand's freshwater eel stocks, Maori, New Zealand's indigenous people, were allocated 20% of the commercial quota, with an additional quota set for customary take [76]. In the case of European eels, quotas have been used since 2004, and apart from small catches of glass eels for research purposes, it has not been legal in the EU to import and export eels since 2010 [54].

Author Contributions: Conceptualization, F.S. and J.Y.; methodology, validation, investigation, F.S.; data curation, F.S. and S.Y.; writing—original draft preparation, F.S.; writing—review and editing, J.L. and J.Y.; visualization, F.S.; supervision, J.L. and J.Y.; funding acquisition, F.S. and J.L. All authors have read and agreed to the published version of the manuscript.

Funding: This research was funded by the National Natural Science Foundation of China (31870527), the Science and Technology Program of Guangzhou, China (202201010761), the China-ASEAN Maritime Cooperation Fund (CAMC–2018F) and the Project of Innovation Team of Survey and Assessment of the Pearl River Fishery Resources (2020TD-10).

Institutional Review Board Statement: The collection and use of glass eels complied with the Laws of the People's Republic of China on the Protection of Wildlife. It is approved by Laboratory Animal Ethics Committee Pearl River Fisheries Research Institute, CAFS. (No. LAEC-PRFRI-2023-03- 14).

Data Availability Statement: The data that support the findings of this study are available through the Pearl River Fisheries Research Institute, Chinese Academy of Fishery Sciences (contact: Fangmin Shuai; sfm@prfri.ac.cn).

Acknowledgments: We are deeply grateful to Zhuhai Detachment of the Guangdong Fishing Administrative Brigade for their assistance in the field sample collection.

Conflicts of Interest: The authors declare no conflict of interest.

References

1. Sala, O.E.; Chapin, F.S.; Armesto, J.J.; Berlow, E.; Bloomfield, J.; Dirzo, R.; Huber-Sanwald, E.; Huenneke, L.F.; Jackson, R.B.; Kinzig, A.; et al. Global biodiversity scenarios for the year 2100. *Science* **2000**, *287*, 1770–1774. [CrossRef] [PubMed]
2. Dulvy, N.K.; Sadovy, Y.; Reynolds, J.D. Extinction vulnerability in marine populations. *Fish Fish.* **2003**, *4*, 25–64. [CrossRef]
3. Stenseth, N.C.; Rouyer, T. Ecology: Destabilized fish stocks. *Nature* **2008**, *452*, 825–826. [CrossRef] [PubMed]
4. Anderson, C.N.; Hsieh, C.H.; Sandin, S.A.; Hewitt, R.; Hollowed, A.; Beddington, J.; May, R.M.; Sugihara, G. Why fishing magnifies fluctuations in fish abundance. *Nature* **2008**, *452*, 835–839. [CrossRef]
5. Mangel, M.; Brodziak, J.; DiNardo, G. Reproductive ecology and scientific inference of steepness: A fundamental metric of population dynamics and strategic fisheries management. *Fish Fish.* **2010**, *11*, 89–104. [CrossRef]
6. Righton, D.; Piper, A.; Aarestrup, K.; Amilhat, E.; Belpaire, C.; Casselman, J.; Castonguay, M.; Díaz, E.; Dörner, H.; Faliex, E.; et al. Important questions to progress science and sustainable management of anguillid eels. *Fish Fish.* **2021**, *22*, 762–788. [CrossRef]
7. Tesch, F.W. *The Eel*, 5th ed.; Wiley-Blackwell: Oxford, UK, 2003; p. 436.
8. Tsukamoto, K. Discovery of the spawning area for Japanese eel. *Nature* **1992**, *356*, 789–791. [CrossRef]

9. Kurogi, H.; Okazaki, M.; Mochioka, N.; Jinbo, T.; Hashimoto, H.; Takahashi, M.; Tawa, A.; Aoyama, J.; Shinoda, A.; Tsukamoto, K.; et al. First capture of post-spawning female of the Japanese eel *Anguilla japonica* at the southern West Mariana Ridge. *Fish. Sci.* **2011**, *77*, 199–205. [CrossRef]
10. Cheng, P.; Tzeng, W. Timing of metamorphosis and estuarine arrival across the dispersal range of the Japanese eel *Anguilla japonica*. *Mar. Ecol. Prog. Ser.* **1996**, *131*, 87–96. [CrossRef]
11. Tsukamoto, K. Oceanic migration and spawning of Anguillid eels. *J. Fish. Biol.* **2009**, *74*, 1833–1852. [CrossRef]
12. Sullivan, M.C.; Able, K.W.; Hare, J.A.; Walsh, H.J. *Anguilla rostrata* glass eel ingress into two, US east coast estuaries: Patterns, processes and implications for adult abundance. *J. Fish Biol.* **2006**, *69*, 1081–1101. [CrossRef]
13. Tanaka, E. Stock assessment of Japanese eels using Japanese abundance indices. *Fish. Sci.* **2014**, *80*, 1129–1144. [CrossRef]
14. Kaifu, K.; Yokouchi, K. Increasing or decreasing?—Current status of the Japanese eel stock. *Fish. Res.* **2019**, *220*, 105348. [CrossRef]
15. MARA (Ministry of Agriculture and Rural Affairs of the People's Republic of China). *Chinese Fishery Statistical Yearbook*; China Agriculture Press: Beijing, China, 2018–2022.
16. Masuda, Y.; Imaizumi, H.; Oda, K. Artificial completion of the Japanese eel, *Anguilla japonica*, life cycle: Challenge to mass production. *Bull. Fish. Res. Agency* **2012**, *35*, 111–117.
17. Tanaka, H. Progression in artificial seedling production of Japanese eel *Anguilla japonica*. *Fish. Sci.* **2015**, *81*, 11–19. [CrossRef]
18. Shuai, F.M.; Li, Z.Q.; Liu, G.W.; Li, X.H.; Li, Y.F.; Yang, J.P.; Li, J. Resource status of Japanese eel (*Anguilla japonica*) in the Pearl River Estuary. *South China Fish. Sci.* **2015**, *11*, 85–89. (In Chinese)
19. Tatsukawa, K. *Eel Resources in East Asia. Eel Biology*; Springer: Berlin/Heidelberg, Germany, 2003; pp. 293–298.
20. Harrison, A.J.; Walker, A.M.; Pinder, A.C.; Briand, C.; Aprahamian, M.W. A review of glass eel migratory behaviour, sampling techniques and abundance estimates in estuaries: Implications for assessing recruitment, local production and exploitation. Rev. *Fish Biol. Fish.* **2014**, *24*, 967–983. [CrossRef]
21. Jacoby, D.M.P.; Casselman, J.M.; Crook, V.; DeLucia, M.B.; Ahn, H.; Kaifu, K.; Kurwie, T.; Sasal, P.; Silfvergrip, A.M.C.; Smith, K.G.; et al. Synergistic patterns of threat and the challenges facing global anguillid eel conservation. *Glob. Ecol. Conserv.* **2015**, *4*, 321–333. [CrossRef]
22. Lu, K.X. *Fishery Resources in the Pearl River System*; Guangdong Science and Technology Press: Guangzhou, China, 1990.
23. Tesch, F.W.; White, R.J. *The Eel*; Blackwell Publishing: London, UK, 2008; pp. 9–24.
24. Fukuda, N.; Miller, M.J.; Aoyama, J.; Shinoda, A.; Tsukamoto, K. Evaluation of the pigmentation stages and body proportions from the glass eel to yellow eel in *Anguilla japonica*. *Fish. Sci.* **2013**, *79*, 425–438. [CrossRef]
25. Kim, J.Y.; Jeong, H.C.; Kim, H.; Kang, S. Forecasting the monthly abundance of anchovies in the South Sea of Korea using a univariate approach. *Fish. Res.* **2015**, *161*, 293–302. [CrossRef]
26. Box, G.E.; Jenkins, G.M.; Reinsel, G.C.; Ljung, G.M. *Time Series Analysis: Forecasting and Control*, 5th ed.; John Wiley & Sons: Hoboken, NJ, USA, 2015.
27. Tsitsika, E.V.; Maravelias, C.D.; Haralabous, J. Modeling and forecasting pelagic fish production using univariate and multivariate ARIMA models. *Fish. Sci.* **2007**, *73*, 979–988. [CrossRef]
28. R Development Core Team. *R: A Language and Environment for Statistical Computing*; R Foundation for Statistical Computing: Vienna, Austria, 2021; Available online: http://www.r-project.org/ (accessed on 1 January 2022).
29. Tzeng, W.N. Immigration timing and activity rhythms of the eel, *Anguilla japonica*, elvers in the estuary of northern Taiwan, with emphasis on environmental influences. *Bull. Jap. Soc. Fish. Oceanogr.* **1985**, *47*, 11–28.
30. Hwang, S.D.; Lee, T.W.; Choi, I.S.; Hwang, S.W. Environmental factors affecting the daily catch levels of *Anguilla japonica* glass eels in the Geum River estuary, South Korea. *J. Coast. Res.* **2014**, *297*, 954–960. [CrossRef]
31. Jellyman, D.; Lambert, P. Factors affecting recruitment of glass eels into the Grey River, New Zealand. *J. Fish Biol.* **2003**, *63*, 1067–1079. [CrossRef]
32. Tsukamoto, K.; Otake, T.; Mochioka, N.; Lee, T.W.; Fricke, H.; Inagaki, T.; Aoyama, J.; Ishikawa, S.; Kimura, S.; Miller, M.J.; et al. Seamounts, new moon and eel spawning: The search for the spawning site of the Japanese eel. *Environ. Biol. Fish.* **2003**, *66*, 221–229. [CrossRef]
33. Han, Y.S.; Wu, C.R.; Iizuka, Y. Batch-like Arrival Waves of Glass Eels of Anguilla japonica in Offshore Waters of Taiwan. *Zool. Stud.* **2016**, *55*, e36.
34. Cresci, A.A. Comprehensive hypothesis on the migration of European glass eels (*Anguilla anguilla*). *Biol. Rev.* **2020**, *95*, 1273–1286. [CrossRef]
35. Yamamoto, T.; Mochioka, N.; Nakazono, A. Seasonal occurrence of anguillid glass eels at Yakushima Island, Japan. *Fish. Sci.* **2001**, *67*, 530–532. [CrossRef]
36. Helme, H.; Bertucci, F.; Moussa, R.M.; Wolff, Y.; Sasal, P. Temporal dynamics of the recruitment of glass eels in two valleys of French Polynesia (Tahiti and Moorea Islands). *Cybium* **2018**, *42*, 341–348.
37. Chow, S.; Masuda, Y.; Satomi, M.; Kamoshida, M.; Takahashi, M. A method to separate eel leptocephalus larvae from turbid water by controlling the light environment. *Nippon Suisan Gakkaishi* **2019**, *85*, 585–590. [CrossRef]
38. Yamada, Y.; Okamura, A.; Mikawa, N.; Utoh, T.; Horie, N.; Tanaka, S.; Miller, M.J.; Tsukamoto, K. Ontogenetic changes in phototactic behavior during metamorphosis of artificially reared Japanese eel *Anguilla japonica* larvae. *Mar. Ecol. Prog. Ser.* **2009**, *379*, 241–251. [CrossRef]

39. Laffaille, P.; Caraguel, J.M.; Legault, A. Temporal patterns in the upstream migration of European glass eels (*Anguilla anguilla*) at the Couesnon estuarine dam. *Estuar. Coast. Shelf Sci.* **2007**, *73*, 81–90. [CrossRef]
40. Guo, H.Y.; Zhang, X.G.; Zhang, Y.; Tang, W.Q.; Wu, J.M. Effects of environmental variables on recruitment of *Anguilla japonica* glass eels in the Yangtze Estuary, China. *Fish. Sci.* **2017**, *83*, 333–341. [CrossRef]
41. August, S.M.; Hicks, B.J. Water temperature and upstream migration of glass eels in New Zealand: Implications of climate change. *Environ. Biol. Fishes* **2008**, *81*, 195–205. [CrossRef]
42. Jellyman, D.; Booker, D.; Watene, E. Recruitment of *Anguilla* spp. glass eels in the Waikato River, New Zealand. Evidence of declining migrations? *J. Fish Biol.* **2009**, *74*, 2014–2033. [CrossRef] [PubMed]
43. Zhang, H.; He, W.; Tong, C.; Lu, J. The effect of fishing the anguillid elver (*Anguilla japonica*) on the fishery of the Yangtze estuary. *Estuar. Coast. Shelf Sci.* **2008**, *76*, 902–908. [CrossRef]
44. Belkin, I.M. Rapid warming of large marine ecosystems. *Prog. Oceanogr.* **2009**, *81*, 207–213. [CrossRef]
45. Yao, Y.; Wang, C. Variations in summer marine heatwaves in the South China Sea. *J. Geophys. Res. Oceans* **2021**, *126*, e2021JC017792. [CrossRef]
46. Rozenfeld, C.; García-Carpintero, V.; Pérez, L.; Gallego, V.; Herranz-Jusdado, J.G.; Tveiten, H.; Johnsen, H.K.; Fontaine, R.; Weltzien, F.; Cañizares, J.; et al. Cold seawater induces early sexual developmental stages in the BPG axis of European eel males. *BMC Genom.* **2019**, *20*, 597. [CrossRef]
47. Sudo, R.; Yada, T. Anguillid eels as a model species for understanding endocrinological influences on the onset of spawning migration of fishes. *Biology* **2022**, *11*, 934. [CrossRef]
48. Desaunay, Y.; Guerault, D. Seasonal and long-term changes in biometrics of eel larvae: A possible relationship between recruitment variation and North Atlantic ecosystem productivity. *J. Fish. Biol.* **1997**, *51*, 317–339. [CrossRef]
49. Kawakami, Y.; Mochioka, N.; Kimura, R.; Nakazono, A. Seasonal changes of the RNA/DNA ratio, size and lipid contents and immigration adaptability of Japanese glass-eels, Anguilla japonica, collected in northern Kyushu, Japan. *J. Exp. Mar. Biol. Ecol.* **1999**, *238*, 1–19. [CrossRef]
50. Kearney, M.; Jeffs, A.; Lee, P. Seasonal differences in the quality of shortfin glass eel, Anguilla australis, and subsequent effects on growth and survival in captivity. *N. Z. Nat. Sci.* **2011**, *36*, 45–55.
51. Belpaire, C.G.J.; Goemans, G.; Geeraerts, C.; Quataert, P.; Parmentier, K.; Hagel, P.; De Boer, J. Decreasing eel stocks: Survival of the fattest? *Ecol. Freshw. Fish* **2009**, *18*, 197–214. [CrossRef]
52. Arai, T. Do we protect freshwater eels or do we drive them to extinction? *Springerplus* **2014**, *3*, 534. [CrossRef]
53. Kahn, D.M. Trends in abundance and fishing mortality of american eel. *Fisheries* **2018**, *44*, 129–136. [CrossRef]
54. Sonne, C.; Peng, W.X.; Alstrup, A.K.O.; Lam, S.S. European eel population at risk of collapse. *Science* **2021**, *372*, 1271. [CrossRef]
55. Feunteun, E. Management and restoration of european eel population (*Anguilla anguilla*): An impossible bargain. *Ecol. Eng.* **2002**, *18*, 575–591. [CrossRef]
56. Dekker, W. Did lack of spawners cause the collapse of the European eel, *Anguilla anguilla*? *Fish. Manag. Ecol.* **2003**, *10*, 365–376. [CrossRef]
57. Leone, C.; Zucchetta, M.; Capoccioni, F.; Gravina, M.F.; Franzoi, P.; Ciccotti, E. Stage-specific distribution models can predict eel (*Anguilla anguilla*) occurrence during settlement in coastal lagoons. *Estuar. Coast. Shelf Sci.* **2016**, *170*, 123–133. [CrossRef]
58. Shuai, F.M.; Liu, Q.F.; Liu, Y.Q.; Wu, Z.; Lek, S.; Chen, Y.S. Spatial Distribution Patterns of Japanese Eel (*Anguilla japonica*) in a Large Subtropical River (Pearl), China. *N. Am. J. Fish. Manag.* **2021**, *41*, 929–938. [CrossRef]
59. Tzeng, W.N.; Lin, H.R.; Wang, C.H.; Xu, S.N. Differences in size and growth rates of male and female migrating Japanese eels in Pearl River, China. *J. Fish Biol.* **2000**, *57*, 1245–1253. [CrossRef]
60. Podda, C.; Palmas, F.; Pusceddu, A.; Sabatini, A. When the Eel Meets Dams: Larger Dams' Long-Term Impacts on *Anguilla anguilla* (L., 1758). *Front. Environ. Sci.* **2022**, *10*, 876369. [CrossRef]
61. Acou, A.; Laffaille, P.; Legault, A.; Feunteun, E. Migration pattern of silver eel (*Anguilla anguilla*, L.) in an obstructed river system. *Ecol. Freshw. Fish* **2008**, *17*, 432–442. [CrossRef]
62. Trancart, T.; Teichert, N.; Lamoureux, J.; Gharnit, E.; Acou, A.; Oliveira, E.D.; Roy, R.; Feunteun, E.; Feunteun, E. A possible strong impact of tidal power plant on silver eels' migration. *Estuar. Coast. Shelf Sci.* **2022**, *278*, 108116. [CrossRef]
63. Robinet, T.T.; Feunteun, E.E. Sublethal effects of exposure to chemical compounds: A cause for the decline in atlantic eels? *Ecotoxicology* **2002**, *11*, 265–277. [CrossRef]
64. Corsi, I.; Mariottini, M.; Badesso, A.; Caruso, T.; Borghesi, N.; Bonacci, S.; Iacocca, A.; Focardi, S. Contamination and sub-lethal toxicological effects of persistent organic pollutants in the European eel (*Anguilla anguilla*) in the Orbetello lagoon (Tuscany, Italy). *Hydrobiologia* **2005**, *550*, 237–249. [CrossRef]
65. Sancho, E.; Fernandez-Vega, C.; Sanchez, M.; Ferrando, M.D.; Andreu-Moliner, E. Alterations on ache activity of the fish *Anguilla anguilla* as response to herbicide-contaminated water. *Ecotoxicol. Environ. Saf.* **2000**, *46*, 57–63. [CrossRef] [PubMed]
66. Pierron, F.; Baudrimont, M.; Bossy, A.; Bourdineaud, J.; Brèthes, D.; Elie, P.; Massabuau, J. Impairment of lipid storage by cadmium in the european eel (*Anguilla anguilla*). *Aquat. Toxicol.* **2007**, *81*, 304–311. [CrossRef]
67. Belpaire, C.; Reyns, T.; Geeraerts, C.; Loco, J.V. Toxic textile dyes accumulate in wild european eel *Anguilla anguilla*. *Chemosphere* **2015**, *138*, 784–791. [CrossRef]

68. Maes, G.E.; Raeymaekers, J.A.M.; Pampoulie, C.; Seynaeve, A.; Goemans, G.; Belpaire, C.; Volckaert, F.A.M. The catadromous European eel *Anguilla anguilla* (L.) as a model for freshwater evolutionary ecotoxicology: Relationship between heavy metal bioaccumulation, condition and genetic variability. *Aquat. Toxicol.* **2005**, *73*, 99–114. [CrossRef] [PubMed]
69. Tao, W.; Niu, L.; Dong, Y.; Fu, T.; Lou, Q. Nutrient Pollution and Its Dynamic Source-Sink Pattern in the Pearl River Estuary (South China). *Front. Mar. Sci.* **2021**, *8*, 713907. [CrossRef]
70. Niu, L.; Cai, H.; Jia, L.; Luo, X.; Tao, W.; Dong, Y.; Yang, Q. Metal pollution in the Pearl River Estuary and implications for estuary management: The influence of hydrological connectivity associated with estuarine mixing. *Ecotoxicol. Environ. Saf.* **2021**, *225*, 112747. [CrossRef] [PubMed]
71. Garcia-Vazquez, S.; Linde, A.R.; Ayllon, F.; Garcia-Vazquez, E. Induction of micronuclei in eel (*Anguilla anguilla* L.) by heavy metals. *Ecotox Environ. Saf.* **2001**, *49*, 139–143.
72. Pujolar, J.M.; Marino, I.A.; Milan, M.; Coppe, A.; Maes, G.E.; Capoccioni, F.; Ciccotti, E.; Bervoets, L.; Covaci, A.; Belpaire, C.; et al. Surviving in a toxic world: Transcriptomics and gene expression profiling in response to environmental pollution in the critically endangered European eel. *BMC Genom.* **2012**, *13*, 507. [CrossRef]
73. Palstra, A.P.; Ginneken, V.; Murk, A.J.; Thillart, G.E.E.J.M. Are dioxin-like contaminants responsible for the eel (*Anguilla anguilla*) drama? *Naturwissenschaften* **2006**, *93*, 145–148. [CrossRef] [PubMed]
74. Dekker, W.; Casselman, J.M.; Cairns, D.K.; Tsukamoto, K.; Jellyman, D.; Lickers, H. Worldwide decline of eel resources necessitates immediate action: Québec Declaration of Concern. *Fisheries* **2003**, *28*, 28–30.
75. Shuai, F.M.; Li, H.Y.; Li, J.; Jiang, T.; Yang, J.; Yang, W.L. Unravelling the life-history patterns and habitat preferences of the Japanese eel (*Anguilla japonica*) in the Pearl River, China. *J. Fish Biol.* **2023**. *early view*. [CrossRef]
76. Jellyman, D.J. Status of New Zealand fresh-water eel stocks and management initiatives. *ICES J. Mar. Sci.* **2007**, *64*, 1379–1386. [CrossRef]

fishes

MDPI

Article

Biomass Quantification of the Critically Endangered *European eel* from Running Waters Using Environmental DNA

Sara Fernandez [1], Álvaro Gutiérrez [1], Dumas Deconinck [1], Jose Luis Martinez [2], Almudena Alvarez [3], Isabel Marquez [3], Gonzalo Machado-Schiaffino [1,*,†] and Eva Garcia-Vazquez [1,*,†]

1 Department of Functional Biology, University of Oviedo, 33006 Oviedo, Spain
2 DNA Analysis Unit, Scientific-Technical Services, University of Oviedo, 33071 Oviedo, Spain
3 Centro de Experimentación Pesquera, Dirección General de Pesca Marítima, Consejería de Medio Rural y Cohesión Territorial del Principado de Asturias, Escuela de Formación Profesional Naútico Pesquera 2ª Planta, Avda, Príncipe de Asturias 74, 33212 Gijón, Spain
* Correspondence: machadogonzalo@uniovi.es (G.M.-S.); egv@uniovi.es (E.G.-V.)
† These authors contributed equally to this work.

Abstract: The *European eel Anguilla anguilla* is a critically endangered catadromous species. There is an urgent need for close surveillance of the populations that are still viable in European rivers. The species is difficult to observe in freshwater because of its bottom-dwelling behavior; the currently employed methods of eel monitoring in Europe based on the physical capture of individuals are stressful and may cause mortality. Here, we present a new highly sensitive method based on an *A. anguilla*-specific qPCR marker designed within the cytochrome oxidase I mitochondrial gene for application on environmental DNA (eDNA). Since the detectability of eDNA depends on the hydrographic conditions, we applied correction for altitude and a linear model and were able to predict the eel biomass from the eDNA in the different rivers of northern Spain still holding wild populations. The method was validated by electrofishing surveys. This novel eDNA-based marker allows for estimating the European eel biomass in running waters from small 1.5 L water samples and could complement, or replace in some cases, current eel surveys without disturbing wild populations.

Keywords: *Anguilla anguilla*; biomass prediction; environmental DNA; hydrographic correction; qPCR; specific marker

Key Contribution: This is a new molecular marker to estimate *Anguilla anguilla* biomass from water samples in running rivers by using qPCR. The quantity of environmental eel DNA measured is corrected by the river hydrography for a more accurate estimation.

Citation: Fernandez, S.; Gutiérrez, Á.; Deconinck, D.; Martinez, J.L.; Alvarez, A.; Marquez, I.; Machado-Schiaffino, G.; Garcia-Vazquez, E. Biomass Quantification of the Critically Endangered *European eel* from Running Waters Using Environmental DNA. *Fishes* **2023**, *8*, 279. https://doi.org/10.3390/fishes8060279

Academic Editor: Hyun-Woo Kim

Received: 28 March 2023
Revised: 17 May 2023
Accepted: 22 May 2023
Published: 24 May 2023

1. Introduction

European eel (*Anguilla anguilla*) populations have been threatened by anthropogenic activities during the last decades. Populations were reduced by 80% across the species' distribution range, from Norway to North Africa, in the 1980s [1], principally due to habitat losses for the construction of river barriers and overexploitation [2] as well as due to the presence of *Anguillicola crassus*, a nematode that parasites the species, causing injuries in the swimbladder that impede migration [3]. Catalogued as critically endangered by the IUCN since 2007 [4] and declared out of safe biological limits by the International Council for the Exploration of the Sea (ICES), actions are needed to maintain the still-viable populations [5].

European eel has a complex life cycle that involves a challenging 5000 km migration from the Sargasso Sea, where they spawn at a depth of 250 m, to the European rivers where they feed until maturity to a return to reproduce back in the Sargasso Sea [6]. Thus, they face a variety of threats from the sea to the rivers. Given the wide latitudinal range of the species' distribution, management involves different countries. Of enormous ecological, economic, and cultural importance [5], the European eel is a target of professional and recreational fishing in Europe and Northern Africa [7]. Mitigation actions to avoid stock depletion include reducing fishing quotas, restocking, river habitat improvement, and the removal of river barriers [2]. However, the conservation measures in practice as well as the monitoring techniques vary a lot depending on the local stakeholders [8].

European legislation mandates that eel mortality is reduced and that new sampling techniques are incorporated to monitor and facilitate an efficient management of the species [9]). However, the lack of the standardization of monitoring methods across countries makes the management of the European eel very complex [10]. In Spain and other countries, the most-employed technique to monitor eel populations is electrofishing [11], while other countries use fyke nets, mark and recapture, or tagging and telemetry for population surveys (as reviewed by [12]). All these methods require the fish to be extracted from the water, which is problematic because, like the majority of wild fish, this species is sensitive to human manipulation. Practices such as catch-and-release may encompass mortalities of 60% or higher [13], and electrofishing and fyke nets cause stress and increase post-survey mortality [12,14]. Therefore, non-intrusive monitoring techniques are urgently needed. However, it is not easy to find a method that is useful for all the life stages in different hydrological conditions. For example, high-frequency multi-beam sonar can be used to estimate the biomass of the large eels abandoning rivers during the escapement period [12], and video surveillance can serve to visualize fish and understand their behavior [15]; however, these techniques rely on good river conditions, so they are inaccurate in turbid waters with suspended material, and are not adequate to estimate the biomass of small juveniles.

In addition to the problems mentioned above, conventional methods such as electrofishing and fyke nets have limitations in detecting species at low densities [16]. This is the case of endangered species such as *Anguilla anguilla*, for which many populations remain at a low density after population declines, and, even in relatively large populations, the density is very low upstream [17]. In these cases, the use of environmental DNA (eDNA) for species detection has advantages over individual sampling [18]. The eDNA-based methodology is a noninvasive technique that does not disturb endangered populations [19], is highly efficient in detecting scarce or evasive species [20], and can be used in river areas where electrofishing is not feasible such as those under a high current speed or in deep ponds [21]. The use of eDNA was proposed as a complement to or substitute for traditional methods such as electrofishing [22]. Some countries, such as Japan, several EU countries, the UK, and the US, incorporated this technology into their governmental monitoring programs [23]. However, it is still necessary to test the tool before its incorporation into a new program, in order to determine its limitations and to know how to interpret the obtained results. In order to perform this, it is highly advisable to follow the recommendations by [24] as much as possible, such as by reporting the primer sequences and design, detecting the target species from the environmental DNA, and disclosing the filter type and pore size.

In the North American congeneric *Anguilla rostrata*, which has a similar life cycle and latitudinal distribution range, the use of eDNA was encouraged to assist with population assessments and guide conservation efforts [25]. A quantitative PCR marker for application on eDNA was further described for the assessment of North American eel populations [26]. Regarding the European eel, two qPCR markers were published for the detection of *Anguilla anguilla* from eDNA. Halvorsen et al. [27] targeted a DNA fragment within the mitochondrial cytochrome b gene and used the marker to study the occurrence of eels along rivers with different types and numbers of barriers in Norway; however, they employed the marker for eel detection and not to quantify eel biomass. Weldon et al. [9,10,27] conducted a survey in Irish lakes with eel populations of different sizes, comparing the results of a qPCR marker that was also developed within the cytochrome b gene with those obtained from fyke net surveys. The probability of the detection of eels using the eDNA marker was 83%, with eDNA concentrations generally associated with population size, which were lower in lakes with low eel populations [9]—although with no clear difference between high and medium eel populations. It is worth noting that this marker was not tested in rivers.

Halvorsen et al. [27] encouraged the development of eDNA methods that could quantify populations from running waters, for application in conservation efforts; Halvorsen et al. [28] developed a method to estimate haplotypes from water samples, although they are not yet able to quantify individuals. For community inventories from the presence/absence of metabarcoding data, eDNA has a spatial signal (upstream–downstream species distribution) comparable to that of local-capture-based methods [29]. However, using eDNA concentrations to estimate biomass in rivers cannot be accomplished in a straightforward way because the river hydrography may have a strong effect on the amount of eDNA that can be detected from water samples. Working with crayfish, Rice et al. [30] found a strong relationship between eDNA detection probability and upstream river distance, which was interpreted as the downstream transport of eDNA from upstream locations. River discharge is positively associated with the average length of downstream eDNA transport [31]; therefore, in streams with high discharges, eDNA rapidly runs downstream. Another intervening factor is the average flow velocity, which is inversely proportional to the eDNA concentration at a river point [32]. The influence of these factors is likely intertwined with the habitat preferences of the European eel in the river because the probability of eel occurrence increases with river size and water temperature (both being generally higher downstream than upstream) and decreases with the distance to the river mouth [17]. Correction for river hydrography is yet to be considered in European eel eDNA studies in running waters.

To fill the gap detected by Halvorsen et al. [27], in this study we developed a qPCR assay to detect and quantify eels' eDNA from running waters and tested it in experimental tanks and in real river samples from the Asturias region in northern Spain (Figure 1), where electrofishing surveys are conducted as part of the governmental annual monitoring program. We validated this method in the field by comparing the eDNA results with electrofishing surveys, applying a simple correction to the eDNA concentration for the dependence of eDNA capture on river hydrography [30–32].

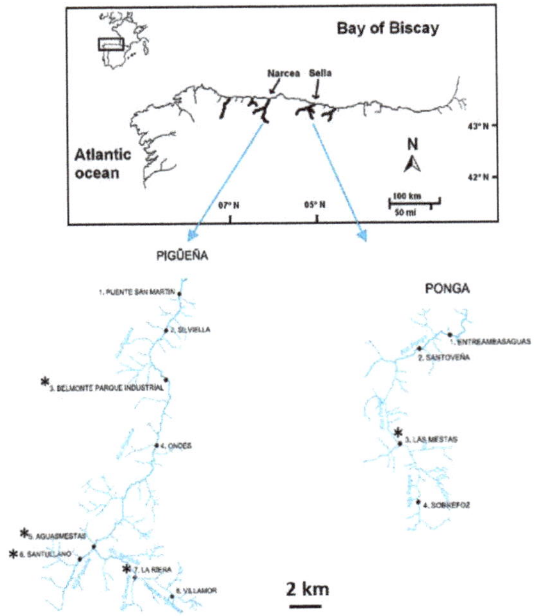

Figure 1. Map showing the rivers where the marker was validated in situ in the region of Asturias (southern Bay of Biscay, Spain). Sampling locations are marked with an asterisk.

2. Material and Methods

2.1. Real-Time PCR Marker Design and qPCR Procedures

2.1.1. Primer Design

TaqMan primers and probe were designed to amplify a fragment from the mitochondrial gene cytochrome oxidase subunit 1 (COI) of *Anguilla anguilla*. National Center for Biotechnology Information (NCBI) nucleotide database was employed to download the sequences from *A. anguilla* and other related and cohabitant species in order to find a specific region allowing for the specific amplification of the target *A. anguilla*. PrimerExpress 3.0 (Applied Biosystems, Waltham, MA, USA) was the software employed for the design. BlastN tool (Basic Local Alignment Search Tool-N) was then used to check possible in silico cross-amplification. There was no in silico cross-amplification against the species present in the database. The primers and probe designed were as follows: ANG-Forward-5′-GGA GCT GGT ACA GGC TGA ACT G-3′; ANG-Reverse-5′-AGT GAG AAA ATT GTC AGG TCA ACA GA-3′; ANG-Probe: 5′-6-FAM TGG CTG GAA ACT TAG CCC ACG CC BHQ1-3′.

Amplification was assayed in vitro using tissue from *A. anguilla*. Other species that can be found in these and other European rivers, such as the fish *Salmo trutta, Salmo salar, Carasius auratus,* and *Squalius caroliterti* and the invasive gastropod *Potamopyrgus antipodarum*, were also assayed to check for possible cross-amplification.

2.1.2. qPCR

Amplification was performed using a 7900 HT Fast Real-Time PCR System (Life Technologies, Inc., Applied Biosystems, Carlsbad, CA, USA). It was carried out in a total volume of 25 μL containing 1X TaqMan® Environmental Master Mix 2.0 (Life Technologies, Inc., Applied Biosystems, Carlsbad, CA, USA), 6 μL environmental DNA, 0.5 μL of each primer, and the probe (10 μM). PCR standard conditions were 50 °C for 2 min and 95 °C for 10 min, followed by 40 cycles of 95 °C 15 s and 60 °C 1 min. Negative controls to monitor for contamination as well as extraction and filtration negative controls were included in the amplification. Three PCR technical replicates were amplified per sample. The standard curve was performed by serial dilution of DNA from European eel, of known concentration (72.8 ng/μL, from 10^{-1} to 10^{-6} serial dilution).

2.2. Validation of the qPCR Marker

2.2.1. Experimental Validation in Aquaculture Tanks

To determine the ability of the new qPCR assay as a quantitative method for the inventory of eel biomass from eDNA, an experiment using tanks with different biomass of eels was developed. A linear relation between the biomass of eels in a tank and the amount of eel eDNA is expected for low and moderate eel densities. In conditions of very high eel density and biomass, we expect a saturation of extraction and quantification methods due to excessive eDNA amounts. The experiment consisted of nine 60 L tanks that were filled with freshwater and supplied with oxygen. *Anguilla anguilla* individuals were acquired from Marina Eel company, weighed, and distributed among tanks.

The tanks were kept at 18 °C. Eels were left for a week and were not fed to minimize contamination from fishmeal. Water samples of 1.5 L were taken from each tank (three samples per tank as sampling replicates). For this, a sterile bottle was introduced 10 cm from the bottom, avoiding touching eels. Sterile gloves were worn and changed between tanks to avoid contamination. The water samples were preserved at 4 °C before filtration. All samples were filtered within 24 h after sampling.

2.2.2. Validation in Natural Rivers

The field validation in running waters was conducted in rivers from Asturias (northwestern Spain). The region is located in the center of the southern Bay of Biscay (Figure 1), which is one of the original distribution areas of *Anguilla anguilla* still holding eel populations [5]. The information was found from the Hydrographic Confederation of North Spain, https://www.chcantabrico.es/las-cuencas-cantabricas/marco-fisico/hidrologia/rios/piguena and https://

www.chcantabrico.es/organismo/las-cuencas-cantabricas/marco-fisico/hidrologia/rios/dhc-occidental/-/asset_publisher/OFa1sWDJLb6J/content/rio-ponga for Pigüeña and Ponga, respectively, accessed on 1 March 2023). Pigüeña River's head is at 1700 m over sea level, the basin is much larger (47.6 km until discharging in the main Narcea River, 404.46 km^2), and the regime is regulated by three hydroelectric facilities, one of them upstream at La Riera site (Figure 1). The upstream part of the river is located within Somiedo Natural Park. Ponga River is smaller (head at 1500 m, 28.9 km) and runs free until discharging in the main Sella River. Data of discharge and flow velocity are not available for these rivers. The upper part where the sample was collected belongs to the Natural Park "Cabecera del río Ponga". Relatively large populations of native salmonids (Atlantic salmon *Salmo salar* and brown trout *Salmo trutta*) occur in the two rivers.

In contrast to the experimental tanks, low amount of eel eDNA, mixed with eDNA from other species, is expected to be found in wild samples. Water samples from the same sampling points where eel biomass was previously estimated from electrofishing were employed for both qPCR and eel-specific end-time PCR validation.

The validation consisted of three steps.

1. Correction for river hydrography

In our case study, we considered stream order (level of branching in a river system: 2–6 in the streams where this marker was validated, based on Strahler, 1957) as a proxy for river discharge and altitude—also associated with temperature and implying a steeper slope in these rivers—as a proxy for mean flow velocity. These two hydrographic features are inversely proportional to eDNA concentration at a river point [33] and are especially important here because the probability of eel occurrence is higher at lower altitudes and higher-order streams [18]. We checked the association of these variables with indicators of the amount of eDNA occurring at the sampling points: the total eDNA quantity (measured from QubitHS methodology as described above); positive PCR amplification using end-time PCR with universal primers; end-time PCR with eel-specific primers [21] that show the presence of specific eel eDNA. Quantitative eDNA data (in terms of Ct values) are corrected by the factor showing a higher correlation with total eDNA quantity.

2. Correlational validation

The quantity of *Anguilla anguilla* eDNA from water samples is expected to be significantly correlated with eel biomass measured from the individuals captured in electrofishing sampling (see below), applying the correction defined in 1. This was accomplished in Pigüeña River.

3. Predictive value of the new marker

Proof of concept to explore the applicability of the marker for routine surveys in different waters, where the eel eDNA from Ponga River is predicted from eel biomass using the curve equation estimated from Pigüeña River.

A regression with significant slope in (2), with the values corrected for hydrography according to results in (1), and eel eDNA quantity within the range estimated from that regression in (3) indicate that the new marker can be considered reliable for its use in the field.

In the river samples, additional markers amplified with end-time PCR (PCR) were assayed to check the effect of river hydrography (see below Section 2.4.1) on total eDNA detectability and to confirm lack of PCR inhibition.

2.2.3. Sampling Procedures in the Field

Water sample collection took place, combined with electrofishing surveys, during September 2021. The Pigüeña River (Asturias, Spain) was selected (Figure 1), where eels are annually monitored by electrofishing by the competent authority (Government of Asturias Principality). Pigüeña is a 46 km long mountainous stream tributary of Narcea River within Nalón-Narcea basin. Samples for the proof of concept were taken from Ponga River, a 25 km long tributary of Sella River.

The government of Asturias region has carried out periodical electrofishing at predetermined sites within a regional river monitoring network since 2011. Water samples for eDNA analysis were taken immediately before electrofishing (Table 1). Electrofishing was conducted by applying the same sampling effort in all the sites. Zippin's method was used to determine eel density [34,35]. Three-pass electrofishing covering the whole river from bank to bank was completed at each sampling point, without replacement. Eels were taken out from the river with a handle net and placed in containers with water and air supply until the end of the electrofishing passes. The individuals were counted, measured, and weighed, before being released back into the same river area.

Table 1. Sampling sites' locations and results of the electrofishing surveys. Altitude in meters over sea level and stream order [36] are indicated. Total eel biomass is expressed in grams.

River	Location Name	Coordinates	Altitude (m)	Electrofishing Surface	Stream Order	Eel Biomass	Eel Number
Pigüeña	La Riera	43°09′09.0″ N; 6°15′12.3″ O	480	280 m^2	3	1789	74
Pigüeña	Santullano	43°09′57.4″ N; 6°19′01.1″ O	430	210 m^2	2	418	17
Pigüeña	Aguasmestas	43°10′31.2″ N; 6°18′00.5″ O	400	385 m^2	4	1910	72
Pigüeña	Belmonte P. Industrial	43°17′17.6″ N; 6°13′16.94″ O	200	455 m^2	6	310	11
Ponga	Las Mestas	43°10′10.8″ N; 5°10′37.9″ O	350	329 m^2	4	290	6

From each sampling point, three sampling replicates of 1.5 L water were collected using sterile plastic bottles. After collection, water samples were frozen until filtration. Researchers wore disposable gloves that were changed between sampling points to avoid cross-contamination.

2.3. *eDNA Analysis Procedures*

2.3.1. Water Samples Filtration

Each sample of 1.5 L water was vacuum-filtered using filter membranes with a 0.22 μm pore size and a 47 mm diameter (Pall Corporation, Life Sciences, Ann Arbor, MI, USA). Filtering process was performed in a specialized laboratory separated from the main molecular facility to avoid contamination. Filtration apparatus, forceps, and surfaces were cleaned using 10% bleach solution between samples to avoid cross-contamination as much as possible. Filtration blanks of 1.5 L distilled water were filtered under the same conditions after each filtration process.

2.3.2. eDNA Purification

Power Water DNA Isolation Kit (Qiagen, Hilden, Germany), in accordance with the protocol of the manufacturer, was used to isolate eDNA from filters. Each filter was divided into two halves to perform the extraction. Then, both extractions were combined after the elution step. Extractions were quantified using Qubit™ dsDNA HS Assay Kit (Invitrogen, Carlsbad, CA, USA).) (Supplementary Table S1).

2.3.3. Inhibition Testing and Detection Probability

To check for potential presence of enzymatic inhibitors in the DNA extracts, a quantitative polymerase chain reaction (qPCR) assay targeting an internal positive control was carried out using Applied Biosystems® TaqMan® Exogenous Internal Positive Control Reagents (Thermo Fisher Scientific, Waltham, MA, USA). Amplification values for cycle threshold (CT; the first PCR cycle at which DNA is detected, with smaller CT values corresponding to higher DNA quantities) were compared between reactions containing 2 μL PCR grade water and reactions containing 2 μL of template eDNA. To enable an appropriate assessment of potential inhibition during subsequent steps, Taqman® environmental master mix 2.0 (ThermoFisher Scientific, Waltham, MA, USA) was used in all the performed qPCRs. Each qPCR was carried out in a 25 μL reaction volume containing 2.5 μL of 10× Exo

IPC Mix and 0.5 μL of 50× Exo IPC DNA. The thermal cycle profile consisted of a hot start at 94 °C for 2 min, a denaturing step at 95 °C for 5 s, and an annealing step at 60 °C for 15 s for 40 cycles.

Detection probability (ρ) was calculated as demonstrated by Laramie et al. (2015). It is calculated based on the number of positive sampling replicates (η = 0–3) divided by the total number of replicates per sampling point (N = 3).

2.4. End-Time PCR Markers Assayed on Field Samples

In addition to the new qPCR marker, two more markers were assayed in the field water samples. A universal marker was employed to confirm the presence of DNA in each sample, and an eel-specific marker detectable by end-time PCR [21] was used for comparison of its performance with the new marker.

2.4.1. Universal Marker

Universal primers were chosen to check for amplification success, as they can amplify DNA from a broad range of organisms and discard inhibition issues in the end-time PCR. The primers mlCOIintF (5′ GWACWGGWTGAACWGTWTAYCCYCC-3′) and jgHCO2198: 5′-TANACYTCNGGRTGNCCRAARAAYCA-3′) from Leray et al. [37], amplifying a 313 bp fragment within the cytochrome oxidase subunit 1 gene (COI), were also employed. End-time PCR reaction was performed with the same reagents using the following conditions: 95 °C for 5 min; 35 cycles of 95 °C for 1 min, 48 °C for 1 min, and 72 °C for 1 min; 72 °C for 5 min extension.

2.4.2. End-Time PCR Eel-Specific Marker

European eel species-specific primers from Burgoa Cardás et al. [21] were employed in end-time PCR using the eDNA samples (Forward-5′-GCT GTA TTA GTA ACC GCC GTT TT-3′, Reverse-5′-GCA GGA TCA AAG AAG GTC GT-3′). End-time PCR amplifications were performed in a total volume of 20 μL, including Green GoTaq® Buffer (1X), $MgCl_2$+ (2.5 mM), dNTPs (0.25 mM), forward and reverse primers (1 μM), BSA (200 ng/μL), 0.65 U of TaqMan Polymerase (Promega®), and 4 μL eDNA. Amplification conditions were 95 °C for 5 min, 95 °C for 30 s, and 65 °C for 30 s and a final extension step at 72 °C for 7 min.

2.5. Data Analysis and Statistics

Variables employed in the analyses were as follows.

- Stream order: The number defining the level of river branching at the considered sampling point. It is a discontinuous quantitative variable.
- Altitude: Meters above sea level at each sampling point. It is a continuous quantitative variable.
- Total eDNA quantity: The amount of eDNA in a sample, measured using HS Qubit Fluorometer. It is a continuous quantitative variable.
- End-time PCR eel-specific marker: Using the primers described above from Burgoa Cardás et al. [21], it measures presence/lack of presence of eels' eDNA (positive amplification/no amplification); thus, the primary variable is binary (0/1). It may be secondarily transformed into a discontinuous quantitative variable using the number of sampling replicates with positive end-time PCR amplification per sampling site (variation range 0–3).
- End-time PCR universal marker: As in the previous case but using Leray et al. [37] universal primers instead of an eel-specific marker.
- Eel eDNA quantity: CT values obtained when amplifying eel eDNA using the species-specific primers developed in the current study. It is a continuous quantitative variable.
- Adjusted eel eDNA quantity: Amount of eel eDNA estimated from field samples as an extrapolation from the standard curve. It is a continuous quantitative variable. In the analysis of field results, CT values were preferred as a proxy to eDNA quantity estimations for being a fitter measurement for regression analysis.

To check the effect of river hydrography on eDNA detectability in the field case study, a multivariate multiple regression was performed with stream order and altitude as independent variables and the total eDNA quantity and positive PCR amplification using end-time PCR universal marker as dependent variables. Independent variables predicting the dependent ones were employed for the correction of eel eDNA quantity in field samples.

Best-fit model for the relation between eel biomass and eel eDNA as CT values (in experimental tanks and in field samples) was tested using Akaike information criterion procedure, and the corresponding equation was determined. Ordinary least squares regression was used to express linear relationships between pairs of variables, such as eel biomass and CT or eDNA quantity measured using Qubit. Homoscedasticity was tested using Breusch–Pagan statistics. Durbin–Watson test was employed to check for the presence of autocorrelation in the errors of regression models, in order to ensure the dataset met the conditions for regression analysis.

Pearson's r was used to test for linear correlations between pairs of variables. Statistical analyses were performed using PAST software [38].

3. Results

3.1. Primers Validation In Vitro and in Experimental Tanks

In vitro, no cross-amplification of the new marker was found with the DNA of any species that was tested.

In experimental tanks, the eel DNA quantity increased with the eel biomass up to a point where a saturation effect appeared (Table 2). Considering the nine tanks, the best-fit model for the adjusted DNA quantity means (Akaike IC = 11.45) corresponded to a power curve (positive with saturation) with the following equation:

$$y = 63.46x^{0.003} - 63.7$$

For CT means (Akaike IC = 11.48) the best-fit model was Michaelis–Menten (negative with saturation), with the following equation:

$$Y = 26.72x/(-1.976 + x)$$

Table 2. Tank experiment results. DNA quantity: amount of DNA (ng) measured by qPCR using the new markers. Eel biomass is expressed in grams.

Tank Number	Biomass	Number of Individuals	Tank Sampling Replicate	CT Mean	Eels' DNA Quantity Per Tank (ng)
			A	28.27	0.275
T1	34.5	3	B	28.81	0.176
			C	28.49	0.197
			A	27.04	0.508
T2	80.57	4	B	26.52	0.723
			C	26.70	0.651
			A	28.23	0.344
T3	81.15	4	B	28.29	0.286
			C	27.91	0.310
			A	26.31	0.832
T4	196.3	8	B	27.31	0.465
			C	24.93	1.902
			A	26.18	0.896
T5	211.7	14	B	26.74	0.691
			C	26.04	1.027
			A	26.17	1.619
T6	358.6	20	B	26.10	1.772
			C	26.44	1.496
			A	25.13	1.750
T7	399.5	24	B	26.86	0.608
			C	27.15	0.945
			A	28.60	0.459
T8	745.5	43	B	28.68	0.392
			C	27.41	0.857
			A	27.12	0.935
T9	748.6	43	B	27.58	0.778
			C	28.19	0.490

However, for further validation and the objective of applying this method in wild populations, we used the linear part of the curve before saturation, because it is extremely unlikely to find densities like those of tanks T8 and T9 (43 eels summing 750 g of biomass in only 60 L of water) in the wild.

Considering the part of the curve before saturation, that is, excluding the two tanks with the highest biomass T8 and T9 (Figure 2), a positive significant correlation between the eels' DNA quantity in the tanks and the eels' biomass was found ($r^2 = 0.485$, $p = 0.001$). The Durbin–Watson statistic was 2.05 with $p = 0.54$ n.s. (no autocorrelation of errors), and the Breusch–Pagan statistic was 4.58 with $p = 0.03$ (no homoscedastic); thus, the conditions for regression analysis were not totally met. The equation of the linear regression was $y = 0.003x + 0.239$. For the estimation of eel eDNA from CT values, a significant negative correlation was found ($r^2 = 0.399$, $p = 0.002$). In this case, the Durbin–Watson was 1.44 with $p = 0.09$ n.s., and the Breusch–Pagan was 0.00005 with $p = 0.994$ n.s., confirming homoscedasticity for this estimator. The equation was $y = -0.005x + 27.919$. For its adjustment to homoscedasticity, we used CT values as estimators of the eel eDNA quantity, with the higher CT value for the smaller eel eDNA amount.

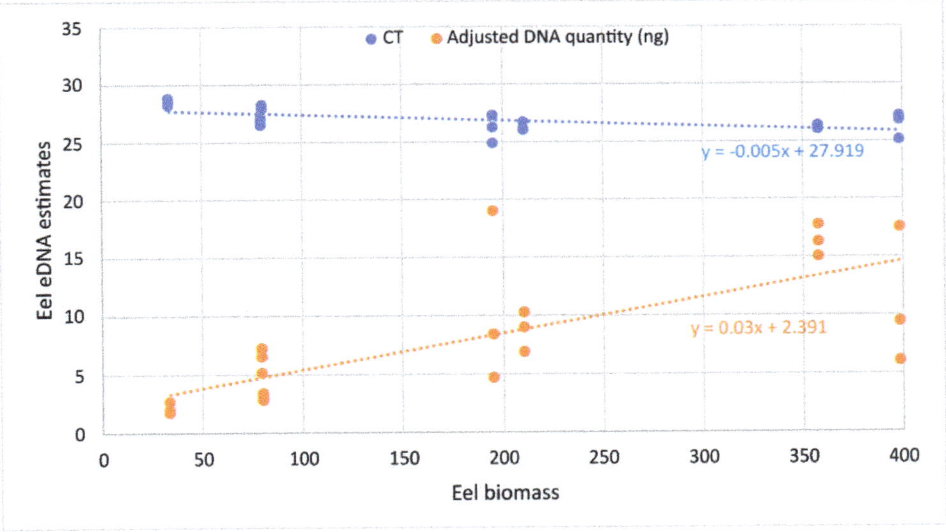

Figure 2. Linear regression showing the relationship between the eel biomass (g) per tank and the eel eDNA measured by qPCR, with the new marker in the linear part of the curve (before saturation) in experimental tanks. Eel eDNA estimates are CT and adjusted DNA quantity (ng*10 in the plot for better visualization). The equations are shown.

3.2. *Field Validation*

3.2.1. eDNA Detection and River Hydrography

The three end-time PCR technical replicates of each sample gave the same result (positive or negative) for the two assayed end-time PCR markers.

DNA extracted from water samples (total eDNA) was quantifiable using Qubit for at least one water sample per site (one sampling replicate) in the Pigüeña River locations. Positive amplification from universal primers [37] was achieved for those water samples and two more (Table 3).

Table 3. Results of eDNA quantification using Qubit fluorometer and end-time PCR amplification using universal [37] or *Anguilla anguilla*-specific primers [21] from the water samples analyzed in Pigüeña River. PCR amplification: 1 = positive; 0 = negative. The amount of total eDNA is expressed in ng/µL. ND, not detected.

Location	Sample	Total eDNA Quantity	Universal Primers	Eel-Specific Primers
Aguasmestas	A1	ND	0	0
	A2	ND	0	0
	A3	0.046	1	0
Belmonte	B1	ND	1	0
	B2	ND	0	1
	B3	0.052	1	1
La Riera	LR1	0.032	1	1
	LR2	0.026	1	1
	LR3	ND	1	0
Santullano	S1	0.158	1	0
	S2	0.114	1	1
	S3	0.255	1	1

Positive amplification from end-time PCR eel-specific primers [21] was obtained from six samples, corresponding to three of the locations analyzed (none from Aguasmestas); in one of these samples that was positive for eel DNA, PCR amplification was not achieved using universal primers (Table 3). This means that eDNA was present in all the samples where PCR amplification was obtained with any of the two markers, even in samples where the amount of DNA was below the detection threshold.

Regarding the effect of the hydrographic river profile on the total eDNA, multivariate multiple regression indicated that both altitude ($t = -4.28$, $p = 0.002$) and stream order ($t = -5.66$, $p = 0.0003$) significantly predicted the total eDNA quantity detected from the water samples. The association was negative in the two cases, as expected given the higher water speed at higher altitudes and the higher water flow in the lower river reaches with a higher stream order; both water speed and flow are physical constraints for eDNA detection. This means that the eDNA results should be corrected by any of those hydrographic indicators that, at the same time, are normally correlated in free-flowing rivers such as the zone considered here ($r = -0.93$, $p << 0.001$).

As predicted, the positive amplification using the end-time PCR universal marker (binary variable) was significantly correlated with the amount of eDNA detected using Qubit in a sampling replicate ($r = 0.68$, $p = 0.014$). The samples with positive amplification from the end-time PCR eel-specific marker were not significantly correlated with the total eDNA quantity ($r = 0.30$, $p = 0.35$), as expected because eDNA belongs to the different species inhabiting a location, not only to eels.

3.2.2. Relationship between Eel Biomass and Eel eDNA

In the river water samples, amplification was not detected in any of the negative controls, so no evidence of contamination was found. Furthermore, there was no evidence of inhibition (Supplementary Table S1).

The standard curves for *A. anguilla* fitted the equation $y = -3.33x + 23.07$, with $R^2 = 0.995$ and $y = -3.49x + 21.33$, and with $R^2 = 0.986$ for the two replicates (Figure 3). The limit of detection obtained was 2.42×10^{-5} ng/µL (LOD_6).

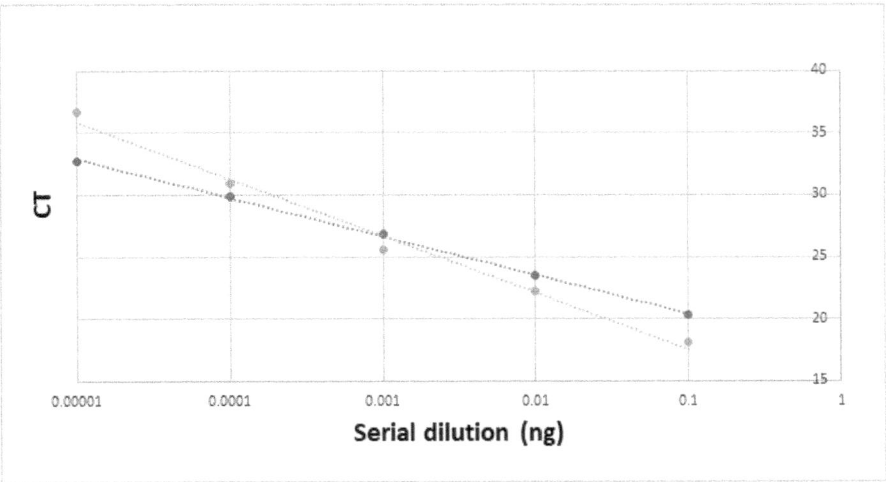

Figure 3. Standard curve performance. Serial dilution and CT values as the axes. Regression lines are shown.

European eel DNA was detected from qPCR at all sampling points, although not all the PCR technical replicates from each sample were positive (Supplementary Table S1; positive results shown in Table 4). Following the criteria already employed in similar studies [10,27,39], a sample was considered positive if at least one of the PCR technical replicates was positive. All sampling points in the Pigüeña River (n = 4) and Las Mestas in the Ponga River showed positive amplifications, reaching 100% detection rate for the locations and 73.3% for the water samples.

Table 4. River water positive results obtained from qPCR employing the new marker. Adjusted DNA quantity: ng, estimated from the standard curve calibration.

Location	Sampling Replicate	PCR Replicate	CT	Adjusted Eel eDNA Quantity
Aguasmestas	A2	A2.3	36.45	4.71×10^{-5}
	A3	A3.1	36.71	8.08×10^{-5}
Belmonte	B1	B1.1	36.58	8.84×10^{-5}
	B2	B2.1	34.37	1.85×10^{-4}
		B2.3	34.96	2.71×10^{-4}
	B3	B3.1	35.32	9.92×10^{-5}
		B3.2	34.22	2.06×10^{-4}
		B3.3	34.07	5.02×10^{-4}
La Riera	LR1	LR1.1	36.11	6.22×10^{-4}
		LR1.3	37.13	6.05×10^{-5}
Santullano	S1	S1.3	35.22	1.06×10^{-4}
	S2	S2.1	35.14	2.38×10^{-4}
	S3	S3.2	35.25	1.00×10^{-4}

As expected from the results obtained from the total eDNA quantity and real-time PCR markers, multiple regression (adjusted multiple $r^2 = 0.49$, $p = 0.014$) showed that the eel eDNA quantity measured with CT values was significantly predicted from the stream order ($t = 2.45$, $p = 0.03$, $r^2 = 0.12$) and from the altitude ($t = 3.27$, $p = 0.008$, $r^2 = 0.32$). From the stronger correlation, we chose altitude to correct the raw CT values, which were accordingly divided by altitude for further analysis.

A significant correlation ($r = -0.64$, $p = 0.018$) was found between the altitude-corrected CT, as the eel eDNA quantity estimator, and the biomass of the eels found by electrofishing (Figure 4). The relationship between the corrected CT as the dependent variable and the eel biomass was:

$$y = -4 \times 10^{-5}x + 0.1596$$

With 95% bootstrapped (N = 1999) confidence intervals (-6.288×10^{-5}, -2.401×10^{-5}) for the slope and (0.128, 0.196) for the intercept, the correlation between the corrected CT and the number of eels caught by electrofishing was also statistically significant ($r = -0.667$, $p = 0.013$; regression formula $y = -0.001x + 0.159$), as expected given the strong correlation between the number of eels and the biomass in this case study ($r = 0.997$, $p << 0.001$).

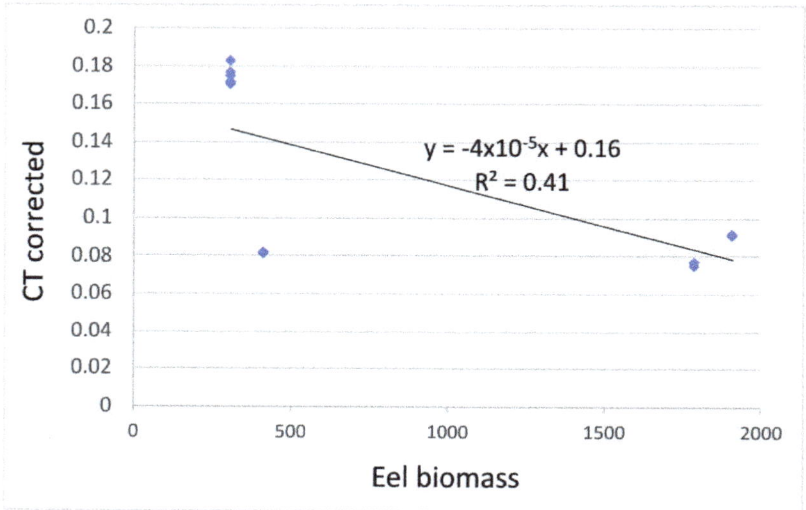

Figure 4. Linear regression showing the relationship between the biomass (in grams) of eels sampled by electrofishing and the CT obtained in qPCR-eDNA assay on Pigüeña River water samples. The equation and r^2 are given.

3.2.3. Predictive Value of the New Marker

In the Las Mestas (LM) location within the Ponga River (350 m above sea level), one sampling replicate of water sample LM2 (total eDNA quantity 0.162 ng/µL) and one of water sample LM3 (total eDNA 0.232 ng/µL) provided significant amplification in the qPCR, with CT values of 36.39 and 37.25, respectively. In these samples, no positive amplification was found from the end-time PCR using eel-specific primers, although the end-time PCR with universal primers was positive. The altitude-corrected CT values were 0.104 and 0.11 for LM2.3 and LM3.1, respectively.

From the eel biomass found in Las Mestas (x = 290 g), by applying the formula obtained for the Pigüeña River, we expected a corrected CT in the 95% range (0.11, 0.189). The values obtained from the water samples were in the lower part of the range.

4. Discussion

4.1. Overview of the Innovations of the New qPCR Marker

The qPCR assay designed here was successfully validated in controlled conditions and applied in river samples where *Anguilla anguilla* eDNA was detected and quantified. Despite the limited number of river sampling points, the detection probabilities found using the assay were very high, especially considering the small water volume that was analyzed (100% detection rate by location, based on only three water samples of 1.5 L). The tool was, therefore, useful for the detection of the species of interest. The sensitivity of the marker was similar in Weldon et al. [10], but the probability of detection was higher for the current assay (being 83% for Weldon et al. [10]); however, they employed lake samples, and the markers were not tested in mesocosm conditions. No limit of detection was reported in Halvorsen et al. [27]; thus, the sensitivity of the markers cannot be compared.

In addition to detection, it was possible to quantify the eels' eDNA. Based on the experimental tanks, the linearity between the eels' eDNA quantity and the eels' biomass is lost once it reaches a certain level of DNA quantity. In any case, the density of eels and biomass per liter in Tanks 7 and 9 were much higher than any expected value in the field. This type of experiment provides information about the thresholds of new assays, but river conditions are very dissimilar to mesocosms, so other variables influence the DNA detection and quantification [24]. With the obtained formula for the controlled experiment in aquaculture tanks, we expected a range of raw CT values of (25.05, 28.14), if the 290 g of eels were living in 60 L tanks for several days. The obtained CT values were much higher, as expected for a wild population of eels living in running waters, with a much lower biomass density per liter of water. However, they were still detectable, highlighting the sensitivity of the marker. Moreover, one of the main novelties of this study was the capacity of eDNA concentration to predict eel biomass. Even if this case study was limited to a few sampling points (those determined by the resource managers in the network of eel surveys in the region), and a very simple correction for altitude was made, the linear regression was sufficient, principally due to the fact that the eel eDNA quantities in the field were far smaller than the saturation values found from the experimental tanks. After further refinement, considering other hydrographic factors and more sampling points, this method would meet the need of the quantitative population assessment from eDNA highlighted by Halvorsen et al. [27].

Local hydrographic conditions could explain the results of qPCR in Santullano, which seem to not fit the regression line shown in Figure 4 (with corrected CT values close to 0.08). Located upstream of the Pigüeña River (Figure 1), it was the field point over 400 m of altitude with the lowest density of eels measured by electrofishing (1.99 g of biomass and 0.08 individuals per square meter). However, at that particular point, the total amount of eDNA directly measured from Qubit was much higher than at the other points (Table 3), suggesting a somewhat lower flow velocity or discharge—thus, an easier eDNA capture—at that point in the moment of sampling.

Differences between the end-time PCR and qPCR results for eel-specific markers were found in this study, with a positive end-time PCR for six sampling replicates and a positive qPCR for nine. Although it is not possible to properly estimate the relative sensitivities of Burgoa Cardas et al.'s [21] marker and this qPCR marker in the field with the current data, the results from Aguasmestas in the Pigüeña River and Las Mestas in the Ponga River would suggest that the new qPCR marker is more sensitive, since end-time PCR was not positive in any sample from those sites, while qPCR was positive in two samples from each site. Similar results were found in rivers of the same region for trout eDNA, which was better detected from qPCR than from end-time PCR [40]. More data are needed to confirm this point, but the results obtained from the different species and markers would point to a higher sensitivity for qPCR.

4.2. Limitations of This Study

One of the limitations of the eDNA-based population surveys highlighted in the current study is the strong effect of river hydrography on the recovered DNA quantity. Both the altitude and stream order influenced the capacity to recover sufficient eDNA for the detection and quantification of eels, as water discharge and velocity did [37]. In population inventories based on electrofishing, this fact was also observed [17,41]. Different electrofishing efforts would be necessary to obtain representative samples depending on the altitude, flow, and other river features as well as on the density of eels. Therefore, eDNA sampling strategies must be adapted to this issue: a higher sampling effort and, thus, higher water volumes and more replicates, would be necessary when monitoring areas with lesser eel abundance and higher water flow and speed.

Another limitation of this study was the small number of sampled field sites, as commented above. Although we found a significant correlation between the eDNA and biomass estimated from electrofishing, by applying a simple correction of the CT values by altitude, it is worth noting that it could be much more accurate with more sampling points and larger water volumes [10,42–44]. Individual hauls are nearly impossible to replicate [45]; thus, not only a higher number of points but also replicates from the same location would be recommended.

5. Management Recommendations

The results obtained in this study reveal the potential utility of eDNA and the developed assay to monitor European eel (*Anguilla anguillla*) populations. It may replace some of the intrusive electrofishing and fyke net surveys, since it gives numerous advantages, especially because it is harmless and does not disturb any fish. Conventional monitoring using netting, trapping, etc., requires specific skills and resources that can be overcome by using eDNA.

More expensive eDNA-based techniques such as metabarcoding were previously employed to monitor fish species, for example in the assessment of barriers removal [46]. We propose including the novel eel-specific qPCR technique developed here in the monitoring programs of the European eel, as part of the toolkit for endangered species surveys. This allows for not only the detection but also relative comparisons of eel abundance in different river zones, once corrected for altitude, stream order, and/or any other indicator of river hydrography. This is crucial for management purposes, especially in hard-to-reach areas where electrofishing is not efficient.

For application of this new method in other European regions, further research would be needed, taking into account the enormous diversity of European rivers. The method was validated here in relatively small rivers in southern Europe. Although the marker seems to be highly sensitive, it should still be validated in large rivers and lakes. Moreover, given the high variation of biotic communities in Europe, the local biota should be considered for additional cross-amplification tests.

6. Conclusions

Here, we describe and validate, using experiments in tanks and in the field, a new tool for the quantification of European eel biomass from environmental DNA in running waters. After correction for river hydrography, this qPCR marker enables the prediction of *Anguilla anguilla* biomass from water samples. The newly validated eDNA-based approach could complement or even replace conventional surveys, as a new molecular toolkit for European eel monitoring programs.

Supplementary Materials: The following supporting information can be downloaded at https://www.mdpi.com/article/10.3390/fishes8060279/s1, Table S1: Raw data of this study showing variables measured from each sampling point. CT, values for the new eel-specific marker. CT-IPC means the CT values in the qPCR when amplifying the inhibition control.

Author Contributions: S.F.: conceptualization, methodology, formal analysis, and writing—original draft. Á.G.: methodology and review and editing. D.D.: methodology and review and editing. J.L.M.: methodology, review and editing, I.M.: methodology and review and editing. A.A.: methodology and review and editing. E.G.-V.: conceptualization, formal analysis, writing—review and editing, and funding acquisition. G.M.-S.: conceptualization, methodology, writing—review and editing, supervision, project administration, and funding acquisition. All authors have read and agreed to the published version of the manuscript.

Funding: This study was funded by the Government of the Principado de Asturias, Spain (contract numbers FUO-18-106, FUO-21-342, and IDI/AYUD2021/50967). S.F. holds a Margarita Salas Grant, with reference number AYUD/2021/58385.

Institutional Review Board Statement: The study was approved by the Research Ethics Committee of the Asturias Principality (Spain). Approval code: 101/16. Approval date: 15 July 2016.

Data Availability Statement: All the data employed in this study are available upon approval.

Acknowledgments: We would like to thank the technicians and staff of the Centro de Experimentacion Pesquera (CEP-Gijon).

Conflicts of Interest: The authors declare no conflict of interest.

References

1. Feunteun, E. Management and Restoration of European Eel Population (*Anguilla anguilla*): An Impossible Bargain. *Ecol. Eng.* **2002**, *18*, 575–591. [CrossRef]
2. Bevacqua, D.; Melià, P.; Di Milano, P. Assessing Management Plans for the Recovery of the European Eel: A Need for Multi-Objective Analyses. *Am. Fish. Soc. Symp.* **2009**, *69*, 637–647.
3. Székely, C.; Palstra, A.; Molnár, K.; van den Thillart, G. Impact of the Swim-Bladder Parasite on the Health and Performance of European Eels. In *Spawning Migration of the European Eel*; Springer: Berlin/Heidelberg, Germany, 2009; pp. 201–226. [CrossRef]
4. Jacoby, D.; Gollock, M. *Anguilla anguilla. The IUCN Red List of Threatened Species*; IUCN: Gland, Switzerland, 2014.
5. Clavero, M.; Hermoso, V. Historical Data to Plan the Recovery of the European Eel. *J. Appl. Ecol.* **2015**, *52*, 960–968. [CrossRef]
6. Aarestrup, K.; Økland, F.; Hansen, M.M.; Righton, D.; Gargan, P.; Castonguay, M.; Bernatchez, L.; Howey, P.; Sparholt, H.; Pedersen, M.I.; et al. Oceanic Spawning Migration of the European Eel (*Anguilla anguilla*). *Science* **2009**, *325*, 1660. [CrossRef] [PubMed]
7. Dekker, W. The History of Commercial Fisheries for European Eel Commenced Only a Century Ago. *Fish. Manag. Ecol.* **2019**, *26*, 6–19. [CrossRef]
8. Dekker, W. A Conceptual Management Framework for the Restoration of the Declining European Eel Stock. *Am. Fish. Soc. Symp.* **2009**, *58*, 3–19.
9. European Union. *Establishing Measures for the Recovery of the Stock of European eel in E Union. 1100/2007*; European Union: Maastricht, The Netherlands, 2007.
10. European Union. *Official Journal of the European Union*; European Union: Maastricht, The Netherlands, 2007; Volume 248, pp. 281–304.
11. Weldon, L.; O'Leary, C.; Steer, M.; Newton, L.; Macdonald, H.; Sargeant, S.L. A Comparison of European Eel *Anguilla anguilla* eDNA Concentrations to Fyke Net Catches in Five Irish Lakes. *Environ. DNA* **2020**, *2*, 587–600. [CrossRef]
12. Lobón-Cerviá, J.; Iglesias, T. Long-Term Numerical Changes and Regulation in a River Stock of European Eel *Anguilla anguilla*. *Freshw. Biol.* **2008**, *53*, 1832–1844. [CrossRef]
13. Bilotta, G.S.; Sibley, P.; Hateley, J.; Don, A. The Decline of the European Eel *Anguilla anguilla*: Quantifying and Managing Escapement to Support Conservation. *J. Fish Biol.* **2011**, *78*, 23–38. [CrossRef]
14. Weltersbach, M.S.; Strehlow, H.V.; Ferter, K.; Klefoth, T.; de Graaf, M.; Dorow, M. Estimating and Mitigating Post-Release Mortality of European Eel by Combining Citizen Science with a Catch-and-Release Angling Experiment. *Fish. Res.* **2018**, *201*, 98–108. [CrossRef]
15. Nielsen, J.L. Scientific Sampling Effects: Electrofishing California's Endangered Fish Populations. *Fisheries* **1998**, *23*, 6–12. [CrossRef]
16. Ellender, B.R.; Becker, A.; Weyl, O.L.F.; Swartz, E.R. Underwater Video Analysis as a Non-Destructive Alternative to Electrofishing for Sampling Imperilled Headwater Stream Fishes. *Aquat. Conserv.* **2012**, *22*, 58–65. [CrossRef]
17. Kanno, Y.; Vokoun, J.C.; Dauwalter, D.C.; Hughes, R.M.; Herlihy, A.T.; Maret, T.R.; Patton, T.M. Influence of Rare Species on Electrofishing Distance When Estimating Species Richness of Stream and River Reaches. *Trans. Am. Fish. Soc.* **2011**, *138*, 1240–1251. [CrossRef]
18. Degerman, E.; Tamario, C.; Watz, J.; Nilsson, P.A.; Calles, O. Occurrence and Habitat Use of European Eel (*Anguilla anguilla*) in Running Waters: Lessons for Improved Monitoring, Habitat Restoration and Stocking. *Aquat. Ecol.* **2019**, *53*, 639–650. [CrossRef]
19. Fernandez, S.; Sandin, M.M.; Beaulieu, P.G.; Clusa, L.; Martinez, J.L.; Ardura, A.; García-Vázquez, E. Environmental DNA for Freshwater Fish Monitoring: Insights for Conservation within a Protected Area. *PeerJ* **2018**, *2018*, e4486. [CrossRef]
20. Thomsen, P.F.; Willerslev, E. Environmental DNA—An Emerging Tool in Conservation for Monitoring Past and Present Biodiversity. *Biol. Conserv.* **2015**, *183*, 4–18. [CrossRef]
21. Jerde, C.L.; Mahon, A.R.; Chadderton, W.L.; Lodge, D.M. "Sight-Unseen" Detection of Rare Aquatic Species Using Environmental DNA. *Conserv. Lett.* **2011**, *4*, 150–157. [CrossRef]
22. Burgoa Cardás, J.; Deconinck, D.; Márquez, I.; Peón Torre, P.; Garcia-Vazquez, E.; Machado-Schiaffino, G. New eDNA Based Tool Applied to the Specific Detection and Monitoring of the Endangered European Eel. *Biol. Conserv.* **2020**, *250*, 108750. [CrossRef]
23. Lacoursière-Roussel, A.; Côté, G.; Leclerc, V.; Bernatchez, L. Quantifying Relative Fish Abundance with EDNA: A Promising Tool for Fisheries Management. *J. Appl. Ecol.* **2016**, *53*, 1148–1157. [CrossRef]

24. Minamoto, T.; Miya, M.; Sado, T.; Seino, S.; Doi, H.; Kondoh, M.; Nakamura, K.; Takahara, T.; Yamamoto, S.; Yamanaka, H.; et al. An Illustrated Manual for Environmental DNA Research: Water Sampling Guidelines and Experimental Protocols. *Environ. DNA* **2021**, *3*, 8–13. [CrossRef]
25. Thalinger, B.; Deiner, K.; Harper, L.R.; Rees, H.C.; Blackman, R.C.; Sint, D.; Traugott, M.; Goldberg, C.S.; Bruce, K. A Validation Scale to Determine the Readiness of Environmental DNA Assays for Routine Species Monitoring. *Environ. DNA* **2021**, *3*, 823–836. [CrossRef]
26. Cairns, D.K.; Benchetrit, J.; Bernatchez, L.; Bornarel, V.; Casselman, J.M.; Castonguay, M.; Charsley, A.R.; Dorow, M.; Drouineau, H.; Frankowski, J.; et al. Thirteen Novel Ideas and Underutilised Resources to Support Progress towards a Range-Wide American Eel Stock Assessment. *Fish. Manag. Ecol.* **2022**, *29*, 516–541. [CrossRef]
27. Moyer, G.R.; Bartron, M.L.; Galbraith, H.S.; Grassi, J.; Rees, C.B. Development and Validation of Two Environmental DNA Assays for American Eel (Anguilla rostrata). *Environ. DNA* **2023**, *5*, 175–190. [CrossRef]
28. Halvorsen, S.; Korslund, L.; Gustavsen, P.; Slettan, A. Environmental DNA Analysis Indicates That Migration Barriers Are Decreasing the Occurrence of European Eel (*Anguilla anguilla*) in Distance from the Sea. *Glob. Ecol. Conserv.* **2020**, *24*, e01245. [CrossRef]
29. Halvorsen, S.; Korslund, L.; Mattingsdal, M.; Slettan, A. Estimating Number of European Eel (*Anguilla anguilla*) Individuals Using Environmental DNA and Haplotype Count in Small Rivers. *Ecol. Evol.* **2023**, *13*, e9785. [CrossRef]
30. Cantera, I.; Decotte, J.B.; Dejean, T.; Murienne, J.; Vigouroux, R.; Valentini, A.; Brosse, S. Characterizing the Spatial Signal of Environmental DNA in River Systems Using a Community Ecology Approach. *Mol. Ecol. Resour.* **2022**, *22*, 1274–1283. [CrossRef]
31. Rice, C.J.; Larson, E.R.; Taylor, C.A. Environmental DNA Detects a Rare Large River Crayfish but with Little Relation to Local Abundance. *Freshw. Biol.* **2018**, *63*, 443–455. [CrossRef]
32. Jo, T.; Yamanaka, H. Meta-Analyses of Environmental DNA Downstream Transport and Deposition in Relation to Hydrogeography in Riverine Environments. *Freshw. Biol.* **2022**, *67*, 1333–1343. [CrossRef]
33. Carraro, L.; Hartikainen, H.; Jokela, J.; Bertuzzo, E.; Rinaldo, A. Estimating Species Distribution and Abundance in River Networks Using Environmental DNA. *Proc. Natl. Acad. Sci. USA* **2018**, *115*, 11724–11729. [CrossRef] [PubMed]
34. Zippin, C. The Removal Method of Population Estimation. *J. Wildl. Manag.* **1958**, *22*, 82–90. [CrossRef]
35. Zippin, C. An Evaluation of the Removal Method of Estimating Animal Populations. *Biometrics* **1956**, *12*, 163–189. [CrossRef]
36. Strahler, A.N. Quantitative Analysis of Watershed Geomorphology. *Eos Trans. Am. Geophys. Union* **1957**, *38*, 913–920. [CrossRef]
37. Leray, M.; Yang, J.Y.; Meyer, C.P.; Mills, S.C.; Agudelo, N.; Ranwez, V.; Boehm, J.T.; Machida, R.J. A New Versatile Primer Set Targeting a Short Fragment of the Mitochondrial COI Region for Metabarcoding Metazoan Diversity: Application for Characterizing Coral Reef Fish Gut Contents. *Front. Zool.* **2013**, *10*, 34. [CrossRef] [PubMed]
38. Hammer, Ø.; Harper, D.A.T.; Ryan, P.D. PAST: Paleontological Statistics Software Package for Education and Data Analysis. *Paleontol. Stat. Softw. Stat* **2013**, *105*, 1352–1357.
39. Laramie, M.B.; Pilliod, D.S.; Goldberg, C.S. Characterizing the Distribution of an Endangered Salmonid Using Environmental DNA Analysis. *Biol. Conserv.* **2015**, *183*, 29–37. [CrossRef]
40. Fernández, S.; Rodríguez, S.; Martínez, J.L.; Borrell, Y.J.; Ardura, A.; García-Vázquez, E. Evaluating Freshwater Macroinvertebrates from EDNA Metabarcoding: A River Nalón Case Study. *PLoS ONE* **2018**, *13*, e0201741. [CrossRef]
41. Lobón-Cerviá, J.; Rincón, P. Variations in the Populationdynamics of the European Eel *Anguilla anguilla* (L.) along the Course of a Cantabrian River. *Ecol. Freshw. Fish.* **1995**, *4*, 17–27. [CrossRef]
42. Takahara, T.; Minamoto, T.; Yamanaka, H.; Doi, H.; Kawabata, Z. Estimation of Fish Biomass Using Environmental DNA. *PLoS ONE* **2012**, *7*, 3–10. [CrossRef]
43. Pilliod, D.S.; Laramie, M.B.; MacCoy, D.; Maclean, S. Integration of EDNA-Based Biological Monitoring within the U.S. Geological Survey's National Streamgage Network. *JAWRA J. Am. Water Resour. Assoc.* **2019**, *55*, 1505–1518. [CrossRef]
44. Itakura, H.; Wakiya, R.; Sakata, M.K.; Hsu, H.-Y.; Chen, S.-C.; Yang, C.-C.; Huang, Y.-C.; Han, Y.-S.; Yamamoto, S.; Minamoto, T.; et al. Estimations of Riverine Distribution, Abundance, and Biomass of Anguillid Eels in Japan and Taiwan Using Environmental DNA Analysis. *Zool. Stud.* **2020**, *59*, 17. [CrossRef]
45. Knudsen, S.W.; Ebert, R.B.; Hesselsøe, M.; Kuntke, F.; Hassingboe, J.; Mortensen, P.B.; Thomsen, P.F.; Sigsgaard, E.E.; Hansen, B.K.; Nielsen, E.E.; et al. Species-Specific Detection and Quantification of Environmental DNA from Marine Fishes in the Baltic Sea. *J. Exp. Mar. Biol. Ecol.* **2019**, *510*, 31–45. [CrossRef]
46. Muha, T.P. Using EDNA Metabarcoding to Monitor Changes in Fish Community Composition After Barrier Removal. *Front. Ecol. Evol.* **2021**, *9*, 28. [CrossRef]

fishes

Article

Human Chorionic Gonadotropin Enhancement of Early Maturation and Consequences for Reproductive Success of Feminized European Eel (*Anguilla anguilla*)

Arjan P. Palstra [1,*], Ida van de Ven [1], Pauline Jéhannet [1], Leo Kruijt [1], Henk Schipper [2], William Swinkels [3] and Leon T. N. Heinsbroek [4]

[1] Animal Breeding and Genomics Department, Wageningen University & Research, 6708 PB Wageningen, The Netherlands
[2] Experimental Zoology Group, Wageningen University and Research, 6700 AH Wageningen, The Netherlands
[3] Palingkwekerij Koolen, 5571 XC Bergeijk, The Netherlands
[4] Wageningen Eel Reproduction Experts B.V., 3708 AB Wageningen, The Netherlands
* Correspondence: arjan.palstra@wur.nl

Abstract: To induce oocyte development, eels are weekly injected with salmon or carp pituitary extract (CPE). The weekly handling and hormone peaks result in inferior oocyte quality; therefore, alternative treatments that improve oocyte quality and reproductive success require investigation. The enhancement of early sexual maturation by a single injection with human chorionic gonadotropin (hCG), administered prior to CPE treatment, was investigated. Fifty feminized eels were subjected to simulated migration, after which eels received either a hCG or a sham injection. After two months, the hCG-treated eels showed an increase in eye size, gonadosomatic index (GSI), and plasma 11-ketotestosterone concentration, when compared with the sham-injected controls. The hCG-treated eels showed increases in oocyte diameter and lipid area, and in ovarian expression of aromatase (*cyp19*), follicle stimulating hormone receptor (*fshr*) and lipoprotein lipase (*lpl*). Yolk was present in the oocytes of the hCG-treated eels, not yet in the oocytes of the controls. The hCG-induced deposition of yolk may relate to early-life treatment with 17β-estradiol during feminization. hCG-treated eels required four CPE injections less to mature than the controls. hCG treatment may benefit reproductive success in feminized eels by initiating vitellogenesis and reducing the hypophysation period, although larvae were obtained from most females in both groups.

Keywords: aquaculture; eel reproduction; propagation; feminization; oocyte quality; egg yolk; lipid deposition; 11-ketotestosterone; 17β-estradiol; ultrasound

Key Contribution: This study quantified the many and diverse stimulatory effects of a single injection with human chorionic gonadotropin on the early sexual maturation of feminized European eels, which included the induced deposition of yolk in the oocytes. When administered prior to the regular protocol of induced maturation by weekly injections of pituitary extract, hCG injection shortened the hypophysation period significantly, which may benefit reproductive success.

Citation: Palstra, A.P.; van de Ven, I.; Jéhannet, P.; Kruijt, L.; Schipper, H.; Swinkels, W.; Heinsbroek, L.T.N. Human Chorionic Gonadotropin Enhancement of Early Maturation and Consequences for Reproductive Success of Feminized European Eel (*Anguilla anguilla*). *Fishes* **2023**, *8*, 281. https://doi.org/10.3390/fishes8060281

Academic Editor: Jose Martin Pujolar

Received: 14 April 2023
Revised: 16 May 2023
Accepted: 22 May 2023
Published: 25 May 2023

1. Introduction

Chorionic gonadotropin is produced by the placenta of pregnant mammals and belongs, together with the gonadotropins follicle-stimulating hormone Fsh and luteinizing hormone Lh, to the family of glycoprotein hormones [1]. Fsh and Lh are essential in the regulation of sexual maturation in vertebrates [2], including the European eel. Fsh induces vitellogenesis in females and spermatogenesis in males; Lh plays a major role in oocyte maturation in females and spermiation in males (e.g., [3,4]). Human chorionic gonadotropin (hCG) is an Lh-analog and binds and activates the luteinizing hormone receptor (Lhr) [5–7].

11-Ketotestosterone (11KT) plays a major role during previtellogenesis in female eels [8]. This non-aromatizable androgen acts during the migratory silver stage of the eels and is referred to as the main puberty initiating androgen [9]. 11KT is responsible for ovarian lipid accumulation in eels, since elevated lipoprotein lipase (Lpl) levels were found after an increase in 11KT [10]. Lipid quantity and composition are key determinants of egg quality in marine fishes [11]. 11KT also induces the expression of the ovarian *fshr* [8] and stimulates the production of the phospholipoglycoprotein vitellogenin (Vtg; [12]).

Fsh controls the synthesis of 17β-estradiol (E2) by increasing ovarian aromatase (cyp19) activity during vitellogenesis [13]. Cyp19 is an enzyme that converts testosterone (T) into E2. When released into the blood circulation of the eel, E2 will bind to its hepatic nuclear estrogen receptor (Esr-1) and therewith induces the synthesis of Vtg by hepatocytes [14]. When released into the circulation, Vtg binds to its ovarian vitellogenin receptor (Vtgr) [15] and is cleaved into yolk protein in the vitellogenic oocytes [16]. Yolk is the main nutrient for developing embryos and is essential for early larval survival and development [17].

To induce the sexual maturation of eels in captivity, hypophysation is the most practiced way. Recently, also eel specific recombinant gonadotropins have successfully been applied to mature eels and produce larvae [18]. Hypophysation involves weekly injections with common carp (*Cyprinus carpio*) or Atlantic salmon (*Salmo salar*) pituitary extracts (CPE and SPE, respectively) leading to weekly handling of the eels and peaking hormone levels. Weekly handling causes stress in the animals and increases their susceptibility to diseases [19,20]. The weekly peaks in hormone levels are associated with oocyte developmental abnormalities [21]. Therefore, alternatives for hypophysation are under investigation, such as steroid implants [22] and human chorionic gonadotropin (hCG).

hCG is generally used to mature male eels. Fontaine [23] discovered that urine of pregnant women could be used to mature male eels, and later, it was shown that hCG was the active compound causing the effects. hCG has a longer half-life compared to other gonadotropins because it contains four glycosylated serine residues [24]. Therefore, hCG does not have to be injected weekly, in contrast to pituitary extract. In the study of Nguyen et al. [25], the effects of hCG were tested on ovarian morphology, sex steroid levels and messenger ribonucleic acid (mRNA) levels of genes expressed in the pituitary, in the short-finned eel *A. australis*. The study suggested the involvement of the Fsh receptor (Fshr) in the regulation of previtellogenic oocyte development. Eels that were treated with hCG (20, 100 or 500 IU kg^{-1} BW) showed significant increases in ovarian fshr mRNA and plasma levels of 11KT and E2. The study concluded that hCG stimulates oocyte growth and development during previtellogenesis; yolk deposition, however, was not observed [25].

In this study, we aimed to determine the effects of an hCG injection on the early maturation in feminized European eels and to assess the consequences for reproductive success. Effects of an hCG injection, administered before the regular protocol of CPE injections to mature European eels, and specifically feminized eels, have not been studied before. The controlled production of larvae enabled us to assess the potential consequences for reproductive success. Specifically, effects were investigated by analyses of the changes in external body appearance and biometry; of the gonadosomatic index (GSI) as determined by ultrasound; of the plasma 11 KT levels as determined by ELISA; of the expression patterns of key hepatic and ovarian genes in vitellogenesis; of oocyte histology, and of the relation with indicators of reproductive success.

2. Materials and Methods

2.1. Experimental Setup

Young elvers were transferred from eel farm Palingkwekerij Koolen (Bergeijk, The Netherlands) to the animal experimental facilities of Wageningen University & Research (CARUS, Wageningen, The Netherlands) and feminized by feeding them with E2-coated pellets over a 7 month period [26,27]. The feminizing treatment does not only realize female eels but also accelerates the oocyte development before the vitellogenic stage and shortens the generation time to 2 years from the glass-eel stage [28]. After an additional 6 months of feeding with a custom-made broodstock diet (protein 525 g kg^{-1}, fat 98 g kg^{-1} and ash 76 g kg^{-1}), 50 eels were transferred to seawater (Tropic Marine, 36 ppt) and fed no longer for a period of 7 weeks.

After this period, the 50 eels were anesthetized (phenoxy ethanol 0.2 mL L^{-1}), portrait pictures were taken, and biometric data were collected. Eels were then subjected to a simulated migration (based on [29]) and swam for 83 days in a swim gutter against a flow of 0.5 m s^{-1} equaling a distance of 3514 km. Eels were swimming in the dark at daily alternating temperatures from 10 °C to 15 °C.

After this simulated migration, all 50 eels were anesthetized again, portrait pics were taken, and biometric data were collected. Eels were now also tagged with passive integrated transponders (PITs) and randomly divided into two groups (pre-treatment groups). One group of eels received an hCG injection (3000 IU kg^{-1}; hCG-treated group; N = 25) and the other group received a physiological salt solution injection of the same volume (control or C group; N = 25). After 8 weeks, again all eels were anesthetized, portrait pics taken, biometric data were collected, and also blood samples were extracted (post-treatment groups). N = 10 eels from each group were dissected, and gonad weight (GW), liver weight (LW), gastrointestinal tract weight (GITW) and swim bladder weight (SBW) were determined. Tissue samples were taken from the liver and ovary and stored in RNAlater stabilization solution (Ambion) at −20 °C for RT-PCR. A second ovary sample was fixed in 4% Paraformaldehyde overnight at 4 °C and then stored in 70% ethanol before histological analysis. The remaining N = 15 eels from each group were used for hypophysation to sexually mature the eels and determine reproductive success (described in Section 2.7).

2.2. Biometrical Parameters

Portrait pics were taken of each eel three times (pre-migration; pre-treatment and post-treatment), and the images were scaled and aligned using ImageJ to visualize morphological changes. Eels were measured for biometrical parameters body length (BL), body weight (BW), body girth (BG), horizontal eye diameter (EDh) and vertical eye diameter (EDv). From these measurements, the Fulton's condition factor (K; [30]), body girth index (BGI; [31]) and the eye index (EI; [32]) were calculated.

2.3. Blood Collection and 11KT Plasma Measurements

Blood samples (0.5 mL) were extracted post treatment from the caudal artery with a heparin-flushed syringe and then immediately put on ice. Plasma was separated by centrifugation (10,000 rpm, 5 min, 4 °C) and stored at −80 °C. Plasma 11KT concentrations in pg mL^{-1} were determined in N = 17 samples from the hCG-treated and control groups, as duplicates using the 11KT ELISA kit (Item No. 582751) from the Cayman Chemical Company (Ann Arbor, Michigan, Unites States) according to the manufacturer's protocol. Three outliers (N = 1 of the control group, N = 2 of the hCG-treated group) with values > 2* SD were not considered in the analysis.

2.4. Tissue Indices and Ultrasound

Gonad weight, liver weight, gastrointestinal tract weight and swim bladder weight were used to calculate the GSI, the hepatosomatic Index (HSI), the gastrointestinal tract index (GITI) and the swim-bladder index (SBI), respectively, by dividing the tissue weights (g) by total body weight (g) and multiplying the outcome times 100.

GSI was also determined non-invasively by ultrasonography (MyLabFiveTMVet with a LA435 probe, Esaote, Genoa, Italy) for all eels at pre-migration, pre-treatment and post-treatment timepoints following the methodology developed by Bureau du Colombier et al. [33] and further validated by Palstra et al. [22]. The ultrasound videos were analyzed using ImageJ to determine the surface area of the gonads. The GSI was calculated according to the following formula [33]:

$$GSI = BW/(e\hat{}(3.02753 + 1.32056 \times LN\ ((A1 + A2 + A3)/3))) \times 100$$

BW: body weight (g); A1: area measurement 1 (cm^2); A2: area measurement 2 (cm^2); A3: area measurement 3 (cm^2).

The GSI values as calculated on the basis of ultrasound videos of dissected eels were compared with their real GSI values. GSI values of two eels in the hCG-treated group were not considered as weights were wrongly determined, leaving N = 8 for the hCG group.

2.5. Histology

Gonadal samples were embedded in paraffin wax and cut into 5 µm sections using a motorized rotary microtome (HM350 Microm). Per sample, two slides containing 6 sections, each at least 30 µm apart, were stained for nuclei and cytoplasm using Mayer's hematoxylin–eosin staining method. Sections were imaged using a Leica DM6b upright microscope. For each sample, the 10 largest oocytes with a visible central nucleus were selected. These oocytes were measured for their diameter using the image-processing software ImageJ [34]. The lipid area and the lipid area relative to the oocyte area were also determined using ImageJ.

2.6. RT-PCR

Total RNA of liver and ovary tissue was isolated and purified using an RNeasy Plus Universal Mini Kit (Qiagen) according to the manufacturer's instructions. RNA quantity (>500 ng mL^{-1}) and quality (>2.0 260/280 ratio) were determined by Nanodrop. Total RNA was reverse transcribed (Applied Biosystems™ High-Capacity cDNA Reverse Transcription Kit) according to the manufacturer's instructions, and the resulting cDNA was diluted to a concentration of 1/50 ng cDNA μL^{-1} in MilliQ (250 ng RNA μL^{-1} solution was used to transcribe into 4 µL cDNA, or RNA-equivalent, that was diluted in 196 µL MilliQ). mRNA levels were quantified using the Quantstudio 5 Real-Time RT-PCR system (ThermoFisher Scientific). RT-PCR was performed on a 20 µL mixture containing cDNA (5 µL, 1/50 ng μL^{-1}), primers (1 µL, 5µM each), SensiFAST SYBR Lo-Rox Kit (10 µL; Bioline, Luckenwalde, Germany) and Milli-Q (4 µL). RT-PCR assays were run using a temperature profile starting at the hold stage (95 °C, 2 min), followed by 30 to 40 cycles of denaturation (95 °C, 15 s) and annealing/extension (°C depending on the selected target gene or housekeeping gene—HKG, 30 s and 72 °C, 5 s) and ending with the melt curve stage (95 °C, 1 s followed by 60 °C, 20 s) gradually heating (0.1 °C s^{-1}) to 95 °C. The melting curves were analyzed for reaction specificity and the presence of primer cross-reaction. Primer efficiencies were determined using a series of dilutions to generate the standard curves. The choices for the selected target genes and HKGs were based on recent literature ([18,35–38]; Table 1).

Table 1. Primers for each of the target genes or housekeeping genes. Abv = gene abbreviation; Accession number: the Genbank accession number for *A. anguilla*; G: sequence obtained from the *A. anguilla* genome by Jéhannet et al. [35] or Setiawan and Lokman [37]; size: the Amplicon size, the PCR product size in base pairs (bp) of nucleotides; temp: annealing temperature in °C of target genes or housekeeping genes, and references. Abbreviations: FW = Forward primer; RV = Reverse primer.

Abv.	Gene	Accession Number	Primer Sequence (5′-3′)	Size (bp)	Temp °C	Ref.
18S	18 s ribosomal RNA	FM946133	FW: GTACACACGGCCGGTACAGT RV: GGTAGGCGCAGAAAGTACCA	302	60	[37]
cyp19	Aromatase cytochrome P450	KF990052	FW: CGCACCTACTTTGCTAAAGCTC RV: AGGTTGAGGATGTCCACCTG	137	62	[35]
elf-1	Elongation factor 1	EU407825	FW: CCCCTGCAGGATGTCTACAA RV: AGGGACTCATGG TGCATTTC	152	64	[37]
esr-1	Estrogen receptor 1	LN879034	FW: GGCATGGCCGAGATTTTC RV: GCACCGGAGTTGAGCAGTAT	116	62	[35]
fshr	Follicle-stimulating hormone receptor	LN831181	FW: CCTGGTCGAGATAACAATCACC RV: CCTGAAGGTCAAACAGAAAGTCC	173	63	[38]
l36	60 s ribosomal protein l36	G	FW: CCTGACCAAGCAGACCAAGT RV: TCTCTTTGCACGGATGTGAG	160	62	[37]

Table 1. *Cont.*

Abv.	Gene	Accession Number	Primer Sequence (5′-3′)	Size (bp)	Temp °C	Ref.
lhr-1	Luteinizing hormone 1	LN831182	FW: GCGGAAACACAGGGAGAAC RV: GGTTGAGGTACTGGAAATCGAAG	155	60	[36]
lhr-2	Luteinizing hormone 2	LN831183	FW: TCAACAACCTCACCAATCTCTCT RV: GCAGTGAAGAAATAGCCGACA	162	62	[18]
lpl	lipoprotein lipase	XM035416270	FW: TGATGCTGATTGCTACTTCTGG RV: ATGCTCTCCTGCTGCTTCTT	115	62	This study
vtgr	Vitellogenin receptor	G	FW: TCTGAACGAACCCCAGGA RV: TTTGGGGAGTGCTTGTTGA	140	59	[35]

Estrogen receptor-1 (*esr-1*) was selected as target gene in the liver. Ribosomal protein L36 (*l36*) was used as HKG to normalize *esr-1* expression by calculating the relative fold change (fc) using the 2-ΔΔCt method.

For the ovary, six target genes were selected. These target genes were Aromatase (*cyp19*), Follicle-stimulating hormone receptor (*fshr*), Luteinizing hormone receptor-1 (*lhr-1*), Luteinizing hormone receptor-2 (*lhr-2*), Lipoprotein lipase (*lpl*) and Vitellogenin receptor (*vtgr*). The primers of Lpl and Lhr-2 were tested for their efficiencies, since the Lpl primer was newly designed, and Lhr-2 had an overall low expression. The efficiency of Lpl was tested to be 93%, and the efficiency of Lhr-2 was 106%; thus, both primers had efficiencies within the desired range (90–110%) and were used for further analysis. As HKGs, 18 s ribosomal RNA (*18s*), Elongation factor-1 (*elf-1*) and *l36* were used to normalize the target genes by calculating the relative fc values using the 2-ΔΔCt method. The BestKeeper tool [39] was used to determine Ct values from 18S, *elf-1* and *l36* combined, so no significant differences were found between the Ct values of the hCG group and the C group.

2.7. Reproduction

The eels used for propagation (N = 15 for each of the groups) were fully matured by hypophysation. Eels received multiple weekly CPE injections until a BWI threshold value of 110 was crossed and eels received an extra booster CPE injection. Final ovulation was then induced by 17,20β-dihydroxy-4-pregnen-3-one (DHP) injection [40–42]. For each group, N = 3 eels died before egg release. From the remaining eels (N = 12 from each group), the days to reach a BWI of 110, the number of CPE injections to reach a BWI of 110, the hours after the final DHP injection until egg release, the percentage of floating eggs, the number of batches of eggs that were obtained and the number of batches that gave embryos and larvae, and the average survival days post-hatch (dph) and the maximal survival dph, were determined.

2.8. Data Analysis

Data analyses were performed using R-studio (version 4.1.1). To test for normal distribution of data, the Shapiro test was performed. If the data were normally distributed ($p \geq 0.05$), Welch two sample t-tests were performed. If data were not normally distributed ($p \leq 0.05$), Wilcoxon rank sum tests were performed. Data were compared using paired tests on the same eels between pre- and post-treatment values, and unpaired when comparing eels of the hCG-treated group with eels of the control group. GSI values as calculated on the basis of ultrasound data were compared with the GSI values as determined by dissection, on the same eels post-treatment, by paired two sample t-tests. Data were considered significantly different when $p \leq 0.05$.

3. Results

3.1. Simulated Migration

All experimental eels managed to complete the simulated migration and were included for comparison of post-migration vs. pre-migration biometrical measurements. Significant decreases were apparent for BW, K and BGI ($p < 0.001$). BW decreased by 57 g, from

398 ± 48 g to 341 ± 42 g. K decreased from 0.204 ± 0.015 g to 0.175 ± 0.013, and BGI decreased from 0.200 ± 0.008 g to 0.179 ± 0.008. EI was 7.43 ± 0.17 after migration and not significantly higher than the average value of 7.32 ± 0.16 before migration.

3.2. Portrait Pictures and Biometry

The portrait images of the experimental eels indicated changes in head morphology that were associated with hCG treatment (Figure 1). In eels of the hCG-treated group, eye enlargement was observed that was not observed in the controls. Eels from the hCG-treated group also showed darkening of the pectoral fins and on the dorsal side of the body, in contrast to the controls. The head shape of the hCG-treated eels was less acute than the head shape of the controls. EI ($p < 0.001$; Table 2) was considerably higher in eels of the hCG-treated group as compared to the controls. Additionally, K ($p < 0.01$; Table 2) and BGI ($p < 0.001$; Table 2) values were higher in eels of the hCG-treated group.

Figure 1. Portrait pics of (**a**) a representative eel of the control group and (**b**) a representative eel of the human chorionic gonadotropin (hCG) treated group. Each eel is shown at three moments from left to right: before migration, pre-treatment and post-treatment. Morphological differences such as eye enlargement, darkening of the pectoral fin and a less acute head were observed in the eels after hCG treatment in comparison with the controls.

Table 2. Paired biometric data (AV ± SD) for eels of the control (C) group and the human chorionic gonadotropin (hCG) treated group, pre- and post-treatment. Significant differences between either the C group post-treatment vs. pre-treatment, or for the hCG-treated group post-treatment vs. pre-treatment are indicated by (−) when post-treatment values were significantly smaller ($p < 0.001$) than pre-treatment values, or by (+) when post-treatment values were significantly larger ($p < 0.001$) than pre-treatment values. Abbreviations: BL = body length; BW = body weight; K = Fulton's condition factor; BGI = body girth index; EI = eye index; C = control group; hCG = human chorionic gonadotropin-treated group.

	BL (cm)	BW (g)	K	BGI	EI
C pre-treatment	58 ± 3	347 ± 47	0.17 ± 0.00	0.18 ± 0.00	7.26 ± 1.27
C post-treatment	58 ± 3	323 ± 45 (−)	0.16 ± 0.00 (−)	0.17 ± 0.00 (−)	7.04 ± 1.26
hCG pre-treatment	58 ± 3	336 ± 37	0.18 ± 0.00	0.18 + 0.00	7.59 ± 1.10
hCG post-treatment	58 ± 3	329 ± 37 (−)	0.17 ± 0.00 (−)	0.18 ± 0.00	9.11 ± 0.92 (+)

3.3. Blood Plasma 11-Ketotestosterone

Significantly higher circulatory 11KT levels were found when comparing hCG-treated eels with the controls: 146 ± 64 vs. 46 ± 26 pg mL^{-1}, respectively ($p < 0.001$; Figure 2).

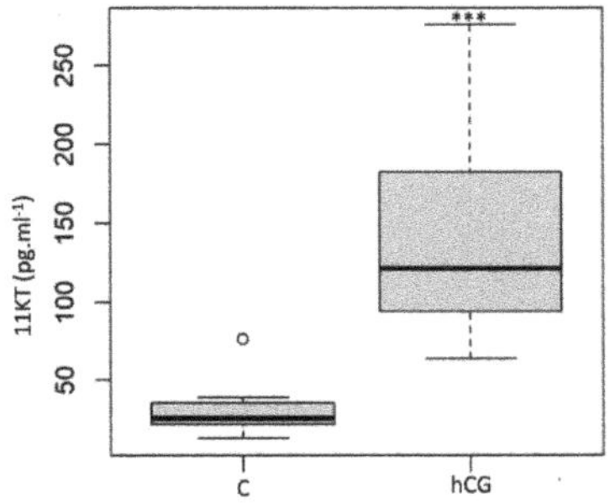

Figure 2. Post-treatment 11-ketotestosterone concentrations (11KT in pg mL^{-1}) as determined by enzyme-linked immunosorbent assay (ELISA) in the blood plasma of eels from the control group (C; N = 16) and from the human chorionic gonadotropin-treated group (hCG; N = 15). The asterisks indicate a significant difference between both groups ($p < 0.001$).

3.4. Ultrasound and Dissection

Ultrasound GSI calculations were validated by GSI determinations on the dissected eels of the hCG-treated and control groups. For eels of the control group (N = 10), GSI as calculated from ultrasound images was 1.34 $\pm$ 0.08 and very similar to the GSI of 1.38 $\pm$ 0.10 as determined on the dissected eels. For eels of the hCG-treated group (N = 8), GSI as calculated from ultrasound images was 2.82 $\pm$ 0.54 and also very similar to the GSI of 2.91 $\pm$ 0.55 as determined on the dissected eels. Ultrasound GSI data could, therefore, be used for further analyses. GSI values for hCG-treated eels (N = 25) were 2.88 $\pm$ 0.57 and significantly higher ($p = 6.422 \times 10^{-15}$) than before treatment (1.14 $\pm$ 0.204; Figure 3). Eels of the control group (N = 25) showed only slightly higher GSI values: 1.16 $\pm$ 0.22 vs. 1.06 $\pm$ 0.15 ($p = 0.0159$; Figure 3). hCG-treated eels had a significantly higher GSI value than the eels of the control group ($p < 0.001$).

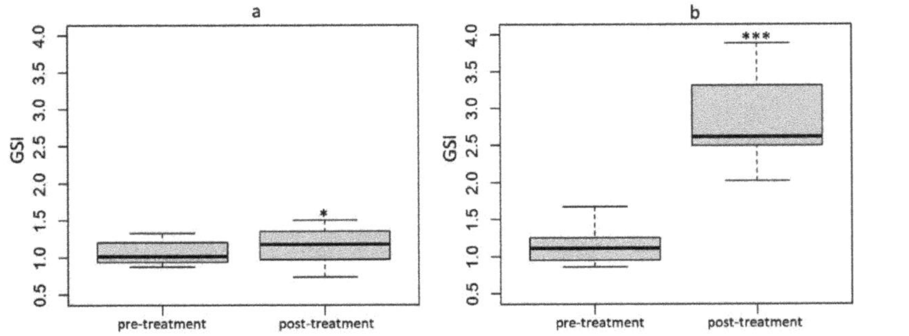

Figure 3. Gonadosomatic index (GSI; AV $\pm$ SD) values, calculated from ultrasound images, for eels pre-treatment and post-treatment of (**a**) the control group (N = 25) and (**b**) the human chorionic gonadotropin (hCG) treated group (N = 25). The asterisks indicate significant differences of * $p < 0.05$ and *** $p < 0.001$.

Eels of the hCG-treated group showed significantly higher HSI values (1.04 $\pm$ 0.16 vs. 0.77 $\pm$ 0.14; $p < 0.01$), lower GITI values (0.53 $\pm$ 0.08 g vs. 1.18 $\pm$ 0.37 g; $p < 0.001$) and higher SBI values (0.29 $\pm$ 0.05 g vs. 0.20 $\pm$ 0.04 g; $p = 0.015$) as compared to eels of the control group.

3.5. Oocyte Histology

Oocytes from the hCG-treated eels were larger (196 ± 23 μm) than the oocytes from eels of the control group (117 ± 12 μm; $p < 0.001$; Figure 4). Yolk was present in the oocytes from the hCG-treated group but absent in the oocytes of the control eels (Figure 5).

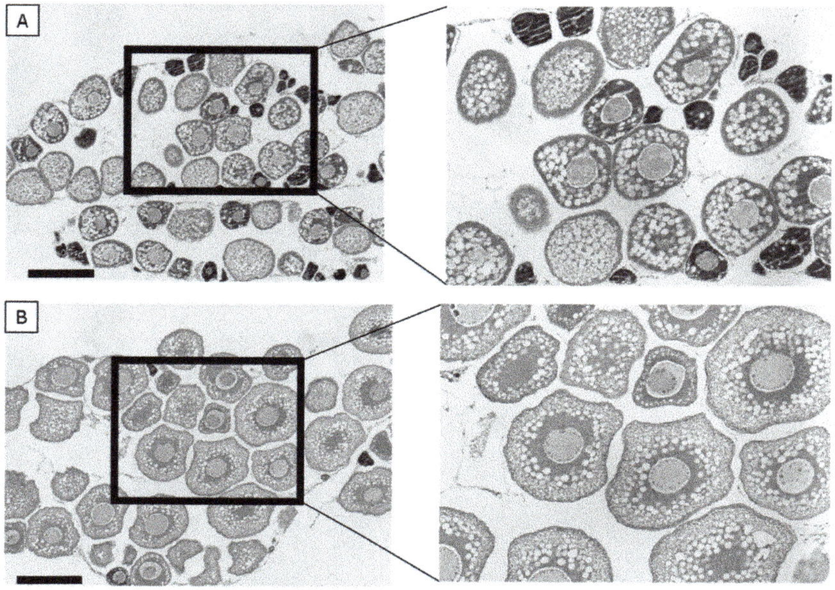

Figure 4. Mayer's hematoxylin eosin (HE) stained oocytes from eels of (**A**) the control group and (**B**) the human chorionic gonadotropin (hCG) treated group. Scale bar = 200 μm.

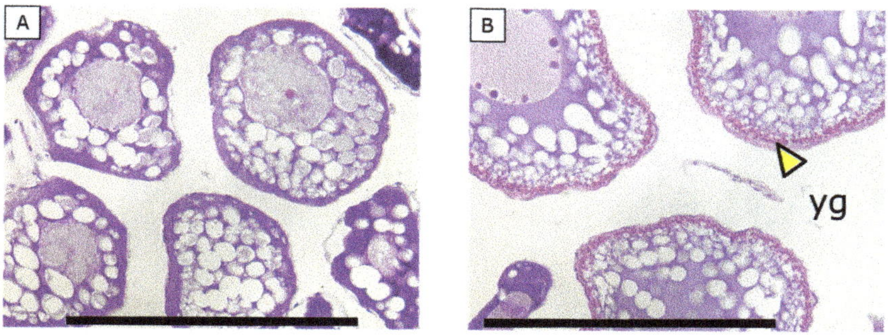

Figure 5. Yolk granules (yg, indicated by the yellow arrowhead), still absent in eels of (**A**) the control group, but present and visible as the peripherally located pink spots (indicated by the yellow arrow head) in eels of (**B**) the human chorionic gonadotropin (hCG) treated group. Scale bar = 200 μm.

The lipid area was much higher in the oocytes from the hCG-treated eels (9414 ± 2034 μm^2) than in oocytes from eels of the control group (3978 ± 694 μm^2; $p < 0.001$). The lipid area relative to the oocyte area was not different between eels from both groups (32 ± 2% vs. 31 ± 2%, hCG group vs. C group, respectively; $p > 0.05$).

3.6. RT-PCR

No significant differences were observed between Ct values of *l36* expression in liver tissue from eels of the hCG-treated group and in liver tissue from the controls ($p > 0.05$). Using Ct values of housekeeping gene *l36* for normalizing, *esr-1* expression was higher in livers of eels of the hCG-treated group than in the controls ($p \leq 0.05$). The relative fold change (fc) between *esr-1* expression in hCG-treated eels and in controls was 2.33 ± 1.13 (Figure 6).

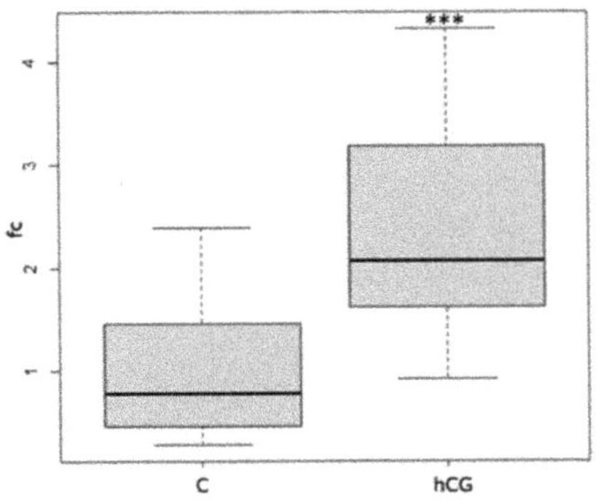

Figure 6. Relative fold change expression of *estrogen-1* (*esr-1*) in the liver using *Ribosomal Protein L36* (*L36*) as housekeeping gene for normalization. Abbreviations: fc = fold change; C = control group; hCG = human chorionic gonadotropin-treated group. The asterisks indicate a significant difference of *** $p < 0.001$.

No significant differences were found between BestKeeper Ct values, of *elf-1*, *l36* and *18s* combined, in ovary tissue from eels of the hCG-treated group and in ovary tissue from the controls ($p > 0.05$). Significantly higher expression was found for *cyp19*, *fshr* and *lpl* with relative fold changes of 2.53 ± 0.93; 6.14 ± 2.13 and 15.75 ± 6.79, respectively, when comparing hCG-treated eels with the controls ($p < 0.001$; Figure 7a–c). Lower expression was found for *vtgr* with a relative fold change of 0.36 ± 0.13 ($p \leq 0.05$; Figure 7d) when comparing hCG-treated eels with the controls. No significant differences were found in the expression of *lhr-1* and *lhr-2* with relative fold changes of 0.89 ± 0.55 and 0.73 ± 0.49, respectively ($p > 0.05$; Figure 7e,f) when comparing hCG-treated eels with the controls.

3.7. Reproduction Parameters

When BWI exceeded the threshold of 110, eels were given a booster CPE injection and then a DHP injection to induce ovulation. For eels of the control group, this was on average after 90 days and 14 CPE injections. Eels ovulated on average 12 h after DHP injection. The percentage of floating eggs was on average 33 ± 24%. For eels of the hCG group, the BWI of 110 was reached after just 64 days and only 10 CPE injections, on average 4 CPE injections less than eels of the control group (Figure 8). Eels of the hCG-treated group also ovulated on average 12 h after DHP injection. The percentage of floating eggs was on average 39 ± 26% (Table 3), slightly higher but not significantly different from eels of the control group.

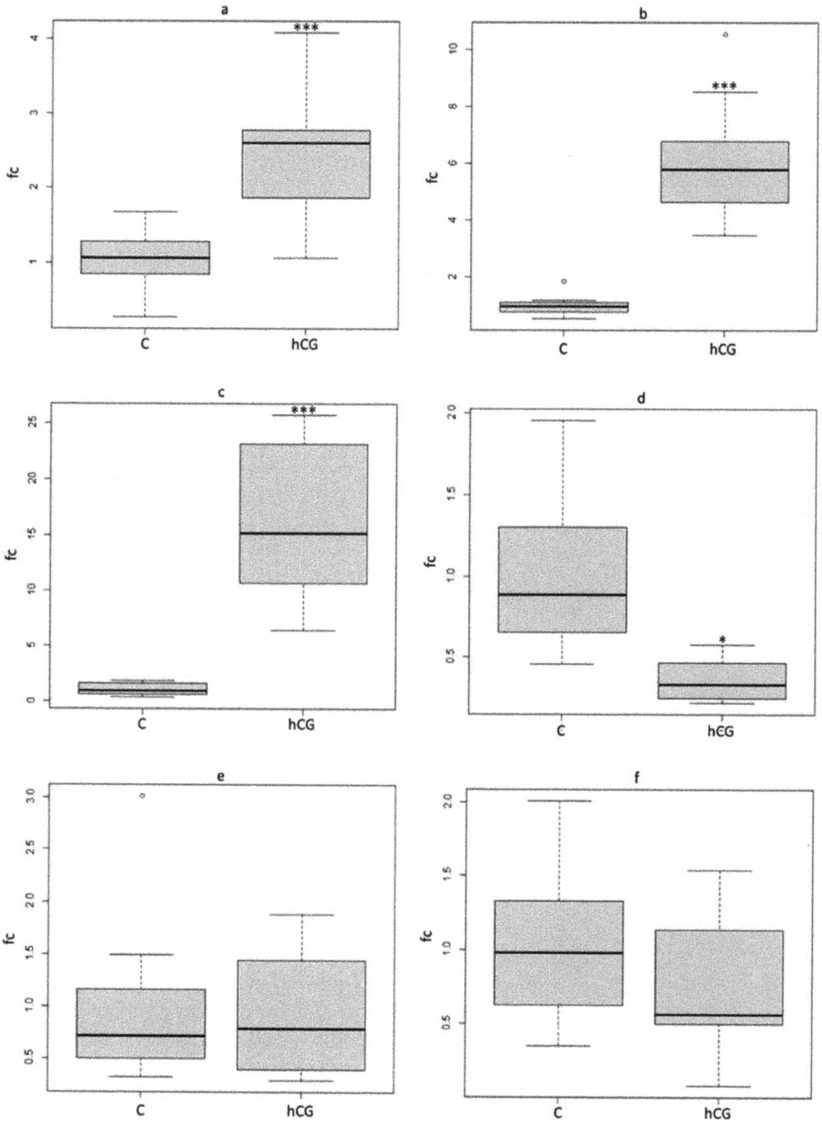

Figure 7. Relative fold change expression of (**a**) aromatase (*cyp19*), (**b**) follicle-stimulating hormone receptor (*fshr*), (**c**) lipoprotein lipase (*lpl*), (**d**) vitellogenin receptor (*vtgr*), (**e**) luteinizing hormone receptor-1 (*lhr-1*) and (**f**) luteinizing hormone receptor-2 (*lhr-2*) expression in the ovary using elongation factor-1 (*elf-1*), ribosomal protein l36 (*l36*) and 18S ribosomal RNAs (*18S*) as housekeeping genes for normalization. Abbreviations: fc = fold change; C = control group; hCG = human chorionic gonadotropin-treated group. The asterisks indicate significant differences of * $p < 0.05$ and *** $p < 0.001$.

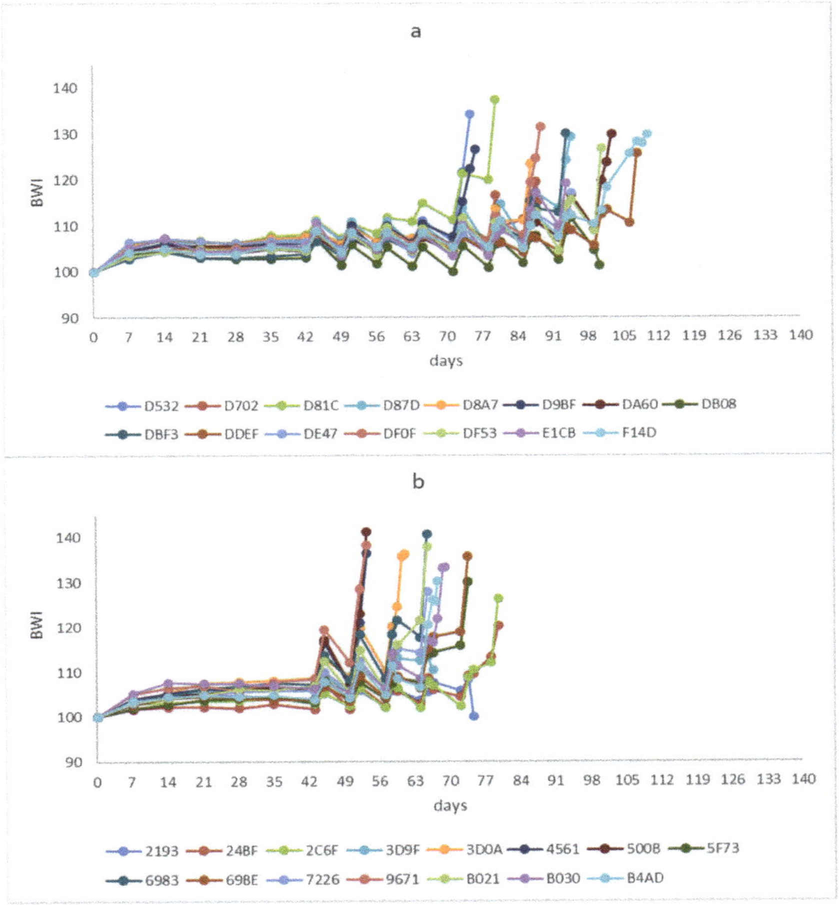

Figure 8. Body weight index evolution of the maturing feminized eels, that released eggs that could be fertilized, versus the number of weekly common carp pituitary extract (CPE) injections for eels from (**a**) the control group and (**b**) the human chorionic gonadotropin-treated group (N = 15 per group). Each experimental eel is represented by PITtag code and colored line connecting the weekly BWI values. Abbreviations: BWI = body weight index; CPE = common carp pituitary extract.

Table 3. Reproduction parameters (AV ± SD) after hypophysation of hCG-treated eels (N = 15) and control eels (N = 15). "Days to reach BWI 110" is the number of days of weekly CPE injections to reach a BWI of 110; "CPE injections mature" is the number of CPE injections to reach final maturation, "Hours after final DHP injection" is the number of hours after DHP injection until egg release and "% Floating eggs" is the average percentage of floating eggs. Abbreviations: BWI = body weight index; CPE = common carp pituitary extract; DHP = 17,20β-dihydroxy-4-pregnen-3-one; C = control; hCG = human chorionic gonadotropin.

	Days to Reach BWI 110	CPE Injections Mature	Hours after Final DHP Injection	% Floating Eggs
C group	90 ± 11	14 ± 2	12 ± 1	33 ± 24
hCG group	65 ± 9	10 ± 1	12 ± 1	39 ± 26

Egg batches were obtained from N = 12 hCG-treated eels and from N = 12 eels of the control group. From nine eels of the control group, larvae batches were obtained. On average, larvae survived for 5 dph, with a maximum survival time of 9 dph. From 11 eels of the hCG-treated group, larvae batches were obtained that had an average survival time of 5 dph and a maximum survival time of 11 dph (Table 4).

Table 4. Reproduction results for hCG-treated and control eels (each group N = 15 eels). "Eggs" is the number of eels that released eggs, "Embryos" is the number of eels that had eggs with developing embryos, "Larvae" is the number of eels from which hatched larvae were produced, and under the header "larval longevity" are shown "average (dph)", which is the average number of days post hatching that the larvae survived, and "max (dph)", which is the maximal number of days post hatching that the larvae survived. Abbreviations: dph= days post hatching; C = control; hCG = human chorionic gonadotropin.

				Larval Longevity	
	Eggs	**Embryos**	**Larvae**	**Average (dph)**	**Max (dph)**
C group	12	9	9	5	9
hCG group	12	11	11	5	11

4. Discussion

This study aimed to investigate the effects of a single human chorionic gonadotropin (hCG) injection on silvering, vitellogenesis and reproductive success of feminized European eels. Here, we will discuss these effects separately.

4.1. hCG Induction of Silvering

When eels become silver, multiple morphological changes occur, such as a significant eye enlargement [32] and darkening of the pectoral fin [43]. After the treatment period, significant eye enlargement was observed when comparing eels from the hCG-treated group with the control eels. Darkening of the pectoral fin was more advanced for the eels treated with hCG than for the controls. Silvering also includes thickening of the swim bladder wall [44], an organ that is of crucial importance for buoyancy control during the daily vertical migrations of the silver eels in the ocean [45]. We observed a higher swim bladder weight in the eels treated with hCG. During silvering, the gastrointestinal tract (GIT) regresses [46], which is in agreement with the lower GIT weight in hCG-treated eels in our study. In yellow eels or "pre-migrant" silver eels, 11KT concentrations are significantly lower than in migrant eels [47]. In the eels treated with hCG, plasma 11KT concentrations were significantly higher than in plasma from eels of the control group. Silvering is primarily under 11KT control [48], and the observed eye enlargement during silvering may be 11KT-mediated [49].

4.2. hCG Induction of Vitellogenesis

In the eels treated with hCG, several vitellogenic changes occurred as compared to the control eels. A significant increase was found in GSI and HSI of the eels treated with hCG. The oocytes in eels treated with hCG were significantly larger than the oocytes in the control eels. In the study of Adachi et al. [21] on *A. japonica*, it was shown that vitellogenesis commenced when the oocytes were about 250 μm in diameter, and Cottrill et al. [50] considered oocytes from *A. rostrata* vitellogenic at 200 μm. The average diameter of oocytes in the feminized eels treated with hCG in this study was 196 ± 23 μm. The higher lipid area and the much higher *lpl* expression in the oocytes of the hCG-treated eels suggests that hCG successfully induced lipid deposition in the oocytes. This is in line with the study of Nguyen et al. [25], where shortfin eels (*A. australis*) were treated with hCG and also showed lipid deposition and upregulated *lpl* expression, relative to the hCG dose. The most outstanding finding with the hCG treatment in our study was the yolk deposition that had occurred in the oocytes, as this has not yet been observed before in eels treated with hCG. The hCG-treated shortfin eels in the study of Nguyen et al. [25] had a significant up-regulation of ovarian *fshr* when increasing the dose of hCG. The eels that received the highest dose of hCG in their study showed advanced oocyte development in the ovaries. The authors concluded that Fsh signaling is essential for previtellogenic oocyte development in shortfin eels. This is in line with the findings of our study where an increase in ovarian *fshr* expression was observed in the eels that were treated with hCG. In the study of Nguyen et al. [25], the treatment with hCG did not affect *cyp19* expression, while increases in plasma E2 and 11KT were observed. In our study, a significant increase in *cyp19* expression was observed after hCG treatment. Cyp19 is a crucial enzyme for vitellogenesis and yolk deposition, since it converts T into E2 in the ovary [13].

Feminization may have had an epigenetic effect, explaining the observed differences in yolk deposition and *cyp19* expression. To feminize eels, the eels were given pellets coated with E2 during a crucial early life period for sex determination. Besides feminization, this procedure may sensitize the eels in their response to hormonal stimulation later in life that may be directed by Cyp19. Additionally, Tzchori et al. [51] found higher *cyp19* expression in the gonads of feminized European eels in comparison with non-feminized individuals.

In eels, hCG binds to the Lhr [5–7]. In our study, no significant differences were found in *lhr-1* or *lhr-2* expression after hCG treatment, similarly to findings in the study of Nguyen et al. [25]. As the Fshr may well be promiscuous and could bind Lh [52,53], also in Japanese eel [54], the question would be if hCG binds to the Fshr. Kazeto et al. [6] concluded that this is not the case for Japanese eel. The increases in *fshr* and *cyp19* expressions mark the start of vitellogenesis. E2 synthesis is moderated by cyp19 activity, and E2 binds to the hepatic nuclear receptor Esr-1 and induces the synthesis of Vtg by the hepatocytes [14]. Vtgs are then released in the circulation of the eels. 11KT does not only induce Vtg and lipid synthesis; it has been suggested that it also plays an important role in Vtg and lipid transport and absorption [55]. The Vtgs are transported to the oocytes, bind to the Vtgrs and are absorbed and reconstructed into yolk protein [16]. In our study, we found a lower *vtgr* expression in hCG-treated eels. This supports the suggestion that the Vtgr is recycled to the oocyte surface during vitellogenic oocyte growth [56].

4.3. hCG Effects on Reproductive Success

The eels treated with hCG were in an advanced stage of maturation in comparison with the controls, which resulted in 4 CPE injections less to mature than the controls. They reached the BWI of 110 threshold 25 days faster than the controls. The presence of yolk and the higher lipid area in the oocytes of the eels treated with hCG may have had consequences for the egg quality. Still, in both groups, 12 out of 15 eels gave batches of larvae that survived on average 5 dph, so an apparent difference in reproductive success was absent.

5. Conclusions

In feminized European eels, pre-treatment with hCG induced silvering, vitellogenic hepatic and ovarian gene expression and lipid and yolk deposition in the oocytes. Yolk deposition and aromatase activation are specific for hCG induction in feminized eels, providing supporting evidence for the view that E2 treatment in early life may sensitize eels to hormonal stimulation later in life. Future research should aim to elucidate the epigenetic effects of early-life E2 treatment on the vitellogenic pathway. hCG also decreased the hypophysation period and could, therefore, be useful as pre-treatment in artificial reproduction protocols.

Author Contributions: Conceptualization, A.P.P. and L.T.N.H.; methodology, A.P.P., I.v.d.V., P.J., L.K., H.S. and L.T.N.H.; software, I.v.d.V.; validation, L.T.N.H.; formal analysis, I.v.d.V.; investigation, A.P.P., I.v.d.V., P.J. and L.T.N.H.; resources, L.K., H.S. and W.S.; data curation, A.P.P., I.v.d.V. and L.T.N.H.; writing—original draft preparation, A.P.P. and I.v.d.V.; writing—review and editing, A.P.P., I.v.d.V. and P.J.; visualization, I.v.d.V.; supervision, A.P.P.; project administration, A.P.P.; funding acquisition, A.P.P. and W.S. All authors have read and agreed to the published version of the manuscript.

Funding: This research was funded as the public–private partnership project "LARVitAAL" by the Dutch Ministry of Agriculture, Nature and Food Quality and the DUPAN foundation, grant number LWV21.061.

Institutional Review Board Statement: Experimental protocols complied with the current laws of The Netherlands and were approved by the CCD (Central Committee for Animal Experiments), project number AVD401002017817, 1 March 2017, and by the DEC (Animal Experiments Committee) and IvD (Authority for Animal Welfare), experiment number 2017.D-0007.005, 15 December 2020, of Wageningen University.

Informed Consent Statement: Not applicable.

Data Availability Statement: Raw data supporting the conclusions of this article will be made available by the authors, without undue reservation, to any qualified researcher.

Acknowledgments: The authors thank H. Komen for his constructive criticisms and zootechnicians of the CARUS facilities. Additionally, we thank our partners of the international EELRIC consortium (www.eelric.eu, accessed on 24 May 2023).

Conflicts of Interest: The authors declare no conflict of interest. The funders had no role in the design of the study; in the collection, analyses, or interpretation of data; in the writing of the manuscript; or in the decision to publish the results.

References

1. Kazeto, Y.; Suzuki, H.; Ozaki, Y.; Gen, K. C-terminal peptide (hCTP) of human chorionic gonadotropin enhances in vivo biological activity of recombinant Japanese eel follicle-stimulating hormone and luteinizing hormone produced in FreeStyle 293-F cell lines. *Gen. Comp. Endocrinol.* **2021**, *306*, 113731. [CrossRef] [PubMed]
2. Gharib, S.D.; Wierman, M.E.; Shupnik, M.A.; Chin, W.W. Molecular biology of the pituitary gonadotropins. *Endocr. Rev.* **1990**, *11*, 177–199. [CrossRef] [PubMed]
3. Molés, G.; Hausken, K.; Carrillo, M.; Zanuy, S.; Levavi-Sivan, B.; Gómez, A. Generation and use of recombinant gonadotropins in fish. *Gen. Comp. Endocrinol.* **2020**, *299*, 113555. [CrossRef] [PubMed]
4. Swanson, P.; Dickey, J.T.; Campbell, B. Biochemistry and physiology of fish gonadotropins. *Fish. Physiol. Biochem.* **2003**, *28*, 53–59. [CrossRef]
5. Aizen, J.; Kowalsman, N.; Kobayashi, M.; Hollander, L.; Sohn, Y.C.; Yoshizaki, G.; Niv, M.Y.; Levavi-Sivan, B. Experimental and computational study of inter- and intra- species specificity of gonadotropins for various gonadotropin receptors. *Mol. Cell. Endocrinol.* **2012**, *364*, 89–100. [CrossRef]
6. Kazeto, Y.; Kohara, M.; Tosaka, R.; Gen, K.; Yokoyama, M.; Miura, C.; Miura, T.; Adachi, S.; Yamauchi, K. Molecular characterization and gene expression of Japanese eel (*Anguilla japonica*) gonadotropin receptors. *Zool. Sci.* **2012**, *29*, 204–211. [CrossRef]
7. Minegishi, Y.; Dirks, R.P.; de Wijze, D.L.; Brittijn, S.A.; Burgerhout, E.; Spaink, H.P.; van den Thillart, G.E.E.J.M. Quantitative bioassays for measuring biologically functional gonadotropins based on eel gonadotropic receptors. *Gen. Comp. Endocrinol.* **2012**, *178*, 145–152. [CrossRef]
8. Lokman, P.M.; Wylie, M.J.; Downes, M.; Di Biase, A.; Damsteegt, E.L. Artificial induction of maturation in female silver eels, *Anguilla australis*: The benefits of androgen pre-treatment. *Aquaculture* **2015**, *437*, 111–119. [CrossRef]
9. Aroua, S.; Schmitz, M.; Baloche, S.; Vidal, B.; Rousseau, K.; Dufour, S. Endocrine evidence that silvering, a secondary metamorphosis in the eel, is a pubertal rather than a metamorphic event. *Neuroendocrinology* **2005**, *82*, 221–232. [CrossRef]
10. Divers, S.L.; McQuillan, H.J.; Matsubara, H.; Todo, T.; Lokman, P.M. Effects of reproductive stage and 11-ketotestosterone on LPL mRNA levels in the ovary of the shortfinned eel. *J. Lipid Res.* **2010**, *51*, 3250–3258. [CrossRef]
11. Rainuzzo, J.R.; Reitan, K.I.; Olsen, Y. The significance of lipids at early stages of marine fish: A review. *Aquaculture* **1997**, *155*, 103–115. [CrossRef]
12. Asanuma, H.; Ohashi, H.; Matsubara, H.; Ijiri, S.; Matsubara, T.; Todo, T.; Adachi, S.; Yamauchi, K. 11-ketotestosterone potentiates estrogen-induced vitellogenin production in liver of Japanese eel (*Anguilla japonica*). *Fish. Physiol. Biochem.* **2003**, *28*, 383–384. [CrossRef]
13. Montserrat, N.; González, A.; Méndez, E.; Piferrer, F.; Planas, J.V. Effects of follicle stimulating hormone on estradiol-17β production and P-450 aromatase (CYP19) activity and mRNA expression in brown trout vitellogenic ovarian follicles in vitro. *Gen. Comp. Endocrinol.* **2004**, *137*, 123–131. [CrossRef]
14. Lafont, A.-G.; Rousseau, K.; Tomkiewicz, J.; Dufour, S. Three nuclear and two membrane estrogen receptors in basal teleosts, *Anguilla* sp.: Identification, evolutionary history and differential expression regulation. *Gen. Comp. Endocrinol.* **2016**, *235*, 177–191. [CrossRef]
15. Nagahama, Y.; Yoshikuni, M.; Yamashita, M.; Tokumoto, T.; Katsu, Y. 4 Regulation of oocyte growth and maturation in fish. *Cur Top. Dev. Biol.* **1995**, *30*, 103–145. [CrossRef]

16. Wallace, R.A. Vitellogenesis and oocyte growth in nonmammalian vertebrates. In *Oogenesis*; Springer: Boston, MA, USA, 1985; pp. 127–177. [CrossRef]
17. Sire, M.-F.; Babin, P.J.; Vernier, J.-M. Involvement of the lysosomal system in yolk protein deposit and degradation during vitellogenesis and embryonic development in trout. *J. Exp. Zool.* **1994**, *269*, 69–83. [CrossRef]
18. Jéhannet, P.; Palstra, A.P.; Giménez Nebot, I.; Swinkels, W.; Heinsbroek, L.T.N.; Komen, H. Recombinant FSH and LH treatment to induce oocyte development in vitro and in vivo in the European eel *Anguilla anguilla*. *Fishes* **2023**, *8*, 123. [CrossRef]
19. Brydges, N.M.; Boulcott, P.; Ellis, T.; Braithwaite, V.A. Quantifying stress responses induced by different handling methods in three species of fish. *Appl. Anim. Behav. Sci.* **2009**, *116*, 295–301. [CrossRef]
20. Schreck, C.B.; Contreras-Sanchez, W.; Fitzpatrick, M.S. Effects of stress on fish reproduction, gamete quality, and progeny. In *Reproductive Biotechnology in Finfish Aquaculture*, 1st ed.; Lee, C.-S., Donaldson, E.M., Eds.; Elsevier: Amsterdam, The Netherlands, 2001; pp. 3–24. [CrossRef]
21. Adachi, S.; Ijiri, S.; Kazeto, Y.; Yamauchi, K. Oogenesis in the Japanese eel *Anguilla japonica*. In *Eel Biology*, 1st ed.; Aida, K., Tsukamoto, K., Yamauchi, K., Eds.; Springer: Tokyo, Japan, 2003; pp. 502–518.
22. Palstra, A.P.; Bouwman, L.J.; Jéhannet, P.; Kruijt, L.; Schipper, H.; Blokland, M.H.; Swinkels, W.; Heinsbroek, L.T.N.; Lokman, P.M. Steroid implants for the induction of vitellogenesis in feminized European silver eels (*Anguilla anguilla* L.). *Front. Genet.* **2022**, *13*, 969202. [CrossRef]
23. Fontaine, M. Sur la maturation complète des organes génitaux de l'anguille mâle et l'émission spontanée de ses produits sexuels. *CR Acad. Sci. Paris* **1936**, *202*, 55.
24. Furuhashi, M.; Shikone, T.; Fares, F.A.; Sugahara, T.; Hsueh, A.J.; Boime, I. Fusing the carboxy-terminal peptide of the chorionic gonadotropin (CG) beta-subunit to the common alpha-subunit: Retention of O-linked glycosylation and enhanced in vivo bioactivity of chimeric human CG. *Mol. Endocrinol.* **1995**, *9*, 54–63. [CrossRef] [PubMed]
25. Nguyen, A.T.; Chia, J.H.Z.; Kazeto, Y.; Wylie, M.J.; Lokman, P.M. Induction of oocyte development in previtellogenic eel, *Anguilla australis*. *Gen. Comp. Endocrinol.* **2020**, *291*, 113404. [CrossRef] [PubMed]
26. Chai, Y.; Tosaka, R.; Sago, K.; Hatanaka, R.; Ijiri, S.; Adachi, S. The relationship between the developmental stage of oocytes in various seasons and the quality of the egg obtained by artificial maturation in the feminized Japanese eel *Anguilla japonica*. *Aquac. Sci.* **2010**, *58*, 269–278. [CrossRef]
27. Böhm, T.; Graziano, M.; Blom, E.; Brittijn, S.A.; Dirks, R.P.; Palstra, A.P. Simulated migration of feminized eels to stimulate and predict the sexual maturation response. In Proceedings of the ICBF2016, San Marcos, TX, USA, 12–18 June 2016.
28. Ijiri, S.; Tsukamoto, K.; Chow, S.; Kurogi, H.; Adachi, S.; Tanaka, H. Controlled reproduction in the Japanese eel (*Anguilla japonica*), past and present. *Aquac. Eur.* **2011**, *36*, 13–17.
29. Mes, D.; Dirks, R.P.; Palstra, A.P. Simulated migration under mimicked photothermal conditions enhances sexual maturation of farmed European eel (*Anguilla anguilla*). *Aquaculture* **2016**, *452*, 367–372. [CrossRef]
30. Fulton, T.W. *Rate of Growth of Sea Fishes*; Neill & Company: Edinburg, UK, 1902.
31. Palstra, A.; van den Thillart, G. Artificial maturation and reproduction of European eel. In *Spawning Migration of European eel. Reproduction Index, a Useful Tool for Conservation Management*, 1st ed.; van den Thillart, G., Dufour, S., Rankin, C., Eds.; Springer: Heidelberg, Germany, 2009; pp. 309–331.
32. Pankhurst, N.W. Relation of visual changes to the onset of sexual maturation in the European eel *Anguilla anguilla* (L.). *J. Fish. Biol.* **1982**, *21*, 127–140. [CrossRef]
33. Bureau du Colombier, S.; Jacobs, L.; Gesset, C.; Elie, P.; Lambert, P. Ultrasonography as a non-invasive tool for sex determination and maturation monitoring in silver eels. *Fish. Res.* **2015**, *164*, 50–58. [CrossRef]
34. Schneider, C.; Rasband, W.; Eliceiri, K. NIH Image to ImageJ: 25 years of image analysis. *Nat. Methods* **2012**, *9*, 671–675. [CrossRef]
35. Jéhannet, P.; Kruijt, L.; Damsteegt, E.L.; Swinkels, W.; Heinsbroek, L.T.N.; Lokman, P.M.; Palstra, A.P. A mechanistic model for studying the initiation of anguillid vitellogenesis by comparing the European eel (*Anguilla anguilla*) and the shortfinned eel (*A. australis*). *Gen. Comp. Endocrinol.* **2019**, *279*, 129–138. [CrossRef]
36. Maugars, G.; Dufour, S. Demonstration of the coexistence of duplicated LH receptors in teleosts, and their origin in ancestral actinopterygians. *PLoS ONE* **2015**, *10*, e0135184. [CrossRef]
37. Setiawan, A.N.; Lokman, P.M. The use of reference gene selection programs to study the silvering transformation in a freshwater eel *Anguilla australis*: A cautionary tale. *BMC Mol. Biol.* **2010**, *11*, 75. [CrossRef] [PubMed]
38. Zadmajid, V.; Falahatimarvast, A.; Damsteegt, E.L.; Setiawan, A.N.; Ozaki, Y.; Shoae, A.; Lokman, P.M. Effects of 11-ketotestosterone and temperature on inhibin subunit mRNA levels in the ovary of the shortfinned eel, *Anguilla australis*. *Comp. Bioch Phys. B* **2015**, *187*, 14–21. [CrossRef] [PubMed]
39. Pfaffl, M.W.; Tichopad, A.; Prgomet, C.; Neuvians, T.P. Determination of stable housekeeping genes, differentially regulated target genes and sample integrity: BestKeeper–Excel-based tool using pair-wise correlations. *Biotechnol. Lett.* **2004**, *26*, 509–515. [CrossRef] [PubMed]
40. Kagawa, H.; Tanaka, H.; Ohta, H.; Okuzawa, K.; Hirose, K. In vitro effects of 17α-hydroxyprogesterone and 17α, 20β-dihydroxy-4-pregnen-3-one on final maturation of oocytes at various developmental stages in artificially matured Japanese eel *Anguilla japonica*. *Fish. Sci.* **1995**, *61*, 1012–1015. [CrossRef]
41. Palstra, A.P.; Cohen, E.G.H.; Niemantsverdriet, P.R.W.; van Ginneken, V.J.T.; van den Thillart, G.E.E.J.M. Artificial maturation and reproduction of European silver eel: Development of oocytes during final maturation. *Aquaculture* **2005**, *249*, 533–547. [CrossRef]

42. Tanaka, H. Progression in artificial seedling production of Japanese eel *Anguilla japonica*. *Fish. Sci.* **2015**, *81*, 11–19. [CrossRef]
43. Han, Y.-S.; Liao, I.-C.; Huang, Y.-S.; He, J.-T.; Chang, C.-W.; Tzeng, W.-N. Synchronous changes of morphology and gonadal development of silvering Japanese eel *Anguilla japonica*. *Aquaculture* **2003**, *219*, 783–796. [CrossRef]
44. Pelster, B. Swimbladder function and the spawning migration of the European eel *Anguilla anguilla*. *Front. Physiol.* **2015**, *5*, 486. [CrossRef]
45. Yamada, Y.; Zhang, H.; Okamura, A.; Tanaka, S.; Horie, N.; Mikawa, N.; Utoh, T.; Oka, H.P. Morphological and histological changes in the swim bladder during maturation of the Japanese eel. *J. Fish. Biol.* **2001**, *58*, 804–814. [CrossRef]
46. Balm, S.P.; Durif, C.; van Ginneken, V.; Antonissen, E.; Boot, R.; van Den Thillart, G.; Verstegen, M. Silvering of European eel (*Anguilla anguilla* L.): Seasonal changes of morphological and metabolic parameters. *Anim. Biol.* **2007**, *57*, 63–77. [CrossRef]
47. Lokman, P.M.; Vermeulen, G.J.; Lambert, J.G.D.; Young, G. Gonad histology and plasma steroid profiles in wild New Zealand freshwater eels (*Anguilla dieffenbachii* and *A. australis*) before and at the onset of the natural spawning migration. I. Females. *Fish. Physiol. Biochem.* **1998**, *19*, 325–338. [CrossRef]
48. Rohr, D.H.; Lokman, P.M.; Davie, P.S.; Young, G. 11-Ketotestosterone induces silvering-related changes in immature female short-finned eels, *Anguilla australis*. *Comp. Biochem. Phys. A* **2001**, *130*, 701–714. [CrossRef] [PubMed]
49. Thomson-Laing, G.; Jasoni, C.L.; Lokman, P.M. The effects of migratory stage and 11-ketotestosterone on the expression of rod opsin genes in the shortfinned eel (*Anguilla australis*). *Gen. Comp. Endocrinol.* **2018**, *257*, 211–219. [CrossRef] [PubMed]
50. Cottrill, R.A.; McKinley, R.S.; Kaak, G.; Dutil, J.-D.; Reid, K.B.; McGrath, K.J. Plasma non-esterified fatty acid profiles and 17beta-oestradiol levels of juvenile immature and maturing adult American eels in the St Lawrence River. *J. Fish. Biol.* **2001**, *59*, 364–379. [CrossRef]
51. Tzchori, I.; Degani, G.; Hurvitz, A.; Moav, B. Cloning and developmental expression of the cytochrome P450 aromatase gene (CYP19) in the European eel (*Anguilla anguilla*). *Gen. Comp. Endocrinol.* **2004**, *138*, 271–280. [CrossRef]
52. Yan, L.; Swanson, P.; Dickhoff, W.W. A two-receptor model for salmon gonadotropins (GTH I and GTH II). *Biol. Reprod.* **1992**, *47*, 418–427. [CrossRef]
53. Miwa, S.; Yan, L.; Swanson, P. Localization of two gonadotropin receptors in the salmon gonad by in vitro ligand autoradiography. *Biol. Reprod.* **1994**, *50*, 629–642. [CrossRef]
54. Kazeto, Y.; Kohara, M.; Miura, T.; Miura, C.; Yamaguchi, S.; Trant, J.M.; Adachi, S.; Yamauchi, K. Japanese eel follicle-stimulating hormone (Fsh) and luteinizing hormone (Lh): Production of biologically active recombinant Fsh and Lh by drosophila S2 cells and their differential actions on the reproductive biology. *Biol. Reprod.* **2008**, *79*, 938–946. [CrossRef]
55. Wang, W.; Zhu, H.; Tian, Z.; Sun, A.; Dong, Y.; Dong, T.; Hu, H. Effects of 11-tetotestosterone on development of the previtellogenic ovary in the Sterlet, *Acipenser ruthenus*. *Front. Endocrinol.* **2020**, *11*, 115. [CrossRef]
56. Perazzolo, L.M.; Coward, K.; Davail, B.; Normand, E.; Tyler, C.R.; Pakdel, F.; Schneider, W.J.; Le Menn, F. Expression and localization of messenger ribonucleic acid for the vitellogenin receptor in ovarian follicles throughout oogenesis in the Rainbow trout, *Oncorhynchus mykiss*. *Biol. Reprod.* **1999**, *60*, 1057–1068. [CrossRef]

fishes

MDPI

Article

Exploring European Eel *Anguilla anguilla* (L.) Habitat Differences Using Otolith Analysis in Central-Western Mediterranean Rivers and Coastal Lagoons from Sardinia

Cinzia Podda [1], Jacopo Culurgioni [2], Riccardo Diciotti [2], Francesco Palmas [1,*], Elsa Amilhat [3,4], Elisabeth Faliex [3,4], Fabien Morat [5,6], Nicola Fois [2] and Andrea Sabatini [1]

1 Department of Life and Environmental Sciences (DiSVA), University of Cagliari, Via T. Fiorelli 1, 09126 Cagliari, Italy; c.podda1@gmail.com (C.P.); asabati@unica.it (A.S.)
2 Ichthyic Products Research Service, Agris–Agricultural Research Agency of Sardinia, Bonassai, S.S. 291 Sassari-Fertilia Km. 18.6, 07100 Sassari, Italy; jculurgioni@agrisricerca.it (J.C.); rdiciotti@agrisricerca.it (R.D.); nfois@agrisricerca.it (N.F.)
3 Université de Perpignan via Domitia, Centre de Formation et de Recherche sur les Environnements Méditerranéens, UMR 5110 F, 66860 Perpignan, France; elsa.amilhat@univ-perp.fr (E.A.); faliex@univ-perp.fr (E.F.)
4 CNRS, Centre de Formation et de Recherche sur les Environnements Méditerranéens, UMR 5110 F, 66860 Perpignan, France
5 Université de Perpignan, PSL Research University: EPHE-UPVD-CNRS, USR 3278 CRIOBE, 66860 Perpignan, France; fabien.morat@univ-perp.fr
6 Laboratoire d'Excellence, CORAIL, 66860 Perpignan, France
* Correspondence: fpalmas@unica.it

Abstract: An otolith shape and morphometric analysis was performed on European eel (*Anguilla anguilla*) subpopulations from five rivers and three coastal lagoons of Sardinia (central-western Mediterranean) to assess the role of different habitats on otolith development. Sagittal otolith shape was described by 11 harmonics from elliptic Fourier descriptors. Comparisons among the harmonics were run through canonical discriminant analyses (CDAs). The CDA reclassification rate (75.7%) demonstrated a spatial environmental discrimination among local eel subpopulations of Sardinia. The Euclidean distance values demonstrated a dissimilarity between the river and lagoon groups. The form factor and roundness shape indices were significantly higher in the river group than in the lagoon group. The distances of the first three rings to the otolith core revealed site-specific otolith development. Moreover, the annual otolith growth rate was faster in the lagoon group than in the river group. The differences among the studied sites in terms of sagittal otolith shape could relate to changes in different local stocks potentially related to environmental peculiarities. Establishing a direct correlation between otolith morphology and environmental factors is challenging, and further studies are needed to investigate the relationship between habitat type/environmental variation and growth/body characteristics of eels. Nevertheless, the achieved results suggest that this method can be considered to be a valuable tool for studying the ontogeny of the European eel.

Keywords: European eel; sagittal otoliths; shape analysis; morphometry; growth; continental waters; Mediterranean

Key Contribution: An otolith shape analysis was performed on European eel subpopulations from five rivers and three lagoons of Sardinia. The obtained differences could lead to changes in different local stocks related to environmental peculiarities.

Citation: Podda, C.; Culurgioni, J.; Diciotti, R.; Palmas, F.; Amilhat, E.; Faliex, E.; Morat, F.; Fois, N.; Sabatini, A. Exploring European Eel *Anguilla anguilla* (L.) Habitat Differences Using Otolith Analysis in Central-Western Mediterranean Rivers and Coastal Lagoons from Sardinia. *Fishes* **2023**, *8*, 386. https://doi.org/10.3390/fishes8080386

Academic Editor: Serena Savoca

Received: 30 June 2023
Revised: 14 July 2023
Accepted: 24 July 2023
Published: 26 July 2023

1. Introduction

Otoliths are biomineralized-crystalline-organic complexes composed mainly of calcium carbonate [1]. With a metabolically inert structure, otoliths are less vulnerable to chemical and structural modifications and grow throughout the life of fish in response to several environmental influences and seasonality [2–5].

Although otoliths typically exhibit a species-specific morphological structure, they may also exhibit intraspecific changes in shape and size in relation to physiological and environmental factors [6]. Variations in otolith morphology have been observed among several populations or stocks of the same species [7–10], as well as within a species depending on factors such as sex [11,12], diet [13,14], and ontogeny [15,16].

For these reasons, otoliths can be defined as among the most useful anatomical structures for studying fish growth [17–19] in the field of ichthyology, ecology [20,21], fisheries biology [22–27], population age structure [28,29], fisheries management [30,31], and the study of fish adaptations to different environmental conditions [32,33].

Otolith shape has widely been used as a phenotypic marker to study variations in the development of populations of the European eel (*Anguilla anguilla* L. 1758) (hereafter the eel), and otoliths are also commonly used for age and growth estimations [34–38]. The pertinence of using otoliths to study development and environmental adaptations in this species may be related to the conservation status of the eel, which is actually in decline and listed as critically endangered, and may support a better understanding of the roles of different habitats to sustain and conserve the species [39,40]. Due to its catadromous life cycle, its wide geographical distribution range, its genomic panmixia, and its ability to live in different aquatic environments, the eel is particularly well suited for understanding how growth is influenced by environmental conditions [28,29,41]. Nevertheless, the body growth of eels can vary significantly within the same subpopulation because of interindividual variation and geographically different habitats [24,29,38,42,43]. Furthermore, the species shows a marked sexual size dimorphism, with the female larger than the male of similar age [24,29,44]. Furthermore, it is also known that eel growth rates can vary across latitudinal gradients of different environmental factors, including, for instance, temperature, photoperiod, hydrology, and productivity [23,24,28,45]. Added to this is the fact that certain aspects, such as site-specific variability that could affect eel growth, have not yet been fully documented and do not provide an overview of the characteristics of the eel's life cycle [24].

The eel also has one of the most complex life cycles in the animal kingdom, which includes two migrations spanning ca. 6000 km from the spawning grounds in the Sargasso Sea to the European and northern African coasts [21,46–48]. Furthermore, the species undergoes a series of metamorphosis to adapt to several aquatic environments throughout its life [28]. Leptocephali (larvae) drift across the Atlantic Ocean and metamorphose into glass eels before entering the continental shelf. During the glass eel stage, the species colonizes continental waters (e.g., rivers, lakes, and lagoons), where it grows and lives from 2 to more than 4–20 years as the yellow eel stage. After this period, eels start to metamorphose into silver eels (adults) and return to their spawning ground [25,28,49–51].

Despite the need to better understand the roles of different habitats in the phenotypic development and growth success of the eel, and considering that an otolith analysis is particularly adapted to investigate this question, only a few studies have examined eels' growth in terms of otolith shape [37,52]. This knowledge gap highlights the need to investigate phenotypic variability in the development of this species in different habitats and across several spatial scales. In this context, the general aim of this study was to analyze otolith shape and growth variations among eel subpopulations inhabiting several rivers and lagoons of Sardinian (central-western Mediterranean).

2. Materials and Methods

2.1. Study Locations

Sampling was conducted in five rivers (the Pramaera, Tirso, Coghinas, Barca, and Mannu di Fluminimaggiore (hereafter UMannu) Rivers) and three lagoons (Calich, Porto Pino, and Sa Praia) in Sardinia. These locations were selected to cover several geographic areas of the island (Figure 1 and Table 1).

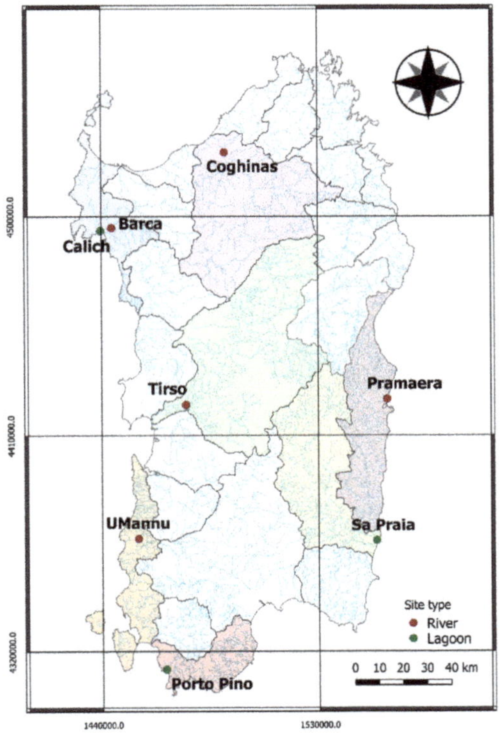

Figure 1. Locations of the eight studied sites within Sardinian continental waters. Rivers are indicated with red dots and lagoons are indicated with green dots.

Table 1. Characterization of the investigated rivers and lagoons, and the number of sampled eels at each site.

Site	River (R) Lagoon (L)	Regime (Only R)	Area (km^2)	Dams (Yes/No) (Only R) [1]	Number of Eels
Pramaera	R	Perennial	184 *	no	26
Tirso	R	Perennial	2043 *	yes	13
Coghinas	R	Perennial	1836 *	yes	10
Barca	R	Intermittent	355 *	yes	10
UMannu (Mannu di Fluminumaggiore)	R	Intermittent	126 *	no	11
Calich	L		0.9		10
Porto Pino	L		0.5		10
Sa Praia	L		0.86		10

[1] [53] * Area of catchment basin.

The Sardinian hydrographic network is characterized by a limited presence of perennial streams and a prevalence of intermittent streams. Most watercourses are located in close proximity to the coast and exhibit irregular flow patterns and significant seasonal hydrological fluctuations [53–58]. These characteristics are further amplified by the presence of steep slopes and short downstream sections. The Pramaera River is a typical Mediterranean small watercourse located in central-eastern Sardinia. This river is characterized by not having fluvial interruptions of anthropogenic origin (i.e., dams or other anthropogenic barriers) [57]. The Tirso River is the main watercourse of the island; it rises in the center of Sardinia and develops from northeast to southwest. Its course differs considerably as it proceeds from its source to the mouth of the river, differentiating in the upstream part with a winding path and considerable slopes, taking on a regular appearance in the central part, and then presenting minimal slopes and large meanders in the downstream section. An important element is the presence of numerous artificial reservoirs that are relevant from the point of view of the quantity of invasable water. Meanwhile, the coastal area has a series of coastal lagoons, some of which dry up completely in the summer. The Coghinas River is the third mainstream in Sardinia; it is located in the northern part of the island. Along its course, the Coghinas River is regulated by two dams, and then it flows into the sea in the Asinara Gulf. The Barca River is found in northwestern Sardinia and is a first-order watercourse. Its downstream trait flows into the Calich lagoon. In the Barca river basin, there are several reservoirs and the natural lake of Baratz. The UMannu River is located in southwestern Sardinia and is a first-order watercourse belonging to the Riu Mannu basin.

Sardinian coastal lagoons extend for a total area of about 120 km^2 and are particularly interesting for their naturalistic value and productivity [59]. The Calich lagoon is located on the northwestern coast of Sardinia. It communicates with the sea through a channel located in the northwestern area of the lagoon. The main tributaries are the Barca and the Calvia Rivers, and the Oruni canal. The continuous tidal flow and the freshwater inputs result in a very variable brackish condition which results in fishing yields that do not exceed 50 kg ha^{-1}. The salinity can vary from 5 in the winter season to 38 in the summer [59]. The Porto Pino lagoon is located on the southern coast of Sardinia; it consists of a series of small basins (Porto Pino, Maestrale, Is Brebeis, Foxi, and Corvo) in communication with each other and used as tanks pre-evaporating from the saline. The salinity can vary from the marine values (ca. 37) and can increase up to 50 [60]. In this lagoon, good integration has been achieved between salt production and fishing activities through the management of bulkheads that regulate the water flow. Fishing activity is carried out using artisanal gill nets, pots, and fyke nets. The Sa Praia lagoon is located on the southeastern coast of Sardinia. It is provisioned by the Gironi River and is connected to the sea by a canal on which a traditional downstream trap called a "lavoriero" is positioned. The salinity ranges from 22.3 to 39.3 (Fish Products Service of Agricultural Research Agency of Sardinia, Agris).

2.2. Eel Samples

Eels were collected from June 2015 to February 2020 during the dry seasons (summer, autumn). In the Pramaera, Barca, and Coghinas Rivers, eels were caught by using experimental fyke nets (2 mm mesh size), while in the Tirso and UMannu Rivers, eels were captured using low-frequency, pulsed DC electrofishing. Lastly, in the Calich, Porto Pino, and Sa Praia lagoons, eels were caught with professional fyke nets (10 mm mesh size).

Individual eel samples were immediately stored in cool and aerated water and anesthetized by immersion in a bath of MS 222 (230 mg L^{-1}) until the termination of opercular movements [58], and then measured for total length (TL, cm) and total weight (TW, g). Subsequently, the eels were sacrificed in situ by decapitation, according to the European Community regulation and Italian legislation for the protection of animals used for scientific purposes (Directive 2010/63/UE L 276 20/10/2010, implemented by Italian Legislative Decree 26/2014). Individuals were kept frozen until head dissection for otolith extraction and gonad dissection for sexual characterization (female, male, undifferentiated).

Sex was determined macroscopically whenever possible, or through histological examination of gonads [61,62]. A preliminary exploratory analysis was carried out on the shape indices of the Pramaera River, which provided the largest number of samples (26 eels) and on which the sex of eels was determined both on a macro- and microscopic basis to test the influence of sex on the shape of sagittal otoliths. In the remaining sites, only macroscopic sex evaluation was possible.

2.3. Otolith Extraction and Shape Analysis

The right and left sagittal otoliths of each eel were extracted for the analysis, cleaned with distilled water to remove remaining adhering tissues, and then placed dry in tubes. Each dried sagittal otolith was observed in the dorsal position under a stereomicroscope equipped with a digital camera (Leica S9i Stereozoom LSR w/TL3000 ergo, Wetzlar, Germany). The digital images were acquired using the Leica LAS 4.12 software to obtain the most highly contrasted images. The extraction and preparation of the sagittal otoliths were developed according to the methodology defined in the *Manual for the Ageing of Atlantic Eel* [36]. For the age reading [36], the sagittal otoliths were prepared by grinding and polishing along the sagittal plane, followed by staining (EDTA and Toluidene Blue, 0.1 M). The sagittal otoliths of eels under 5 years were analyzed without any preparation (in toto), except for their immersion in thyme essential oil to improve the visualization of growth marks.

Different measurements were performed on each sagittal otolith to calculate the shape indices [63] by using the software ImageJ (National Institutes of Health, USA). The following five indices were derived from the area (A), the perimeter (P), the Feret length (L), and the Feret width (w) of otoliths: form factor ($4\pi A/P^2$), circularity (P2/A), roundness ($4A/\pi L^2$), ellipticity $(L - w)/(L + w)$, and rectangularity (A/Lw). All indexes ranged from 0 to 1.

The sagittal otolith shape was first compared between males and females from the Pramaera River, which had the largest number of samples, and on which were conducted a microscopic analysis of the gonads to better assess the sex. As no significant differences were observed between sexes (K–W, $p > 0.05$), shape analyses were conducted on both the males and females together for all investigated sites. The five shape indices as well as the measured distances of the first three rings to the core and the annual otolith growth rate (cm year^{-1}) were compared between the river and lagoon groups. The assumption of linearity (normality and homoscedasticity) was rejected, and therefore a nonparametric Kruskal–Wallis (K–W) test followed by a pairwise comparison post hoc Dunn's (Z) test were performed to test for differences in the median values between the studied sites.

According to the methodology of a shape analysis previously described in Morat et al. [7] and in Mérigot et al. [64], an elliptic Fourier analysis [7,64–66] and comparisons of the shape indices [60] were conducted. The elliptic Fourier analysis describes the outline of sagittal otoliths as several components named harmonics. Each harmonic is characterized by 4 coefficients, derived from the projection of each point along the x- and y-axes. The Fourier coefficients were calculated using the software Shape 1.3 (Tokyo, Japan) [67]. The Fourier power spectrum was calculated for each sagittal otolith to determine the best number of harmonics for the optimal reconstruction of the otolith outline [68,69] for both the right and left sagittal otoliths of the same individual separately, as well as combined. In order to define the suitable number of harmonics to be considered in the analyses, the minimum number of harmonics was set up to obtain a threshold of 99.99% of the outline. For this reason, a total of 11 harmonics of the right sagittal otoliths were selected. Because the first harmonic was not considered (representing a simple ellipse), a total of 40 Fourier coefficients were used to describe each sagittal otolith.

The shape differences between each river and lagoon were determined using a canonical discriminant analysis (CDA) performed with the 40 Fourier coefficients. This classification method investigates the groups' integrity (each river and lagoon) by finding a linear combination of the descriptors that maximizes Wilk's lambda (λ) obtaining values ranging from 0 (low discrimination) to 1 (high discrimination) [70]. The Cohen kappa statistic was used to estimate the global reclassification rate of all groups [71]. The dissimilarity between groups was evaluated by using the Euclidean distance (d) between the barycenters of each group.

For the analysis of the shape indices, pairwise collinearity was investigated by examined scatter plots, excluding redundancy between paired variables using Spearman's $\rho > 0.7$. Shape indices were discarded from the pairwise combination based on the variance inflation factor (VIF) discarding observation with VIF > 3 [72]. Then, differences in shape indices between the river and lagoon groups were analyzed to describe and compare the sagittal otoliths in the different study sites (each river and lagoon). In the Pramaera river, comparisons were also carried out between sexes.

Significance was set at $p < 0.05$. All statistical and shape analyses were performed with the software R 4.3.1 (R Development Core Team) [73].

3. Results

A total of 100 eels were collected for the sagittal otolith analysis (Tables 1 and 2).

Table 2. Biometric ranges (minimum–maximum) of sampled eels per sex (female, male, and undifferentiated eels) for each studied river and lagoon. Total length (TL) and total weight (TW).

	Females		Males		Undifferentiated	
Site	**TL (cm)**	**TW (g)**	**TL (cm)**	**TW (g)**	**TL (cm)**	**TW (g)**
Pramaera	49.50–65.00	190.00–497.70	31.10–41.90	50.80–152.20	6.80–31.60	0.21–51.00
Tirso	/	/	/	/	16.30–32.30	6.70–71.90
UMannu	/	/	26.40–34.20	20.0–58.40	/	/
Barca	77.20	945.65	29.30–43.20	43.78–143.50	28.00	31.00
Coghinas	/	/	/	/	11.00–20.50	1.22–10.66
Calich	47.30–56.00	229.97–450.51	33.40–38.40	68.70–101.95	27.50–30.60	40.02–50.40
Porto Pino	26.50–75.50	24.50–636.80	/	/	/	/
Sa Praia	56.00–56.50	326.30–342.80	32.00–39.50	51.18–104.40	32.1	43.58

Comparisons of shape indices between male and female eels from the Pramaera River revealed no significant sex-dependent differences in sagittal otolith morphology from eels of the same study site (K–W test, $p > 0.05$).

Based on this result, the Fourier coefficients of the right sagittal otoliths were used in the CDA to assess the relative classification of the eight study sites in terms of otolith shape (Figure 2). The CDA revealed Wilks λ values equal to 0.06 and 0.17 for the x- and y-axis, respectively, indicating a relatively low discrimination between groups, while the percentage of reclassification assessed with Cohen's kappa test was 75.7%.

The Euclidean distance values between the barycenters of each group (rivers and lagoons), obtained from the CDAs, showed a distinct clustering pattern. The Pramaera, Tirso, and UMannu Rivers exhibited a close grouping with values of $d < 0.9$. Similarly, the three lagoons (Calich, Porto Pino, and Sa Praia) formed a distinct grouping with d values less than 0.4. The Barca River displayed intermediate characteristics, sharing similarities with both the river and the lagoon groups. In contrast, the Coghinas River did not group with any other site showing a high dissimilarity with d values greater than 2.4 compared to all other sites (Table 3).

Shape indices were analyzed through Spearman's rank correlation ($\rho < 0.7$) (Figure 3) and a VIF score threshold of 3. Based on these criteria, only form factor (VIF = 1.70), roundness (VIF = 1.57), and rectangularity (VIF = 1.18) were considered for the subsequent analysis. Area, perimeter, and the remaining shape indices (Feret length, Feret width, and circularity) showed a correlation higher than 0.7 and VIF greater than 3, and thus were discarded from subsequent analyses.

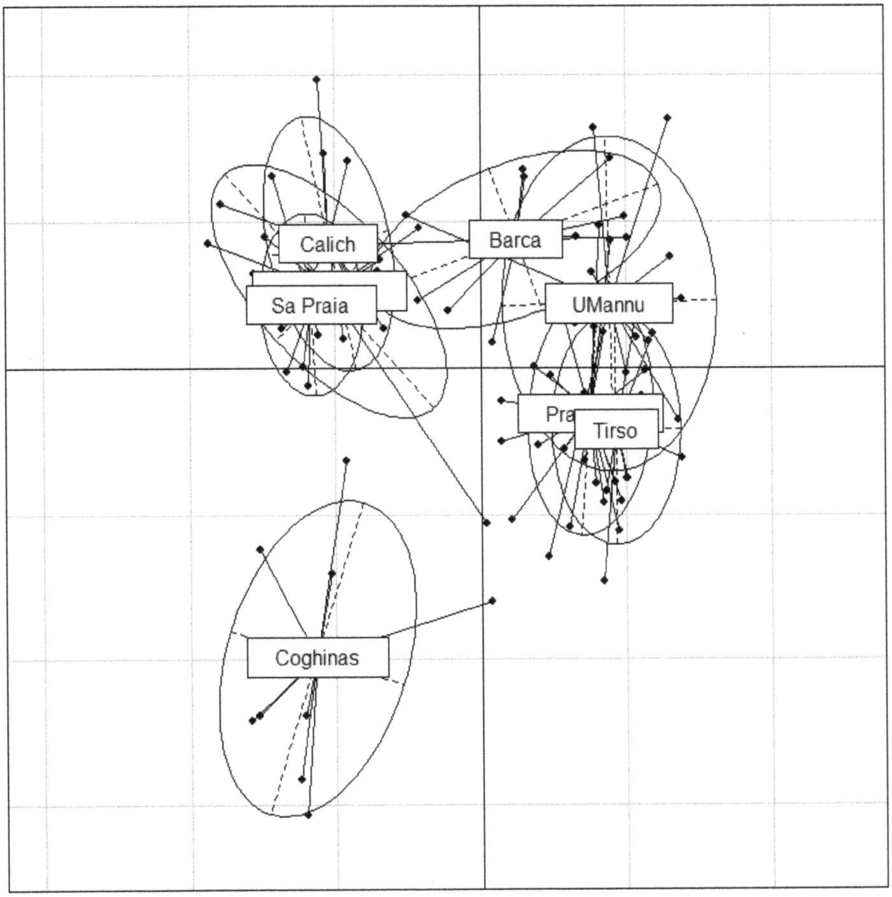

Figure 2. Canonical discriminant analysis (CDA) output achieved with Fourier coefficients for the five rivers (Pramaera, Tirso, UMannu, Barca, and Coghinas) and three lagoons (Calich, Porto Pino, and Sa Praia) investigated in the present study.

Table 3. Euclidean distance values between the barycenters of the study sites resulting from the CDAs performed with right sagittal otoliths (Euclidean distance values < 1, shown in bold, represent strong clustering between sites).

Site	Pramaera	Tirso	UMannu	Barca	Coghinas	Calich	Porto Pino	Sa Praia
Tirso	**0.21**							
UMannu	**0.76**	**0.86**						
Barca	1.29	1.47	**0.78**					
Coghinas	2.52	2.59	3.16	3.18				
Calich	2.14	2.34	1.97	1.28	2.84			
Porto Pino	1.98	2.18	1.93	1.33	2.52	**0.32**		
Sa Praia	2.06	2.26	2.05	1.47	2.42	**0.43**	**0.16**	

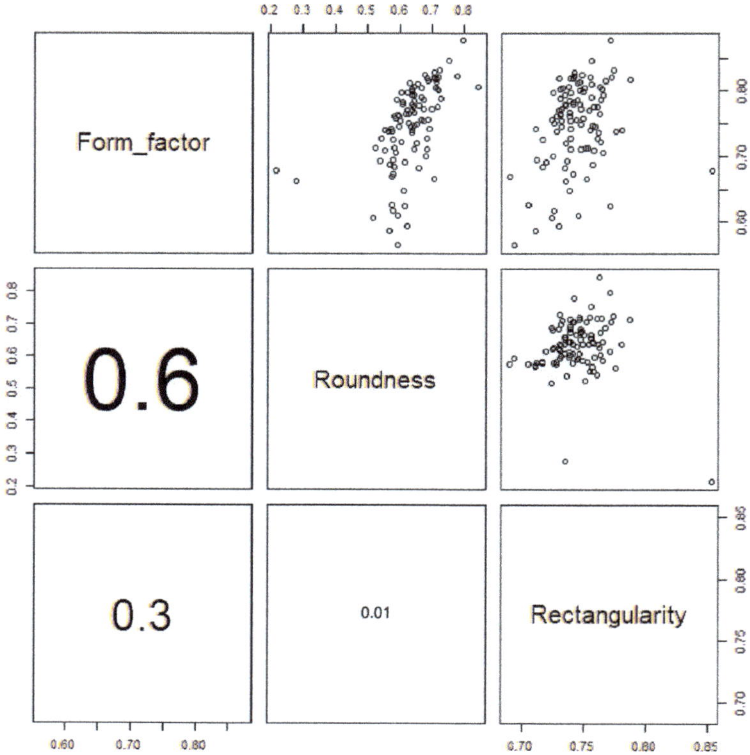

Figure 3. Spearman rank correlation and ρ values for shape otoliths' indexes.

The form factor showed significant differences between the river and lagoon groups (K–W = 29.34, $p = 0.00012$) (Figure 4). Specifically, the form factor values were found to be higher in the river group than in the lagoon group (Table 4).

This result was also confirmed by the post hoc Dunn's test revealing significant differences, especially between the Calich and Porto Pino lagoons compared to the Coghinas, Pramaera, and Tirso Rivers (Z tests, $p < 0.05$).

A similar outline was obtained for the roundness index (Figure 4), highlighting statistical differences detected between sites (K–W = 42.11, $p < 0.0001$) (Figure 4), with greater values of roundness in rivers than in lagoons, especially the Coghinas River (Table 5).

Table 4. In the grey boxes, median values ± standard deviation (SD) of the form factor shape index are described. Below the grey boxes, the *p* values are reported with significant values ($p < 0.05$) shown in bold, and above the grey boxes, asterisks indicate the significance level, i.e., $p < 0.05$ = *, and $p > 0.05$ = ns (not significant).

	Pramaera	Tirso	UMannu	Barca	Coghinas	Calich	Porto Pino	Sa Praia
Pramaera	0.767 ± 0.063	ns	ns	ns	ns	ns	*	ns
Tirso	1	0.777 ± 0.044	ns	ns	ns	ns	*	ns
UMannu	1	1	0.794 ± 0.041	ns	ns	ns	ns	ns
Barca	1	1	1	0.763 ± 0.067	ns	ns	ns	ns
Coghinas	1	1	1	1	0.794 ± 0.038	*	*	ns
Calich	0.14	0.08	0.19	1	**0.04**	0.704 ± 0.055	ns	ns
Porto Pino	**0.03**	**0.02**	0.06	0.90	**0.01**	1	0.695 ± 0.055	ns
Sa Praia	0.44	0.24	0.49	1	0.14	1	1	0.720 ± 0.071

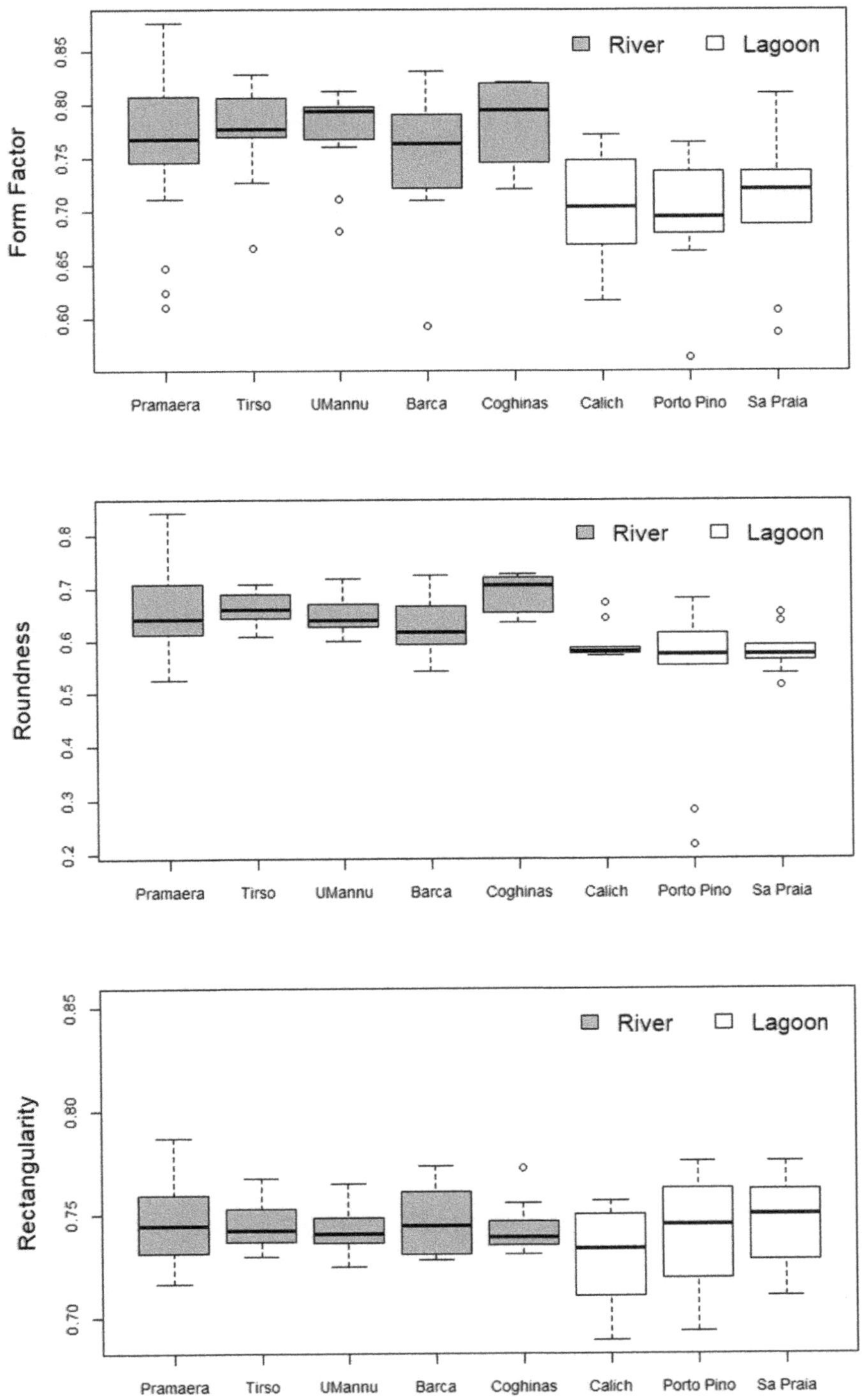

Figure 4. Boxplots for form factor, roundness, and rectangularity shape indices for the rivers (grey) and for the lagoons (white).

Table 5. In the grey boxes, median values ± standard deviation (SD) of the roundness shape index are described. Below the grey boxes, the p values are reported with significant values ($p < 0.05$) shown in bold, and above the grey boxes, asterisks indicate the significance level, i.e., $p < 0.001$ = ***, $p < 0.01$ = **, and $p < 0.05$ = *, and $p > 0.05$ = ns (not significant).

	Pramaera	Tirso	UMannu	Barca	Coghinas	Calich	Porto Pino	Sa Praia
Pramaera	0.642 ± 0.073	ns	ns	ns	ns	ns	*	*
Tirso	1	0.660 ± 0.032	ns	ns	ns	*	*	**
UMannu	1	1	0.639 ± 0.040	ns	ns	ns	ns	ns
Barca	1	1	1	0.616 ± 0.051	ns	ns	ns	ns
Coghinas	1	1	1	0.30	0.705 ± 0.035	**	***	***
Calich	0.07	**0.03**	0.39	1	**0.002**	0.580 ± 0.034	ns	ns
Porto Pino	**0.03**	**0.01**	0.19	1	**<0.001**	1	0.573 ± 0.015	ns
Sa Praia	**0.01**	**0.007**	0.12	1	**<0.001**	1	1	0.574 ± 0.041

On the one hand, significant differences were evident in all three lagoons compared to the Pramaera, Tirso, and Coghinas Rivers (Z tests, $p < 0.05$). On the other hand, the rectangularity index (Figure 4) did not show significant differences in medians between the sites (Table 6) (K–W = 3.88, $p = 0.79$).

Table 6. In the grey boxes, median values ± standard deviation (SD) of the rectangularity shape index are described. Below the grey boxes, the p values are reported with significant values ($p < 0.05$) shown in bold, and above the grey boxes, asterisks indicate the significance level, and $p > 0.05$ = ns (not significant).

	Pramaera	Tirso	UMannu	Barca	Coghinas	Calich	Porto Pino	Sa Praia
Pramaera	0.744 ± 0.018	ns	ns	ns	ns	ns	ns	ns
Tirso	1	0.742 ± 0.012	ns	ns	ns	ns	ns	ns
UMannu	1	1	0.741 ± 0.012	ns	ns	ns	ns	ns
Barca	1	1	1	0.745 ± 0.016	ns	ns	ns	ns
Coghinas	1	1	1	1	0.739 ± 0.013	ns	ns	ns
Calich	1	1	1	1	1	0.733 ± 0.024	ns	ns
Porto Pino	1	1	1	1	1	1	0.745 ± 0.044	ns
Sa Praia	1	1	1	1	1	1	1	0.750 ± 0.021

Analysis of ring distances from the core of the first two years revealed significant differences in otolith growth. The median values were higher in lagoons compared to in rivers (Tables 7 and 8) (K–W = 22.57, $p = 0.0019$ and K–W = 19.30, $p = 0.0073$, respectively) (Figure 5). Pairwise significant differences were observed only between the Calich and Sa Praia lagoons compared to the UMannu River (Z tests, $p < 0.05$).

Table 7. In the grey boxes, the median values ± standard deviation (SD) of the first ring distance from the core of sagittal otoliths are described. Below the grey boxes, the p values are reported with significant values ($p < 0.05$) shown in bold, and above the grey boxes, asterisks indicate the significance level, i.e., $p < 0.05$ = *, and $p > 0.05$ = ns (not significant).

	Pramaera	Tirso	UMannu	Barca	Coghinas	Calich	Porto Pino	Sa Praia
Pramaera	0.326 ± 0.088	ns	ns	ns	ns	ns	ns	ns
Tirso	1	0.347 ± 0.069	ns	ns	ns	ns	ns	ns
UMannu	1	1	0.298 ± 0.061	ns	ns	ns	ns	*
Barca	0.45	1	0.15	0.424 ± 0.060	ns	ns	ns	ns
Coghinas	1	1	1	1	0.347 ± 0.68	ns	ns	ns
Calich	1	1	0.73	1	1	0.414 ± 0.166	ns	ns
Porto Pino	0.39	1	0.07	1	1	1	0.439 ± 0.116	ns
Sa Praia	0.07	0.61	**0.03**	1	1	1	1	0.460 ± 0.086

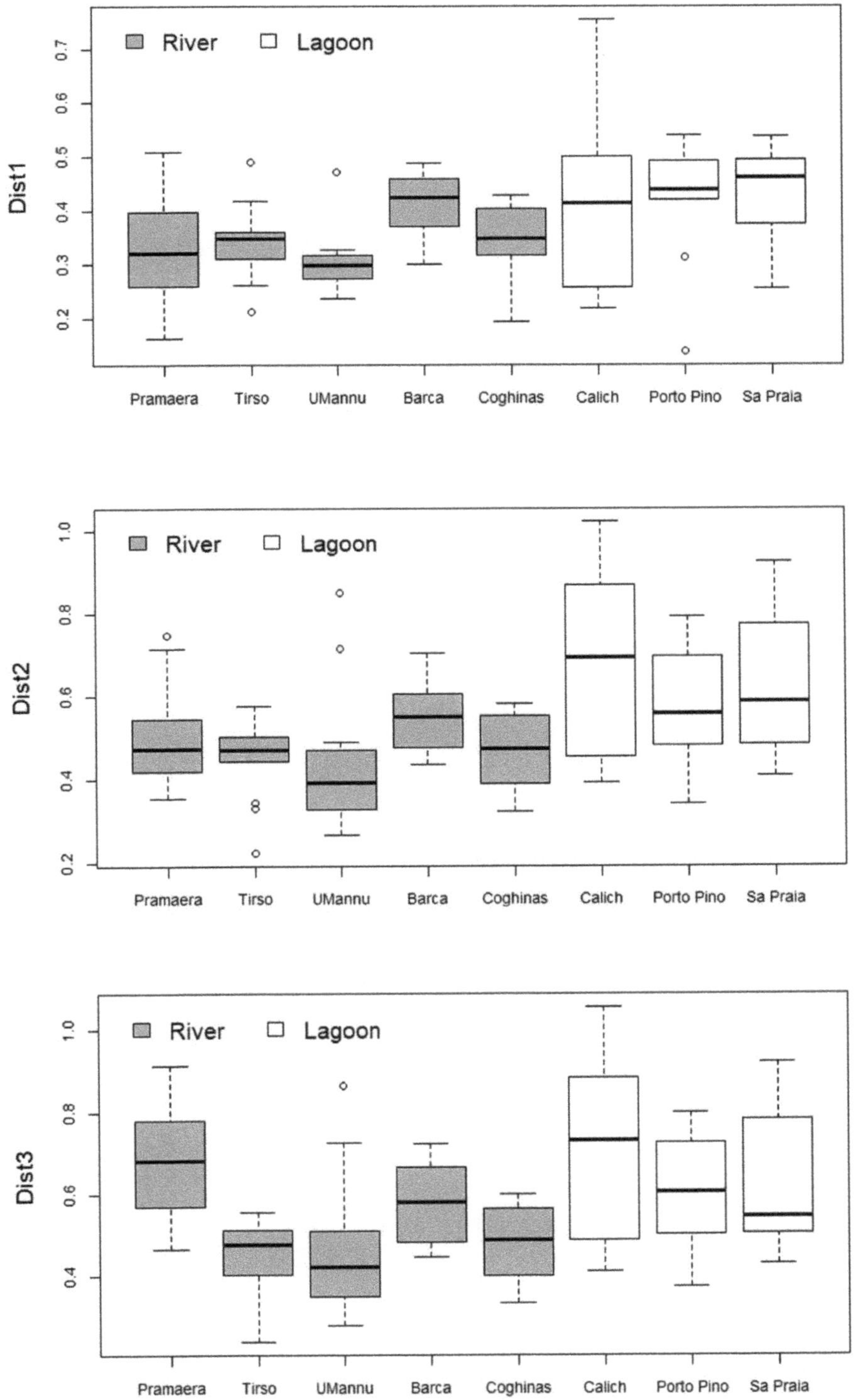

Figure 5. Boxplots for the first (Dist1), second (Dist2), and third (Dist3) ring distances from the sagittal otolith's core for rivers (grey) and for lagoons (white).

Table 8. In the grey boxes, median values ± standard deviation (SD) of the second ring distance from the core of sagittal otoliths are described. Below the grey boxes, the p values are reported with significant values ($p < 0.05$) shown in bold, and above the grey boxes, asterisks indicate the significance level, i.e., $p < 0.05$ = *, and $p > 0.05$ = ns (not significant).

	Pramaera	Tirso	UMannu	Barca	Coghinas	Calich	Porto Pino	Sa Praia
Pramaera	0.496 ± 0.126	ns	ns	ns	ns	ns	ns	ns
Tirso	1	0.469 ± 0.099	ns	ns	ns	ns	ns	ns
UMannu	1	1	0.390 ± 0.184	ns	ns	*	ns	ns
Barca	1	1	0.49	0.549 ± 0.088	ns	ns	ns	ns
Coghinas	1	1	1	1	0.471 ± 0.088	ns	ns	ns
Calich	1	0.36	**0.04**	1	0.89	0.694 ± 0.223	ns	ns
Porto Pino	1	1	0.31	1	1	1	0.556 ± 0.145	ns
Sa Praia	1	0.51	0.07	1	1	1	1	0.587 ± 0.188

For the third ring distances (K–W = 26.33, $p < 0.001$), differences in the distances were found between the Pramaera and the Tirso Rivers, and the Pramaera and the UMannu Rivers (Z tests, $p < 0.05$) (Figure 5 and Table 9).

Table 9. In the grey boxes, median values ± standard deviation (SD) of the third ring distance from the core of sagittal otoliths are described. Below the grey boxes, the p values are reported with significant values ($p < 0.05$) shown in bold, and above the grey boxes, asterisks indicate the significance level, i.e., $p < 0.01$ = ** and $p < 0.05$ = *, and $p > 0.05$ = ns (not significant).

	Pramaera	Tirso	UMannu	Barca	Coghinas	Calich	Porto Pino	Sa Praia
Pramaera	0.693 ± 0.154	**	*	ns	ns	ns	ns	ns
Tirso	**0.007**	0.476 ± 0.095	ns	ns	ns	ns	ns	ns
UMannu	**0.01**	1	0.421 ± 0.180	ns	ns	ns	ns	ns
Barca	1	1	1	0.581 ± 0.100	ns	ns	ns	ns
Coghinas	0.10	1	1	1	0.488 ± 0.094	ns	ns	ns
Calich	1	0.08	0.13	1	0.46	0.729 ± 0.234	ns	ns
Porto Pino	1	0.45	0.66	1	1	1	0.605 ± 0.140	ns
Sa Praia	1	0.52	0.75	1	1	1	1	0.546 ± 0.168

Significant differences in annual otolith growth values were found between the river and lagoon groups (Figure 6) (K–W = 58.27, $p < 0.0001$), with lagoons generally showing higher median values, except for the Pramaera River, which showed variation in annual growth with values close to those for the lagoons. The Coghinas River displayed the lowest median annual otolith growth values (Table 10) (Z tests, $p < 0.05$).

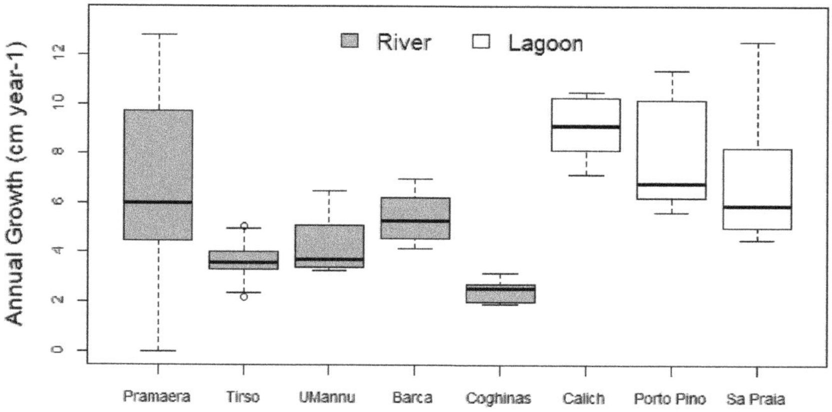

Figure 6. Boxplot for the annual otolith growth (cm year^{-1}) for rivers (grey) and for lagoons (white).

Table 10. In the grey boxes, median values ± standard deviation (SD) of the annual otolith growth (cm year^{-1}) are described. Below the grey boxes, the p values are reported with significant values ($p < 0.05$) shown in bold, and above the grey boxes, asterisks indicate the significance level, i.e., $p < 0.0001$ = ****, $p < 0.001$ = ***, $p < 0.01$ = **, and $p < 0.05$ = *, and $p > 0.05$ = ns (not significant).

	Pramaera	Tirso	UMannu	Barca	Coghinas	Calich	Porto Pino	Sa Praia
Pramaera	6.00 ± 3.50	*	ns	ns	****	ns	ns	ns
Tirso	0.01	3.58 ± 0.84	ns	ns	ns	****	**	ns
UMannu	0.33	1	3.70 ± 1.33	ns	ns	**	ns	ns
Barca	1	0.87	1	5.29 ± 1.03	*	ns	ns	ns
Coghinas	**<0.0001**	1	1	**0.03**	7.60 ± 2.86	****	****	***
Calich	0.72	**<0.0001**	**0.002**	0.31	**<0.0001**	9.15 ± 1.77	ns	ns
Porto Pino	1	**0.002**	0.57	1	**<0.0001**	1	6.81 ± 2.16	ns
Sa Praia	1	0.57	0.65	1	**<0.001**	1	1	5.90 ± 2.64

4. Discussion

In this study, for the first time, to the best of our knowledge, we aimed to investigate the subpopulations of *A. anguilla* from Sardinian continental waters (central-western Mediterranean) using an otolith shape analysis. Our goal was to understand the role of different habitat types (rivers and lagoons) on otolith shape and growth as a marker of eel subpopulation success. The results of the canonical discriminant analysis demonstrated a high value of reclassification of 75.7%, which suggests that the classification analysis and shape index comparisons can represent useful tools for discriminating eel subpopulations that inhabit different habitat typology. Similar findings have already been observed in other regional studies that have focused on Mediterranean eel stocks [37,74]. In particular, Capoccioni et al. [37] found differences in otolith morphology among three Mediterranean eel local stocks, including two brackish lagoons and one river. Additionally, Milošević et al. [75] investigated otolith shape variations between riverine and lacustrine habitats in the Adriatic Basin (Croatia and Montenegro), and their findings aligned with our study, highlighting the role of environmental variability in shaping otolith morphology and size during the growth of the species.

In the present study, we found that sagittal otoliths from rivers were generally rounder and less irregular compared with those from lagoons. This finding corroborated results obtained in previous shape otolith studies, as mentioned above. In addition, we obtained high variability in terms of annual otolith growth rates among several subpopulations as well as in single habitats, which has already been observed in previous studies all over Europe and in the Mediterranean basin, due to the multitude of used habitats by the species [38,76–79]. Our values were in accordance with those obtained in other particular Mediterranean lagoons (about 5 cm year^{-1}) (Valli di Comacchio [80], Vaccarés-Impérieux, Fumemorte [78], Aveiro [81], and Valle Nuova lagoons [80]). Eels from rivers showed lower annual growth rate values that appeared to be similar to values obtained in other studies (between 3 and 4.5 cm year^{-1} (Severn [82], Shannon [83], Frome [84], Koge Lellinge [85], Barrow [86], and Imsa [87] Rivers). We also confirmed that brackish habitats such as coastal lagoons could support faster growth rates than riverine habitats offering probably more suitable conditions to support eel growth and survival [78,88–90].

Furthermore, in our study, the sagittal otoliths from rivers maintained a more circular shape across their entire life, with the only exception of otoliths of the eels from the Coghinas River which showed a shape differentiation resulting in a more circular shape than all otoliths of the other rivers. This peculiar result can be attributed to the size of the sample that was possible to collect, consisting of only small specimens (TL $\leq$ 20.5 cm). Despite their reduced TL values, these eels showed ages ranging from two to five years. Even if coming from the same basin, sagittal otoliths of eels from the Barca River and Calich lagoon were an average distant, according to the Euclidean distance, sharing similarities both with the river and lagoon groups. However, differences found in the shape indices fit well between these two sites. This could be related to the resident behavior of the eels during the trophic stage that showed differences in development according to the habitat typology.

Indeed, several environmental abiotic characteristics (e.g., temperature, salinity, depth, food availability, and ecological niche) create habitats that are more suitable for eels, determining variability in their development that can also be reflected in otolith shape; therefore, more effort should be made to protect them [2,13,64,91–95]. Furthermore, different eels' development strategies in a variety of aquatic habitats located in different geographic areas can reflect the complexity of these environments, and therefore, can help to understand habitat suitability, the success of recruitment, and eel productivity [96,97]. However, these results, allowed the discrimination of local eel subpopulations, corroborating the hypothesis that ecological and morphological differences in otolith shape are influenced by the specific environments inhabited by the species [7]. Eels with rounder otoliths were found in freshwater riverine habitats. These environments generally tend to be less susceptible to variations in salinity or temperature and depth than brackish estuaries or lagoons [98]. However, it remains unknown how environmental abiotic variables may act together, influencing or limiting the growth of the species [29].

Although the European eel is protected according to regional, national, and international regulations, and despite its commercial importance, little has been published on the ecology of this species in Sardinia [53,57,60,98,99]. There are no studies that have analyzed eels' sagittal otolith shape in Sardinian continental waters, including rivers and lagoons. Furthermore, only one study has investigated the relationship between otolith and growth in the European eel in the Porto Pino lagoon [60].

All the differences found in sagittal otolith shape among the studied sites could lead to changes between different local stocks and they could be related to environmental peculiarities. However, establishing a direct correlation between environmental factors and variations in otolith morphology poses challenges. Further studies are necessary to investigate the relationship between habitat type/environmental variability and the growth/body characteristics of eels. Studies should also be conducted taking into consideration intraspecific variability in terms of sex and size, when possible, which is representative of an entire local eel subpopulation. Therefore, because otolith shape has been studied for the European eel in some European areas [37,52], successfully discriminating eels that grow in different habitat types, we also support that this method can be considered to be a valuable tool for studying the species' ontogeny.

Author Contributions: Conceptualization, C.P.; methodology, C.P., E.A., F.M. and E.F.; software, C.P. and F.M.; validation, C.P., F.P., E.A., F.M. and A.S.; formal analysis, C.P.; investigation, C.P., J.C., R.D., F.P. and A.S.; writing—original draft preparation, C.P.; writing—review and editing, J.C., R.D., F.P., E.A., F.M., E.F., N.F. and A.S.; supervision, A.S.; funding acquisition, N.F. and A.S. All authors have read and agreed to the published version of the manuscript.

Funding: This research was financed by the Regional Plan for Eel Management in Sardinia, following the Decree of the Sardinian Regional Minister for Agriculture and Agro-Pastoral Reform No. 3186/DecA/158 of 29 December 2009 and subsequent modifications and additions.

Institutional Review Board Statement: This study makes use of otoliths (or otolithic organs). For the study of otoliths, it is necessary to extract them from the head of dead animals. As reported in Directive 2010/63/EU and in the Italian Legislative Decree 26/2014 in the implementation of the Directive 2010/63/EU, the suppression of animals for the sole purpose of using their organs or tissues, as in our case, is excluded among the actions considered to be procedures. Therefore, we have not applied any procedures on animals to be killed. The ethical approval should be waived.

Informed Consent Statement: Not applicable.

Data Availability Statement: The data presented in this study are available on request from the corresponding author.

Acknowledgments: The authors would like to thank all the fishermen of the cooperatives: Il Golfo e Laguna, San Giuseppe, Compendio Ittico Villaputzu, Alessio Musu, Melissa Serra, Adriano Deiana, Sabina Piras, Giosuè Serreli, and Marco Maxia for their collaboration. C.P. gratefully acknowledges the Sardinian Regional Government for the financial support of her PhD scholarship (P.O.R. Sardegna F.S.E.—Operational Program of the Autonomous Region of Sardinia, European Social Fund 2014–2020—Axis III Education and Training, Thematic Goal 10, Investment Priority 10ii), specific goal 10.5.

Conflicts of Interest: The authors declare no conflict of interest.

References

1. Aydın, İ. *Balık Larvalarında Otolit*; Makale, Sümae Yunus Araştırma Bülteni: Trabzon, Turkey, 2006; Volume 2, p. 6.
2. Campana, E.S.; Thorrold, S.R. Otoliths, increments, and elements: Keys to a comprehensive understanding of fish populations? *Can. J. Fish. Aquat. Sci.* **2001**, *58*, 30–38. [CrossRef]
3. Thorrold, S.R.; Campana, S.E.; Jones, C.M.; Swart, P.K. Factors determining δ13C and δ18O fractionation in aragonitic otoliths of marine fish. *Geochim. Cosmochim. Acta* **1997**, *61*, 2909–2919. [CrossRef]
4. Elsdon, T.S.; Gillanders, B.M. Interactive effects of temperature and salinity on otolith chemistry: Challenges for determining environmental histories of fish. *Can. J. Fish. Aquat. Sci.* **2002**, *59*, 1796–1808. [CrossRef]
5. Lecomte-Finiger, R. Growth history and age at recruitment of European glass eels (*Anguilla anguilla*) as revealed by otolith microstructure. *Mar. Biol.* **1992**, *114*, 205–210. [CrossRef]
6. Mille, T.; Mahe, K.; Villanueva, M.C.; De Pontual, H.; Ernande, B. Sagittal otolith morphogenesis asymmetry in marine fishes. *J. Fish Biol.* **2015**, *87*, 646–663. [CrossRef]
7. Morat, F.; Letourneur, Y.; Nérini, D.; Banaru, D.; Batjakas, I.E. Discrimination of red mullet populations (Teleostean, Mullidae) along multi-spatial and ontogenetic scales within the Mediterranean basin on the basis of otolith shape analysis. *Aquat. Living Resour.* **2012**, *25*, 27–39. [CrossRef]
8. Paul, K.; Oeberst, R.; Hammer, C. Evaluation of otolith shape analysis as a tool for discriminating adults of Baltic cod stocks. *J. Appl. Ichthyol.* **2013**, *29*, 743–750. [CrossRef]
9. Ozpicak, M.; Saygin, S.; Aydın, A.; Hancer, E.; Yilmaz, S.; Polat, N. Otolith shape analyses of *Squalius cephalus* (Linnaeus, 1758) (*Actinopterygii: Cyprinidae*) inhabiting four inland water bodies of the middle Black Sea region, Turkey. *Iran. J. Ichthyol.* **2018**, *5*, 293–302.
10. Zhao, B.; Liu, J.; Song, J.; Cao, L.; Dou, S. Otolith shape analysis for stock discrimination of two *Collichthys* genus croaker (*Pieces: Sciaenidae*) from the northern Chinese coast. *J. Oceanol. Limnol.* **2018**, *36*, 981–989. [CrossRef]
11. Yilmaz, S.; Yazicioglu, O.; Saygin, S.; Polat, N. Relationships of otolith dimensions with body length of European perch, *Perca fluviatilis* L., 1758 from Lake Ladik, Turkey. *Pak. J. Zool.* **2014**, *46*, 1231–1238.
12. Başusta, N.; Khan, U. Sexual dimorphism in the otolith shape of shidrum, *Umbrina cirrosa* (L.), in the eastern Mediterranean Sea: Fish size–otolith size relationships. *J. Fish Biol.* **2021**, *99*, 164–174. [CrossRef]
13. Gagliano, M.; McCormick, M.I. Feeding history influences otolith shape in tropical fish. *Mar. Ecol. Progr. Ser.* **2004**, *278*, 291–296. [CrossRef]
14. Mille, T.; Mahé, K.; Cachera, M.; Villanueva, M.; de Pontual, H.; Ernande, B. Diet is correlated with otolith shape in marine fish. *Mar. Ecol. Prog. Ser.* **2016**, *555*, 167–184. [CrossRef]
15. Campana, S.E. *Photographic Atlas of Fish Otoliths of the Northwest Atlantic Ocean*; NRC Research Press: Ottawa, ON, Canada, 2004. [CrossRef]
16. Bounket, B.; Gibert, P.; Gennotte, V.; Argillier, C.; Carrel, G.; Maire, A.; Logez, M.; Morat, F. Otolith shape analysis and daily increment validation during ontogeny of larval and juvenile European chub *Squalius cephalus*. *J. Fish Biol.* **2019**, *95*, 444–452. [CrossRef]
17. Tsukamoto, K.; Nakai, I.; Tesch, W. Do all freshwater eels migrate? *Nature* **1998**, *396*, 635–636. [CrossRef]
18. Limburg, K.E.; Wickstrom, H.; Svedang, H.; Elfman, M.; Kristiansson, P. Do stocked freshwater eels migrate? Evidence from the Baltic suggests "yes". *Am. Fish. Soc. Symp.* **2003**, *33*, 275–284.
19. ICES. *Report of the Joint EIFAAC/ICES/GFCM Working Group on Eel (WGEEL)*; ICES Document CM 2015/ACOM: 18; ICES: Antalya, Turkey, 2015.
20. Antunes, C.; Tesch, F.-W. A critical consideration of the metamorphosis zone when identifying daily rings in otoliths of European eel, *Anguilla anguilla* (L.). *Ecol. Freshw. Fish* **1997**, *6*, 102–107. [CrossRef]
21. Dekker, W. Status of the European eel stock and fisheries. In *Eel Biology*; Aida, K., Tsukamoto, K., Yamauchi, K., Eds.; Springer: Tokyo, Japan, 2003; pp. 237–254.
22. Bertin, L. *Eels: A Biological Study*; Cleaver-Hume Press: London, UK, 1956.
23. Helfman, G.S.; Facey, D.E.; Hales, L.S.; Bozeman, E.L. Reproductive ecology of the American eel. *Am. Fish. Soc. Symp.* **1987**, *1*, 42–56.
24. Vollestad, L.A. Geographic Variation in Age and Length at Metamorphosis of Maturing European Eel: Environmental Effects and Phenotypic Plasticity. *J. Anim. Ecol.* **1992**, *61*, 41–48. [CrossRef]

25. Thillart, G.v.D.; van Ginneken, V.; Korner, F.; Heijmans, R.; van der Linden, R.; Gluvers, A. Endurance swimming of European eel. *J. Fish Biol.* **2004**, *65*, 312–318. [CrossRef]
26. Edeline, E.; Dufour, S.; Elie, P. Role of glass eel salinity preference in the control of habitat selection and growth plasticity in Anguilla anguilla. *Mar. Ecol. Prog. Ser.* **2005**, *304*, 191–199. [CrossRef]
27. Edeline, E.; Lambert, P.; Rigaud, C.; Elie, P. Effects of body condition and water temperature on *Anguilla anguilla* glass eel migratory behavior. *J. Exp. Mar. Biol. Ecol.* **2006**, *331*, 217–225. [CrossRef]
28. Tesch, F.W. *The eel*; Blackwell Science: Oxford, UK, 2003; p. 408.
29. Daverat, F.; Beaulaton, L.; Poole, R.; Lambert, P.; Wickström, H.; Andersson, J.; Aprahamian, M.; Hizem, B.; Elie, P.; Yalçın-Özdilek, S.; et al. One century of eel growth: Changes and implications. *Ecol. Fresh. Fish* **2012**, *21*, 325–336. [CrossRef]
30. Krueger, W.H.; Oliveira, K. Evidence for environmental sex determination in the American eel, *Anguilla rostrata. Environ. Biol. Fishes* **1982**, *55*, 381–389. [CrossRef]
31. Jessop, B.; Shiao, J.; Iizuka, Y.; Tzeng, W. Variation in the annual growth, by sex and migration history, of silver American eels *Anguilla rostrata. Mar. Ecol. Prog. Ser.* **2004**, *272*, 231–244. [CrossRef]
32. Izzo, C.; Doubleday, Z.A.; Schultz, A.G.; Woodcock, S.H.; Gillanders, B.M. Contribution of water chemistry and fish condition to otolith chemistry: Comparisons across salinity environments. *J. Fish Biol.* **2015**, *86*, 1680–1698. [CrossRef]
33. D'iglio, C.; Natale, S.; Albano, M.; Savoca, S.; Famulari, S.; Gervasi, C.; Lanteri, G.; Panarello, G.; Spanò, N.; Capillo, G. Otolith Analyses Highlight Morpho-Functional Differences of Three Species of Mullet (*Mugilidae*) from Transitional Water. *Sustainability* **2022**, *14*, 398. [CrossRef]
34. Begg, G.A.; Campana, S.E.; Fowler, A.J.; Suthers, I.M. Otolith research and application: Current directions in innovation and implementation. *Mar. Freshw. Res.* **2005**, *56*, 477–483. [CrossRef]
35. Campana, S.E. Otolith science entering the 21st century. *Mar. Freshw. Res.* **2005**, *56*, 485–495. [CrossRef]
36. ICES. *Workshop on Age Reading of European and American Eel (WKAREA)*; ICES CM 2009\ACOM 48; ICES: Bordeaux, France, 2009; p. 66.
37. Capoccioni, F.; Costa, C.; Aguzzi, J.; Menesatti, P.; Lombarte, A.; Ciccotti, E. Ontogenetic and environmental effects on otolith shape variability in three Mediterranean European eel (*Anguilla anguilla*, L.) local stocks. *J. Exp. Mar. Biol. Ecol.* **2011**, *397*, 1–7. [CrossRef]
38. Panfili, J.; Boulenger, C.; Musseau, C.; Crivelli, A.J. Extreme variability in European eel growth revealed by an extended mark and recapture experiment in southern France and implications for management. *Can. J. Fish. Aquat. Sci.* **2022**, *79*, 631–641. [CrossRef]
39. Pike, C.; Crook, V.; Gollock, M. Anguilla anguilla. In *The IUCN Red List of Threatened Species: e.T60344A 152845178*; IUCN: Gland, Switzerland, 2020.
40. ICES. *European eel (Anguilla anguilla) throughout Its Natural Range*; Report; ICES Advice: Copenhagen, Denmark, 2022. [CrossRef]
41. Pujolar, J.M.; Jacobsen, M.W.; Als, T.D.; Frydenberg, J.; Munch, K.; Jónsson, B.; Jian, J.B.; Cheng, L.; Maes, G.E.; Bernatchez, L.; et al. Genome-wide single-generation signatures of local selection in the panmictic European eel. *Mol. Ecol.* **2014**, *23*, 2514–2528. [CrossRef]
42. Panfili, J.; Ximénès, M.-C.; Crivelli, A.J. Sources of Variation in Growth of the European Eel (*Anguilla anguilla*) Estimated from Otoliths. *Can. J. Fish. Aquat. Sci.* **1994**, *51*, 506–515. [CrossRef]
43. Melià, P.; Bevacqua, D.; Crivelli, A.J.; Panfili, J.; De Leo, G.A.; Gatto, M. Sex differentiation of the European eel in brackish and freshwater environments: A comparative analysis. *J. Fish Biol.* **2006**, *69*, 1228–1235. [CrossRef]
44. De Leo, G.A.; Gatto, M.; Mateo, M.; Lambert, P.; Tetard, S.; Castonguay, M.; Ernande, B.; Drouineau, H.; Rigaud, C.; Daverat, F.; et al. A size and age-structured model of the European eel (*Anguilla anguilla* L.). *Can. J. Fish. Aquat. Sci.* **1995**, *52*, 1351–1367. [CrossRef]
45. Daverat, F.; Limburg, K.; Thibault, I.; Shiao, J.; Dodson, J.; Caron, F.; Tzeng, W.; Iizuka, Y.; Wickström, H. Phenotypic plasticity of habitat use by three temperate eel species, *Anguilla anguilla, A. japonica* and *A. rostrata. Mar. Ecol. Prog. Ser.* **2006**, *308*, 231–241. [CrossRef]
46. Schmidt, J. Breeding Places and Migrations of the Eel. *Nature* **1923**, *111*, 51–54. [CrossRef]
47. Miller, M.J.; Bonhommeau, S.; Munk, P.; Castonguay, M.; Hanel, R.; McCleave, J.D. A century of research on the larval distributions of the Atlantic eels: A re-examination of the data. *Biol. Rev.* **2015**, *90*, 1035–1064. [CrossRef]
48. Chang, Y.-L.K.; Feunteun, E.; Miyazawa, Y.; Tsukamoto, K. New clues on the Atlantic eels spawning behavior and area: The Mid-Atlantic Ridge hypothesis. *Sci. Rep.* **2020**, *10*, 15981. [CrossRef]
49. Trancart, T.; Tudorache, C.; Thillart, G.v.D.; Acou, A.; Carpentier, A.; Boinet, C.; Gouchet, G.; Feunteun, E. The effect of thermal shock during diel vertical migration on the energy required for oceanic migration of the European silver eel. *J. Exp. Mar. Biol. Ecol.* **2015**, *463*, 168–172. [CrossRef]
50. Righton, D.; Westerberg, H.; Feunteun, E.; Økland, F.; Gargan, P.; Amilhat, E.; Metcalfe, J.; Lobon-Cervia, J.; Sjöberg, N.; Simon, J.; et al. Empirical observations of the spawning migration of European eels: The long and dangerous road to the Sargasso Sea. *Sci. Adv.* **2016**, *2*, e1501694. [CrossRef] [PubMed]
51. Wright, R.M.; Piper, A.T.; Aarestrup, K.; Azevedo, J.M.N.; Cowan, G.; Don, A.; Gollock, M.; Ramallo, S.R.; Velterop, R.; Walker, A.; et al. First direct evidence of adult European eels migrating to their breeding place in the Sargasso Sea. *Sci. Rep.* **2022**, *12*, 15362. [CrossRef] [PubMed]
52. Moura, A.; Dias, E.; López, R.; Antunes, C. Regional Population Structure of the European Eel at the Southern Limit of Its Distribution Revealed by Otolith Shape Signature. *Fishes* **2022**, *7*, 135. [CrossRef]

53. Podda, C.; Palmas, F.; Pusceddu, A.; Sabatini, A. When the Eel Meets Dams: Larger Dams' Long-Term Impacts on *Anguilla anguilla* (L., 1758). *Front. Environ. Sci.* **2022**, *10*, 876369. [CrossRef]
54. De Waele, J.; Martina, M.L.; Sanna, L.; Cabras, S.; Cossu, Q.A. Flash flood hydrology in karstic terrain: Flumineddu Canyon, central-east Sardinia. *Geomorphology* **2010**, *120*, 162–173. [CrossRef]
55. Sabatini, A.; Podda, C.; Frau, G.; Cani, M.V.; Musu, A.; Serra, M.; Palmas, F. Restoration of native Mediterranean brown trout *Salmo cettii* Rafinesque, 1810 (*Actinopterygii: Salmonidae*) populations using an electric barrier as a mitigation tool. *Eur. Zool. J.* **2018**, *85*, 137–149. [CrossRef]
56. Palmas, F.; Righi, T.; Musu, A.; Frongia, C.; Podda, C.; Serra, M.; Splendiani, A.; Barucchi, V.C.; Sabatini, A. Pug-Headedness Anomaly in a Wild and Isolated Population of Native Mediterranean Trout *Salmo trutta* L., 1758 Complex (Osteichthyes: *Salmonidae*). *Diversity* **2020**, *12*, 353. [CrossRef]
57. Podda, C.; Palmas, F.; Frau, G.; Chessa, G.; Culurgioni, J.; Diciotti, R.; Fois, N.; Sabatini, A. Environmental influences on the recruitment dynamics of juvenile European eels, *Anguilla anguilla*, in a small estuary of the Tyrrhenian Sea, Sardinia, Italy. *Aquat. Conserv. Mar. Freshw. Ecosyst.* **2020**, *30*, 1638–1648. [CrossRef]
58. AA.VV. *Carta Ittica Della Sardegna–D.G.R. N. 2/28, 428. del 20/01/2022;* Regione Autonoma della Sardegna (ADA/STNPF)/Università degli Studi di Cagliari (DISVA): Cagliari, Italy, 2022.
59. AA.VV. *Piano di Salvaguardia e Valorizzazione dei Laghi Salsi;* Università degli Studi di Cagliari: Cagliari, Italy, 2010; p. 317.
60. Rossi, R.; Cannas, A. Eel fishing management in a hypersaline lagoon of Southern Sardinia. *Fish. Res.* **1984**, *2*, 285–298. [CrossRef]
61. Gilderhus, P.A.; Marking, L.L. Comparative efficacy of 16 anesthetic chemicals on rainbow trout. *N. Am. J. Fish. Manag.* **1987**, *7*, 288–292. [CrossRef]
62. Colombo, G.; Grandidr, G. Histological study of the development and sex differentiation of the gonad in the European eel. *J. Fish Biol.* **1996**, *48*, 493–512. [CrossRef]
63. Tuset, V.M.; Lozano, I.J.; González, J.A.; Pertusa, J.F.; García-Díaz, M.M. Shape indices to identify regional differences in otolith morphology of comber, *Serranus cabrilla* (L., 1758). *J. Appl. Ichthyol.* **2003**, *19*, 88–93. [CrossRef]
64. Mérigot, B.; Letourneur, Y.; Lecomte-Finiger, R. Characterization of local populations of the common sole *Solea solea* (*Pisces, Soleidae*) in the NW Mediterranean through otolith morphometrics and shape analysis. *Mar. Biol.* **2007**, *151*, 997–1008. [CrossRef]
65. Stransky, C.; MacLellan, E.S. Species separation and zoogeography of redfish and rockfish (genus *Sebastes*) by otolith shape analysis. *Can. J. Fish. Aquat. Sci.* **2005**, *62*, 2265–2276. [CrossRef]
66. Morat, F.; Gibert, P.; Reynaud, N.; Testi, B.; Favriou, P.; Raymond, V.; Carrel, G.; Maire, A. Spatial distribution, total length frequencies and otolith morphometry as tools to analyse the effects of a flash flood on populations of roach (*Rutilus rutilus*). *Ecol. Freshw. Fish* **2018**, *27*, 421–432. [CrossRef]
67. Iwata, H. Tutorial for SHAPE v.1.3. 2006, p. 21. Available online: http://lbm.ab.a.u--tokyo.ac.jp/~iwata/shape/tutorial.pdf (accessed on 20 September 2021).
68. Kuhl, F.P.; Giardina, C.R. Elliptic Fourier features of a closed contour. *Comput. Graph. Image Process.* **1982**, *18*, 236–258. [CrossRef]
69. Crampton, J.S. Elliptic Fourier shape analysis of fossil bivalves: Some practical considerations. *Lethaia* **1995**, *28*, 179–186. [CrossRef]
70. Ramsay, J.O.; Silveman, B.W. *Functional Data Analysis*, 2nd ed.; Springer: New York, NY, USA, 2005; p. 310.
71. Titus, K.; Mosher, J.A.; Williams, B.K. Chance-corrected Classification for Use in Discriminant Analysis: Ecological Applications. *Am. Midl. Nat.* **1984**, *111*, 1. [CrossRef]
72. Zuur, A.F.; Ieno, E.N.; Elphick, C.S. A protocol for data exploration to avoid common statistical problems. *Methods Ecol. Evol.* **2010**, *1*, 3–14. [CrossRef]
73. R Core Team. *R: A Language and Environment for Statistical Computing;* R Foundation for Statistical Computing: Vienna, Austria, 2012; Available online: http://www.R-project.org (accessed on 28 October 2021).
74. Milošević, D.; Bigović, M.; Mrdak, D.; Milašević, I.; Piria, M. Otolith morphology and microchemistry fingerprints of European eel, *Anguilla anguilla* (Linnaeus, 1758) stocks from the Adriatic Basin in Croatia and Montenegro. *Sci. Total. Environ.* **2021**, *786*, 147478. [CrossRef]
75. Panfili, J.; Ximénès, M.C. Évaluation de l'âge et de la croissance de l'anguille européenne (*Anguilla anguilla* L.) en milieu continental: Méthodologies, validation, application en Méditerranée et comparaisons en Europe. *Bull. Fr. Pêche Piscic.* **1994**, *335*, 43–66. [CrossRef]
76. Pujolar, J.M.; Maes, G.E.; Vancoillie, C.; Volckaert, F.A.M. Growth rate correlates to individual heterozygosity in the European eel, *Anguilla anguilla* L. *Evolution* **2005**, *59*, 189–199.
77. Melia, P.; Bevacqua, D.; Crivelli, A.J.; De Leo, G.A.; Panfili, J.; Gatto, M. Age and growth of Anguilla anguilla in the Camargue lagoons. *J. Fish Biol.* **2006**, *68*, 876–890. [CrossRef]
78. Boulenger, C.; Acou, A.; Trancart, T.; Crivelli, A.J.; Feunteun, E. Length-weight relationships of the silver European eel, *Anguilla anguilla* (Linnaeus, 1758), across its geographic range. *J. Appl. Ichthyol.* **2015**, *31*, 427–430. [CrossRef]
79. Rossi, R.; Colombo, G. Sex Ratio, Age and Growth of Silver Eels, in Two Brackish Lagoons in the Northern Adriatic (Valli of Comacchio and Valle Nuova). *Arch. Oceanogr. Limnol.* **1976**, *18*, 227–310.
80. Gordo, L.S.; Jorge, M.I. Age and growth of the european eel, *Anguilla anguilla* (Linnaeus, 1758) in the Aveiro Lagoon, Portugal. *Sci. Mar.* **1991**, *55*, 389–395.

81. Aprahamian, M.W.; Walker, A.M.; Williams, B.; Bark, A.; Knights, B. On the application of models of European eel (*Anguilla anguilla*) production and escapement to the development of Eel Management Plans: The River Severn. *ICES J. Mar. Sci.* **2007**, *64*, 1472–1482. [CrossRef]
82. O'Connor, W. Biology and Management of European Eel (*Anguilla anguilla*, L.) in the Shannon Estuary, Ireland. Ph.D. Thesis, National University of Ireland, Galway, Ireland, 2003.
83. Mann, R.H.K.; Blackburn, J.H. The biology of the eel *Anguilla anguilla* (L.) in an English chalk stream and interactions with juvenile trout *Salmo trutta* L. and salmon *Salmo salar* L. *Hydrobiologia* **1991**, *218*, 65–76. [CrossRef]
84. Rasmussen, G.; Therkildsen, B. Food, Growth and Production of *Anguilla anguilla* in a Small Danish Stream. Rapp. Procès-Verbaux Réun. *Cons. Int. Pour Explor. Mer.* **1979**, *174*, 32–40.
85. Moriarty, C. Age determination and growth rate of eels, *Anguilla anguilla* (L). *J. Fish Biol.* **1983**, *23*, 257–264. [CrossRef]
86. Vøllestad, L.A.; Jonsson, B. Life-History Characteristics of the European Eel *Anguilla anguilla* in the Imsa River, Norway. *Trans. Am. Fish. Soc.* **1986**, *115*, 864–871. [CrossRef]
87. Daverat, F.; Tomás, J. Tactics and demographic attributes in the European eel *Anguilla anguilla* in the Gironde watershed, SW France. *Mar. Ecol. Prog. Ser.* **2006**, *307*, 247–257. [CrossRef]
88. Simon, J.; Ubl, C.; Dorow, M. Growth of European eel Anguilla anguilla along the southern Baltic coast of Germany and implication for the eel management. *Environ. Biol. Fishes* **2013**, *96*, 1073–1086. [CrossRef]
89. Patey, G.; Couillard, C.M.; Drouineau, H.; Verreault, G.; Pierron, F.; Lambert, P.; Baudrimont, M.; Couture, P. Early back-calculated size-at-age of Atlantic yellow eels sampled along ecological gradients in the Gironde and St. Lawrence hydrographical systems. *Can. J. Fish. Aquat. Sci.* **2018**, *75*, 1270–1279. [CrossRef]
90. Campana, S.E.; Neilson, J.D. Microstructure of Fish Otoliths. *Can. J. Fish. Aquat. Sci.* **1985**, *42*, 1014–1032. [CrossRef]
91. Wilson, J.R.R. Depth-related changes in sagitta morphology in six macrourid fishes of the Pacific and Atlantic Oceans. *Copeia* **1985**, *4*, 1011–1017. [CrossRef]
92. Aguirre, H.; Lombarte, A. Ecomorphological comparisons of sagittae in *Mullus barbatus* and *M. surmuletus*. *J. Fish Biol.* **1999**, *55*, 105–114. [CrossRef]
93. Morales-Nin, B. Influence of environmental factors on microstructure of otoliths of three demersal fish species caught off Namibia. *S. Afr. J. Mar. Sci.* **1987**, *5*, 255–262. [CrossRef]
94. Gonzalez-Salas, C.; Lenfant, P. Interannual variability and intraannual stability of the otolith shape in European anchovy *Engraulis encrasicolus* (L.) in the Bay of Biscay. *J. Fish Biol.* **2007**, *70*, 35–49. [CrossRef]
95. Schiavina, M.; Bevacqua, D.; Melià, P.; Crivelli, A.J.; Gatto, M.; De Leo, G.A. A user-friendly tool to assess management plans for European eel fishery and conservation. *Environ. Model. Softw.* **2015**, *64*, 9–17. [CrossRef]
96. Bevacqua, D.; Melià, P.; Schiavina, M.; Crivelli, A.J.; De Leo, G.A.; Gatto, M. A demographic model for the conservation and management of the European eel: An application to a Mediterranean coastal lagoon. *ICES J. Mar. Sci.* **2019**, *76*, 2164–2178. [CrossRef]
97. Whigham, D.F.; Baldwin, A.H.; Barendregt, A. Chapter 18—Tidal Freshwater Wetlands. In *Coastal Wetlands*, 2nd ed.; Perillo, G.M.E., Wolanski, E., Cahoon, D.R., Hopkinson, C.S., Eds.; Elsevier: Amsterdam, The Netherlands, 2019; pp. 619–640. [CrossRef]
98. Podda, C.; Sabatini, A.; Palmas, F.; Pusceddu, A. Hard times for catadromous fish: The case of the European eel (*Anguilla anguilla*, L. 1758). *Adv. Oceanogr. Limnol.* **2021**, *12*, 9997. [CrossRef]
99. Porceddu, R.; Podda, C.; Mulas, G.; Palmas, F.; Picci, L.; Scano, C.; Spiga, S.; Sabatini, A. Changes in Dendritic Spine Morphology and Density of Granule Cells in the Olfactory Bulb of *Anguilla anguilla* (L., 1758): A Possible Way to Understand Orientation and Migratory Behavior. *Biology* **2022**, *11*, 1244. [CrossRef] [PubMed]

fishes

MDPI

Article

Estuarine-Specific Migration of Glass Eels in the Ems Estuary

Jeroen B. J. Huisman [1,2,*], Henry J. Kuipers [2], Leopold A. J. Nagelkerke [1], Peter Paul Schollema [3] and Inge van der Knaap [2]

1 Aquaculture & Fisheries Group, Wageningen University & Research, 6700 AH Wageningen, The Netherlands; leo.nagelkerke@wur.nl
2 Coastal and Marine Systems Group, Van Hall Larenstein University of Applied Sciences, 8901 BV Leeuwarden, The Netherlands; henry.kuipers@hvhl.nl (H.J.K.); inge.vanderknaap@hvhl.nl (I.v.d.K.)
3 Water Authority Hunze en Aa's, 9640 AD Veendam, The Netherlands; p.schollema@hunzeenaas.nl
* Correspondence: jeroen.huisman@hvhl.nl

Abstract: Understanding recruitment of glass eels in estuaries is crucial for the conservation of the European eel (*Anguilla anguilla*). However, basic knowledge on estuarine-specific glass eel migration, including in estuarine harbours, is mostly lacking. Therefore, we studied glass eel migration in the Dutch–German Ems estuary and the harbour at Delfzijl (The Netherlands) and tagged glass eels with Visual Implant Elastomer tags (VIE tags). We released 2000 tagged glass eels into the Ems estuary itself and 1000 tagged glass eels into the tidal harbour at Delfzijl. At three estuarine locations, i.e., Delfzijl–Duurswold, Termunterzijl, and Nieuwe Statenzijl, glass eel collectors were strategically placed, each location being progressively situated further upstream in the Ems estuary. Most glass eels ($n_{untagged}$ = 97,089, n_{tagged} = 74) were caught at Nieuwe Statenzijl, although this location is much further upstream. Lower numbers of glass eels ($n_{untagged}$ = 1856, n_{tagged} = 31) were caught at Delfzijl–Duurswold and Termunterzijl ($n_{untagged}$ = 1192, n_{tagged} = 7). Glass eels arrived approximately a week earlier at Nieuwe Statenzijl than at the other two locations, and the migration speed of tagged glass eels was highest at Nieuwe Statenzijl (>2 km/day) and lower (<1 km/day) at Delfzijl–Duurswold. Our study highlights that migration and the resulting potential recruitment of glass eels in estuaries and harbours may vary considerably both spatially and temporally. Further research on estuarine-specific factors that influence glass eel migration, such as the (anthropogenically altered) tidal action and flow, will provide valuable information on what influences glass eel migration in estuaries.

Keywords: *Anguilla anguilla*; glass eels; VIE tag; spatial and temporal distribution; tidal migration; estuary; harbour

Key Contribution: Specific characteristics of estuaries may guide glass eel migration towards freshwater entry points. The resulting spatially and temporally variable migration of glass eels can have catchment-wide implications on recruitment. The assessment of factors influencing glass eel migration in estuaries, including anthropogenic impacts, is instrumental for effective prioritization of mitigating measures such as fish passes.

Citation: Huisman, J.B.J.; Kuipers, H.J.; Nagelkerke, L.A.J.; Schollema, P.P.; van der Knaap, I. Estuarine-Specific Migration of Glass Eels in the Ems Estuary. *Fishes* **2023**, *8*, 392. https://doi.org/10.3390/fishes8080392

Academic Editors: Jose Martin Pujolar and Audrey M. Darnaude

Received: 31 May 2023
Revised: 5 July 2023
Accepted: 18 July 2023
Published: 27 July 2023

1. Introduction

Estuaries around the world are affected by a myriad of anthropogenic impacts [1]. In the Wadden Sea, located in the south-eastern part of the North Sea, this has led to reduced numbers and even the extinction of estuarine and diadromous species [2]. The once common European eel (*Anguilla anguilla*) now has an endangered status [3,4]. The eel is a catadromous species and depends on the inland migration of juveniles (glass eels). Estuarine migration is an important bottleneck for glass eel recruitment [5] and thus there is an urgent need to understand the mechanisms driving glass eel migration in estuaries and the resulting glass eel recruitment in tributaries [6]. Additionally, the recruitment indices of glass eels are of major importance in stock assessment [7].

Drouineau et al. [7] modelled glass eel recruitment and assumed that recruitment is a function of the surface area, also raising the question of how glass eels are distributed in estuaries. Some authors report that currents from freshwater entry points and tides are both important directional orientation cues in estuaries [8]. However, Able et al. [9] ascribes the variability in catches of American glass eel (*A. rostrata*) at several tributaries to the difference in the volume of water flowing through both inlets of Barnegat Bay estuary. Glass eel recruitment at freshwater entry points in an estuary could therefore simply be a function of distance, i.e., tributaries furthest from the entrance of the estuary receive fewer glass eels.

Glass eels are poor swimmers and the utilisation of tidal flows is recognized as an important factor guiding glass eel migration [10,11]. Glass eels move up in the water column during flood tides [8] but the precise timing is unknown. Other fish species, such as flounder larvae (*Platichthys flesus*), are present in higher densities just after slack tides and before the middle of flood tides [12], indicating partial use of a single flood tide, and this behaviour may apply to glass eels as well. The available flood tidal flow may therefore function as a conveyer belt for upstream migrating glass eels [6]. In utilising these tidal currents, glass eels are aided by rheotactic orientation and magnetic imprinting to ascertain an upstream direction of migration [13]. Beaulaton et al. [14] demonstrated that glass eels may partially use available flood tides to migrate upstream in the Gironde estuary. The partial use of available tidal flows by glass eels could lead to faster or slower progression depending on estuary-specific tidal flow characteristics. The tidal flow, tidal amplitude, and duration of the different stages of the tide in an estuary are determined by estuary-specific characteristics, such as size, shape, and depth. Indeed, even seemingly similar estuaries like the Elbe, Weser and Jade, and Ems highly differ with regard to flood tide, salinity, and sea level [15].

Like many estuaries around the world, the Ems estuary is regularly dredged. As a result, maximal flood tidal flow velocities in the Ems are now reached just 1 h after slack tides [16], and may exceed 2 m/s [17]. Anthropogenically altered tidal flows may affect glass eel migration (and tidal migrants in general) in estuaries, as they use tidal transport to move upstream in the estuary. Estuarine-specific (high) tidal flows may inhibit the migration of glass eels (or other tidal migrants) towards freshwater entry points that are located perpendicular to the main tidal flow and thus may favour glass eel migration towards more upstream locations. We therefore expected higher glass eel numbers at the end of the estuary. Activities within tidal harbours, such as dredging, may further exacerbate migratory problems for glass eels. Increased silt loads due to dredging could be one of the contributing factors in the mortality of glass eels [18] or provoke avoidance behaviour. However, Pavlov et al. [19] did not find any avoidance reaction to high turbidity in elvers of two eel species.

To understand the distribution and recruitment of glass eels in the Dutch Ems estuary, we performed a mark–recapture experiment in which we tagged glass eels with Visible Implant Elastomer tags (from here on VIE tags). We aimed to determine migration patterns and the timing of the arrival of glass eels at three freshwater entry points in the Dutch Ems estuary: Delfzijl, Termunterzijl, and Nieuwe Statenzijl. The industrial harbour of Delfzijl and recreational harbour of Termunterzijl are situated perpendicular to the main tidal flow and Nieuwe Statenzijl is located at the end of the Ems estuary. We also released VIE-marked glass eels within the Delfzijl harbour channel to determine the migration speed and arrival of glass eels in Delfzijl harbour.

2. Materials and Methods

2.1. Study Area

The Dutch–German Ems estuary (Figure 1a) is used for industrial activities and includes the port of Delfzijl. Dredging and construction measures in harbours greatly altered the physical processes in the Ems river, including an increased tidal range [20]. The tidal amplitude in the Ems estuary is ca. 3 m [21]. The Ems–Dollard is part of the Ems estuary and is designated under the Habitats Directive [22] and is separated from the tidal Ems river by the Geisseleit dam (Figure 1b). There are four freshwater entry points in the industrial harbour of Delfzijl; in this study we focussed on one, the pumping station and sluice complex Delfzijl–Duurswold. The tidal amplitude in the harbour of Delfzijl is also ca. 3 m [23]. Termunterzijl is a small recreational harbour and is situated a few kilometres south of the harbour of Delfzijl (Figure 1b). The opening of both harbours is situated perpendicular to the main tidal flow in the Ems estuary (Figure 1b). The discharge sluice of Nieuwe Statenzijl is located at the end of the Ems–Dollard basin (Figure 1a). The catchments sizes of the freshwater entry points are Nieuwe Statenzijl (89,000 ha), Termunterzijl (18,808 ha), and Delfzijl–Duurswold (21,524 ha). Glass eel fisheries are not allowed in the Ems estuary.

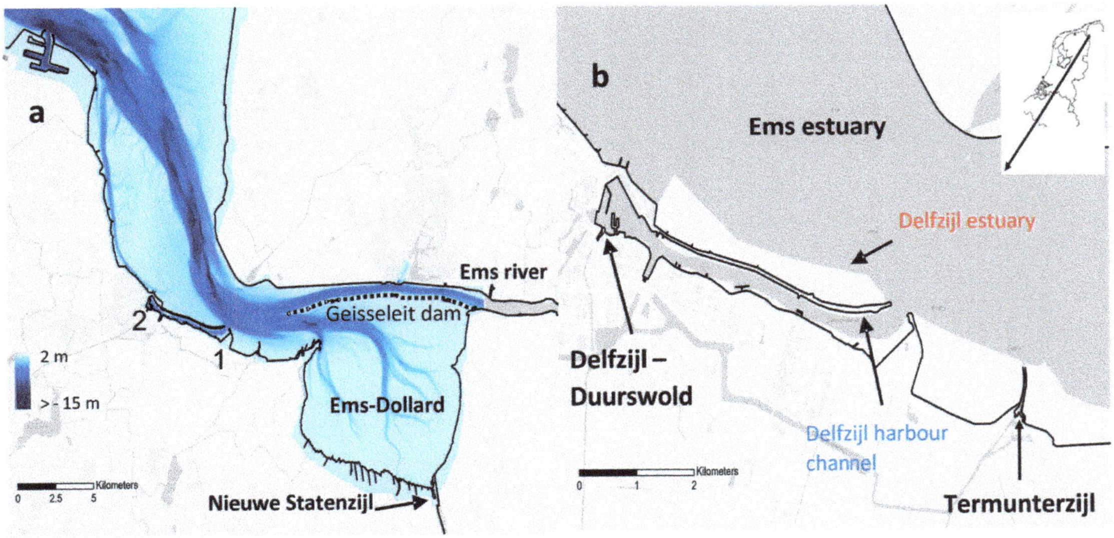

Figure 1. (**a**) Map of the Ems estuary, including tidal gullies, depth (m), and sampling locations. The sampling locations at Termunterzijl (1) and Delfzijl–Duurswold (2) are indicated with numbers. Inlay represents the Netherlands and the general location of the Ems estuary. (**b**) Detailed map of the harbours of Delfzijl and Termunterzijl. Glass eel release sites "Delfzijl estuary" and "Delfzijl harbour channel" are indicated.

2.2. Mark–Recapture of Glass Eels

To assess migration of glass eels towards the freshwater entry points at Delfzijl–Duurswold, Termunterzijl, and Nieuwe Statenzijl (Figure 1a,b), we designed a mark–recapture experiment using VIE tags. VIE tags have been used to mark small fish [24,25]. VIE tags do not affect glass eel survival [26], supporting their suitability for mark–recapture experiments on glass eels in estuaries.

We VIE-tagged a total of 3000 glass eels. Tagged glass eels were caught (between 22nd and 25 April 2022) at the fish pass adjacent to the pumping station of Roptazijl. Roptazijl is located ~100 km west of the Ems estuary (Friesland, The Netherlands). Glass eels were caught on the sea side using a lift net (1 m^2, mesh size 1 mm) that was lowered in front of the inlet of the fish passage and lifted every 10 min for a 6 h period from low to high tide. Caught glass eels were temporally held inside the pumping station in large tubs (65 L) with aerated salt water which was temperature controlled at 5–8 °C for a maximum of 3 days. The glass eels were then transported to the research laboratory at the Van Hall Larenstein

Applied Sciences University, Leeuwarden. On the 25 April 2022, all 3000 glass eels were tagged (Figure 2).

(a) (b)

Figure 2. Glass eel marked with a red VIE tag (**a**) and a glass eel marked with a blue VIE tag (**b**).

VIE tags were inserted on the dorsal side of the glass eel at approximately half of the body length (Figure 2), using a syringe with 0.3 mm needle. Glass eels were anaesthetised using 0.4 mL/L 2-phenoxyethanol and individually tagged with a red or blue VIE tag. After marking, the sedated glass eels were placed in a recovery tub and, after recovery, in large, aerated holding tanks with salt water and a water temperature of 5–8 °C.

On 26 April, the glass eels were transported in aerated tanks with salt water and released at two release sites. To study glass eel migration in the Ems estuary, we released 2000 red-VIE-tagged eels at "Delfzijl estuary" (53.321° N, 6.996° E) (Figure 1b). To study glass eel migration in industrial harbours, we released a batch of 1000 blue-VIE-tagged eels at "Delfzijl harbour channel" (53.315° N, 6.994° E) (Figure 1b) on the inside of the harbour of Delfzijl. The number of VIE-tagged glass eels in the harbour was smaller, as the distance between the release site and the sampling site is smaller (ca. 5 km), and hence the probability of recapture is higher. The two release sites are henceforth named 'Delfzijl estuary' and 'Delfzijl harbour channel'. Glass eels prefer the edges of estuaries [27,28] and were therefore released in the nearshore zone in ca. 0.5 m deep water. In addition, glass eels were released just after low tide and 4 days before spring tide, adhering to natural migration processes [29].

To study recruitment of glass eels and recapture VIE-tagged glass eels, we used eel collectors adapted from an earlier design of Bult et al. [30]. We placed collectors at two locations situated perpendicular to main tidal flow: Delfzijl–Duurswold and Termunterzijl. We placed a glass eel collector in front of the pumping station–sluice complex at Delfzijl–Duurswold, which is located in the industrial harbour of Delfzijl. For practical reasons, it was not possible to place Delfzijl–Duurswold collectors at the other freshwater entry points in the harbour of Delfzijl. We also placed a glass eel collector in the small recreational harbour of Termunterzijl. The glass eel collector was placed adjacent to the pumping station near the opening of the harbour. To determine glass eel migration at the end of the estuary, a collector was placed at the discharge sluice at Nieuwe Statenzijl (Figure 1b).

The eel collector consisted of a 0.6 m × 0.6 m × 0.6 m holding tank in a floating container. A funnel with a 1 × 1 mm mesh size allowed glass eels to enter the collector. The opening (0.1 m diameter) of the collector was ~0.2 m below the surface of the water and freshwater was pumped from the polder side over the dike into the collector, creating a flow of freshwater on the estuary side of ~4 L/s. For practical reasons and because glass eels prefer the sides of water courses [28], the collectors were placed near walls in the vicinity of the discharge points. We used UV light torches to ensure detection of VIE-tagged glass eels [31].

2.3. Data Collection

The collectors were emptied daily at high tide for a total of 25 days between 26 April and 20 May 2022. We recorded daily catches of untagged and VIE-tagged glass eels at Delfzijl–Duurswold, Termunterzijl, and Nieuwe Statenzijl. We also collected data on environmental variables during our data collection period. The average daily water temperature (in °C) was calculated for all three recapture locations using data from the closest measurement buoy in the estuary, 'Randzelgat Noord boei' (waterinfo.rws.nl). The tidal difference was defined as the difference between the daily maximum and minimum of the tide per location as a measure for tidal flow, i.e., the higher the difference, the higher the tidal flow will be. Water level (ca. 15 min) and daily discharge (m^3/day) data of all discharge points were obtained from the Waterschap Hunze en Aa's. The moon phase was included as the daily percentage of moon fullness, in which 0% is a new moon and 100% is a full moon, and was calculated from r-package lunar v0.2-1 [32].

2.4. Data Analysis

2.4.1. Analyses of Total Glass Eels Caught

We used a negative binomial generalized additive model (NB GAM) with a log-link function in R (mgcv version 1.8-35) to test the daily total catches of glass eels at Delfzijl–Duurswold and Nieuwe Statenzijl as a function of the fixed covariates: catch location, daily tidal difference, maximal daily difference between maximum and minimum sea water level (m), day of the year, mean daily water surface temperature (°C), discharge (m^3/day), and moon phase (%). An NB GAM was chosen because of the nonlinear relationship between day of the year and daily catches at Delfzijl–Duurswold and Nieuwe Statenzijl [33]. The log link function in the model ensures positive fitted values. A negative binomial distribution is typically used for over-dispersed count data, which we found when comparing the variance of the catch data to the mean. Visual inspection of the moon phase in relation to the day of the year showed a high correlation ($r = 0.92$, $p < 0.001$), and therefore the moon phase was not included in the model. The initial NB GAM included all main covariates plus a smoother for day of the year for both catch sites, the interaction between daily tidal difference and catch location, and the interaction between daily water temperature and catch location. We kept our focus covariates, catch location and tidal difference, in our final model, and used backwards model selection to select the model that best fitted the data. A control covariate was only included in the model when including it substantially improved the Akaike information criterion (AIC) (i.e., the AIC was reduced by more than 2 units) [34]. Temporal autocorrelation between subsequent days within a catch location did not need to be taken into account based on the low autocorrelation and partial autocorrelation values in the daily catch of glass eels. As a result of the very low numbers of glass eels that were caught at Termunterzijl, this location was not used in the analyses.

2.4.2. Recapture Probability

We used a binomial GLM (r-package stats version 4.1.0) with a logit link function to assess the effect of release location (i.e., Delfzijl estuary or Delfzijl harbour channel) on the recapture probability of the 3000 VIE-tagged glass eels. Per recapture location (i.e., Delfzijl–Duurswold and Nieuwe Statenzijl), one model was made. Binomial GLMs were used because of the binary response variable (1 = recaptured, 0 = not recaptured).

2.4.3. Migration Speed

The migration speed was defined as the time in days between release and recapture location of the VIE-tagged glass eels (105 in total) that were caught in relation to the distance between the release site and sampling location. We used a negative binomial GLM (r package glmmTMB version 1.1.4) [35] with a log link function to assess the effect of release location (i.e., Delfzijl estuary or Delfzijl harbour channel) and recapture location (i.e., Delfzijl–Duurswold and Nieuwe Statenzijl) on the glass eel migration speed (km/day). A negative binomial distribution was used to account for the overdispersion in the migration speed count data, which we found when comparing the variance to the mean. To account for the difference in distance between the two release locations and recapture locations, we included the natural logarithmic of the difference in distance between release and recapture locations (km) as an offset variable. Our initial migration speed NB GLM included covariates: release location (i.e., Delfzijl estuary or Delfzijl harbour channel), recapture location (i.e., Delfzijl–Duurswold and Nieuwe Statenzijl), and the interaction between release and recapture locations. We used backwards model selection, dropping covariates, to find the model with the lowest AIC. Using the r-package emmeans (version 1.7.1-1) [36], we performed a pairwise comparison for the combination of the interaction between release and recapture location.

For all final models, we performed residual diagnostics using the DHARMa R package version 0.4.5 [37]. Plots of the residuals versus the continuous covariates of the analysis of the total catches showed nonlinear patterns, which is why we included a smoother for the day of the year for both collector locations. The residual plots of the recapture probability and migration speed analysis did not show a clear nonlinear pattern, and therefore it was not necessary to model nonlinear patterns.

2.5. Animal Welfare

This research was conducted in accordance with the animal welfare protocol number VD7110020197584.

3. Results

3.1. Spatial and Temporal Arrival of Glass Eels at Freshwater Entry Points in the Estuary

During our research period (26 April 2022–20 May 2022), high numbers of glass eels were caught at Nieuwe Statenzijl (n = 97.089), the furthest location from both release sites. Low numbers of glass eels were caught at Delfzijl–Duurswold (n = 1.856) and Termunterzijl (n = 1.192) (Table 1). The peaks of tagged glass eels mimicked the peaks of the arrival of untagged glass eels (Figure 3). The total recapture rate of VIE-tagged glass eels released in Delfzijl harbour and Delfzijl estuary was 3.5% and 3.9%, respectively.

Table 1. Number (n) of glass eel catches (VIE-tagged and non-VIE-tagged eels) caught at Delfzijl–Duurswold, Termunterzijl, and Nieuwe Statenzijl.

Location	Total Amount of Glass Eels Caught (n)	Total Catch of VIE-Tagged Eels Released at Delfzijl Estuary (n), and Percentage of Recapture (%)	Total Catch of VIE-Tagged Eels Released at Delfzijl Harbour Channel (n), and Percentage of Recapture (%)
Delfzijl–Duurswold	1856	21 (1.1%)	10 (1.0%)
Termunterzijl	1192	6 (0.3%)	1 (0.1%)
Nieuwe Statenzijl	97,089	50 (2.5%)	24 (2.4%)

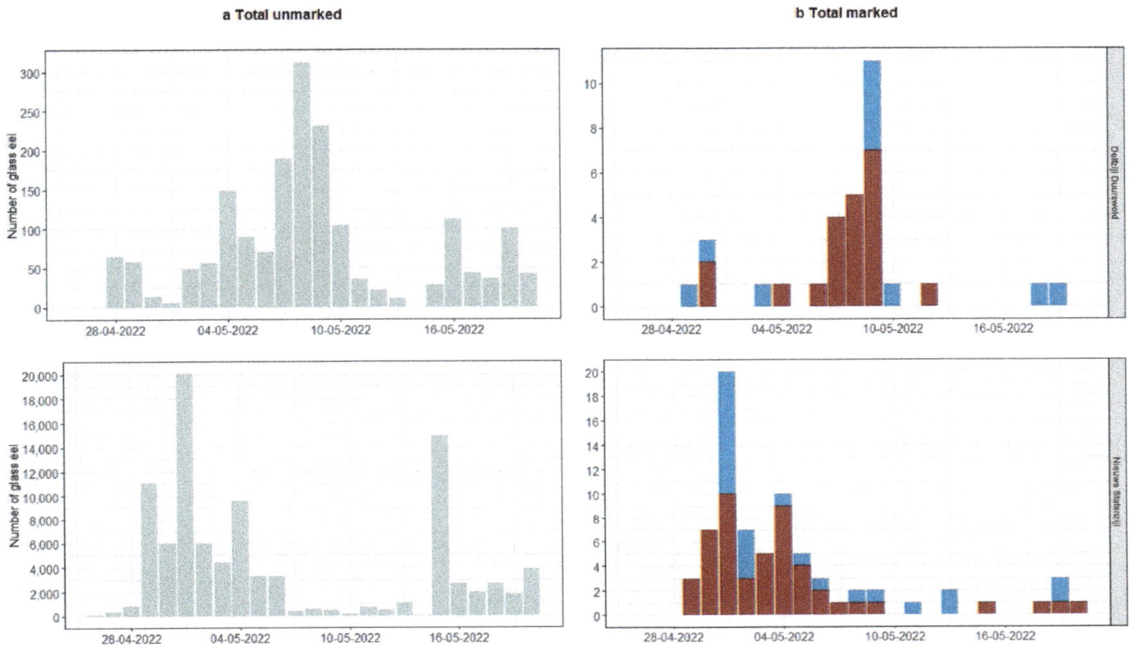

Figure 3. (**a**) Daily total numbers of glass eels that were caught at Delfzijl–Duurswold and Nieuwe Statenzijl and (**b**) daily number of VIE-tagged glass eels that were caught at Delfzijl–Duurswold and Nieuwe Statenzijl. Red bars indicate the number of VIE-tagged glass eels released at Delfzijl estuary and blue bars indicate the VIE-tagged glass eels released at Delfzijl harbour channel.

The final NB GAM to investigate daily catches of glass eels included the recapture location (Delfzijl–Duurswold and Nieuwe Statenzijl), the daily tidal difference per recapture location, and two smoothers for the day of the year by location (Table 2). We found that the daily catches of glass eels were significantly different ($p < 0.001$) between Nieuwe Statenzijl and Delfzijl–Duurswold. At Nieuwe Statenzijl, the catch size was 32 (95% CI [18.31–56.02]) times higher than at Delfzijl–Duurswold. We found no significant influence of the tidal difference on the daily catch size ($p = 0.50$).

Table 2. The table shows the parameters included in the final negative binomial GAM modelling the total daily glass eel catch. Parameters included are catch location (Nieuwe Statenzijl = intercept), daily tidal difference (m), and two smoothers for day of the year at Delfzijl–Duurswold and Nieuwe Statenzijl. The table shows the estimate of the regression coefficient (β) and its standard error (SE), the effective degrees of freedom (Edf), the test statistic (z-value), degrees of freedom (df), exponentiated value of the regression coefficient (Exp(β)), the 95% confidence interval for Exp(β), and *p*-values.

	β ± SE	z-Value	df	Exp(β)	95% IC for Exp(β)	*p*-Value
Intercept	2.654 ± 2.189	1.213		14.22	[0.20, 1037.34]	0.225
Tidal difference (m)	0.004 ± 0.006	0.666	1	1.00	[0.99, 1.02]	0.505
Catch location Delfzijl–Duurswold [a]	3.466 ± 0.285	12.149	1	32.02	[18.31, 56.02]	<0.001
S (day of the year, Delfzijl–Duurswold)	Edf = 2.758	15.31	2.76			0.003
S (day of the year, Nieuwe Statenzijl)	Edf = 5.501	34.34	5.5			<0.001

[a] Reference: Nieuwe Statenzijl.

At Delfzijl–Duurswold, most glass eels were caught on 8 May and at Nieuwe Statenzijl on the 1 May (Figure 3a). This earlier peak in the number of caught glass eels at

Nieuwe Statenzijl was reflected by the modelled predicted catch sizes per day and location (Figure S1). At both locations, the total daily catch of glass eels depended on the day of the year ($p = 0.003$ and $p < 0.001$, respectively) (Figure S1).

3.2. Recapture of VIE-Tagged Glass Eels

Of the 2000 VIE-tagged glass eels released in the estuary, 21 migrated to Delfzijl–Duurswold, 6 were caught at Termunterzijl, and 50 migrated towards Nieuwe Statenzijl (Table 1). Of the 1000 VIE-tagged glass eels released in Delfzijl harbour, 10 were caught at Delfzijl–Duurswold, 1 was caught at Termunterzijl, and 24 were caught at Nieuwe Statenzijl (Table 1). The recapture probability of catching VIE-tagged glass eels from either of the two release sites did not differ significantly for location: Delfzijl–Duurswold (Wald $X^2 = 0.016$, df = 1, $p = 0.90$) and Nieuwe Statenzijl (Wald $X^2 = 0.027$, df = 1, $p = 0.87$) (Figure S2).

3.3. Migration Speed

The final NB GLM to investigate the migration speed included release location with an interaction with recapture location (Table 3). The release site and recapture location had an interactive effect on the migration speed ($p = 0.013$). The average migration speed of the glass eels caught at Delfzijl–Duurswold differed significantly between the two release sites ($p = 0.003$, Figure 4); this was not the case for Nieuwe Statenzijl ($p = 0.83$, Figure 4). Overall, the migration speed of glass eels caught at Nieuwe Statenzijl was higher than glass eels caught at Delfzijl–Duurswold ($p < 0.001$, Table 3, Figure 4).

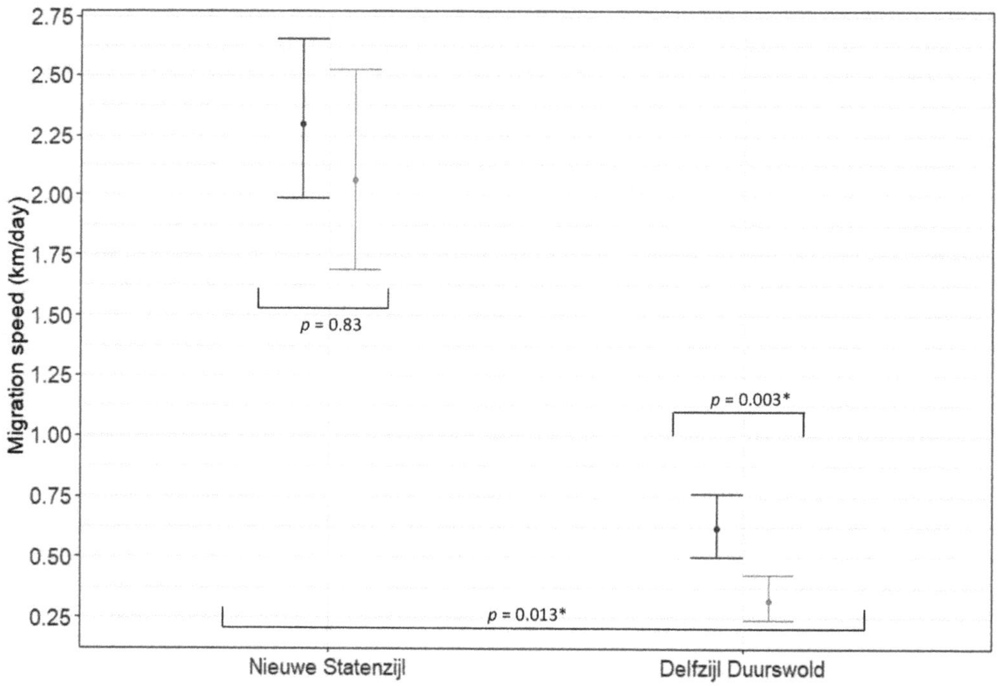

Figure 4. Modelled mean migration speed (in km/day) between release sites and recapture sites of VIE-tagged glass eels released at Delfzijl estuary (red) and those released at Delfzijl harbour channel (blue), including confidence intervals, asterisk indicates statistical significance.

Table 3. The parameters included in the final negative binomial GLM modelling migration speed. Model includes an offset variable of the natural logarithmic of the difference in distance between release and recapture location (km). Parameters included are release location (Delfzijl estuary = intercept), recapture location, and the interaction between release and recapture location. The table shows the estimate of the regression coefficient (β) and its standard error (SE), the test statistic (z-value), degrees of freedom (df), exponentiated value of the regression coefficient (Exp(β)), the 95% confidence interval for Exp(β), and *p*-value.

	β ± SE	z-Value	df	Exp(β)	95% IC for Exp(β)	*p*-Value
Intercept	1.131 ± 0.145	7.779		3.10	[2.33, 4.12]	<0.001
Release location (VIE colour) [a]	−0.654 ± 0.180	−3.634	1	0.52	[0.37, 0.74]	<0.001
Recapture location [b]	−1.855 ± 0.178	−10.406	1	0.16	[0.11, 0.22]	<0.001
Release * Recapture location	0.547 ± 0.220	2.483	1	1.73	[1.12, 2.66]	0.013

[a] Reference: estuary (VIE red). [b] reference: Nieuwe Statenzijl.

4. Discussion

4.1. Spatial Variation of Glass Eels

This study shows that recruitment of glass eel at tributaries in an estuary is not necessarily a function of the distance from the opening of the estuary. On the contrary, our results show that the highest number of glass eels (tagged and untagged) was found at Nieuwe Statenzijl, the tributary furthest from the opening of the estuary. Low catches of glass eels at Termunterzijl and Delfzijl may indicate that anthropogenically altered tidal flows hinder glass eel migration towards locations that are situated perpendicular to the main tidal flow. The glass eel migration speed was the highest at the furthest location, i.e., Nieuwe Statenzijl. We will discuss glass eel catches and migration speed in the light of estuarine-specific tidal flows and anthropogenic impacts in the estuary and harbour of Delfzijl.

Catches of untagged glass eels and tagged glass eels show similar spatial and temporal patterns (Figure 3) per location, resembling the migration patterns of all glass eels in the Ems. In addition, the recapture rate for VIE-tagged glass eels for both release sites was 3.5% and 3.9%, which is similar to other studies. Using dyes, Aprahamian and Wood [38] reported recapture rates of 1.3–4.4% in the Severn estuary and Briand et al. [39] reported recapture rates of 0.4% to 3.7% at several locations in the Vilaine estuary.

The pronounced difference in catches of glass eels between sampled locations was unexpected. Tidal flow is a well-known factor in estuaries with regard to glass eel migration [11,40]. However, most glass eels do not utilize all available flood tidal flows [14], but may skip a tide or partially use a single flood tide. In addition, Gascuel [41] found that only 10% of glass eels use the full flood tides to migrate upstream in a small French estuary. As such, a prime explanation for the pronounced difference in glass eel catches could be the estuarine-specific (tidal) flow conditions. Indeed, the Ems estuary can be classified as a flood-dominated system [42]. The flood tidal flow curve has been altered by human activity, resulting in unnaturally fast flood tidal flows shortly after ebb tides [19]. This may especially affect glass eels, as they move up in the water column after slack tides to utilize flood tidal flows.

Can the tidal flow characteristics of the Ems estuary fully explain the low catches of tagged and untagged glass eels at Delfzijl? Fitri [43] has shown that the angle of approach contributes to nearshore hydro-morphodynamical responses to coastal structures in sediment dynamics, and this could therefore impact tidal migrants such as glass eels as well. As a result of breakwater dams [44], the entrances of the harbours at Delfzijl and Termunterzijl are located perpendicular to the main tidal flow. Therefore, high flood tidal flows (>1 m/s) in the Ems estuary [16] could mean that glass eels are mechanistically hindered in their migration towards both harbours, i.e., washed past the harbour entrances. This notion is supported by the fact that glass eels are considered to be weak swimmers that can sustain a swimming speed of 15 cm/s for 60 min [45].

In addition, two-thirds of the flood tidal flow in the Ems estuary is pushed in to the large Ems–Dollard basin [46], probably due to the shape of the estuary and the Geiseleitdam (Figure 1b). Able et al. [9] suggested that the volume of water exchanged through two inlets between the sea and an estuary could explain spatial variations in glass eel recruitment at tributaries. As such, glass eels may utilize the anthropogenic (high) flood tidal flow and are subsequently pushed by the volume of the (tidal) flow towards Nieuwe Statenzijl.

Some authors point to the discharge of tributaries as directional factors influencing glass eel migration [8,47]. The discharge of freshwater into the harbour of Delfzijl is intermittent and occasionally high (Figure S3), and this may influence [48] the attractiveness of the freshwater entry points for glass eels. However, lower numbers of glass eels were caught in Termunterzijl even though discharge at this location was more continuous (Figure S3). Some models assume that the glass eel distribution is a function of catchment size [7], as catchment size affects discharge. The catchment size of Nieuwe Statenzijl is ca. 4 times larger than the other two catchments. However, the number of glass eels caught at Nieuwe Statenzijl was ca. 32 times higher than at Delfzijl–Duurswold. Furthermore, Nieuwe Statenzijl is located further upstream, and its discharge will be diluted in the large Ems–Dollard basin. Therefore, we believe that the catchment size and/or discharge cannot explain the large difference in glass eel catches at freshwater entry points in the Ems estuary. The sampling locations at Termunterzijl and Delfzijl are in the vicinity of extensively used shipping locks. However, shipping locks are ineffective as a primary fish migration solution [49] and this could be due to the intermittent flow coming from sluicing events. Indeed, glass eels may be less attracted to pumping stations with a low frequency of pumping activity [48].

Some of the glass eels released in the harbour of Delfzijl migrated out into the estuary, pointing towards more local disturbances in the harbour itself. Verwilligen et al. [23] reported that daily dredging of the harbour entrance results in a high sediment load and that the bottom of the harbour channel consists of fluid mud. However, other anthropogenic disturbances such as industrial wastewater discharges or other harbour-related activities cannot be excluded. Importantly, tidal harbours are often situated in a pivotal position in the connection between the estuarine and/or marine environment and the rivers in the hinterland. Further research is needed to determine the possible effects of harbour-related activities on glass eel migration and fish migration in general.

4.2. Temporal Distribution and Migration Speed of Glass Eels

Although Nieuwe Statenzijl is ca. 12 km further from the release sites than Delfzijl, the peak of glass eels reached Nieuwe Statenzijl ca. 8 days (Figure 3) earlier. Glass eels caught at Nieuwe Statenzijl swam with an average speed of 2.08 km/day (release site estuary) and 2.27 km/day (release site harbour channel Delfzijl). This is slightly lower but comparable to reported migration speeds in the Gironde of 3–4 km/day [14] and 3–8 km/day for *A. rostrata* [50]. The maximum daily tidal change influenced (although not significantly) the daily catches at Nieuwe Statenzijl, underpinning the concept that glass eels utilize tidal flows to move to tidal barriers. This notion is further supported by the double peak of glass eel catches at Nieuwe Statenzijl, which seems to adhere to daily tidal action (S4). In contrast, glass eel arrivals at Delfzijl–Duurswold peak at a low daily tidal difference, indicating that other migratory mechanisms are prevailing.

Glass eels released in the estuary migrated almost twice as fast to Delfzijl–Duurswold than those released in the harbour channel. We suggest that glass eels released in the estuary awaited better migration circumstances, for example, a lower turbidity. These results strengthen the notion that local disturbances in the harbour are influencing glass eel migration.

5. Conclusions

Our results show that glass eels migrate in higher numbers and speeds to a tributary at the end of the Ems estuary than towards two other tributaries which are closer to the estuary entrance. Estuarine-specific tidal flows in relation to the angle of approach may guide glass eel migration towards freshwater entry points, especially in anthropogenically altered estuaries. In addition, the low migration speed and tagged glass eels leaving the harbour of Delfzijl point towards harbour-specific migratory problems. Water managers need to be aware that estuary-specific tidal flows may determine recruitment of glass eels (or fish in general) at freshwater entry points, and this may have catchment-wide implications. This knowledge may assist in the assessment of current and future anthropogenic impacts on fish migration in estuaries and understanding the recruitment of fish at freshwater entry points.

Supplementary Materials: The following supporting information can be downloaded at: https://www.mdpi.com/article/10.3390/fishes8080392/s1, Figure S1. Predicted daily catch over days of the year at Delfzijl–Duurswold and Nieuwe Statenzijl. Figure S2. Recapture probability of VIE tagged blue and red eel at Delfzijl (a) and Nieuwe Statenzijl (b). Figure S3. Daily discharge at Delfzijl, Nieuwe Statenzijl, total discharge at Delfzijl and Termunterzijl. Figure S4. Daily number of glass eels caught at Delfzijl (a) and Nieuwe Statenzijl (b), and daily tidal difference (cm).

Author Contributions: Conceptualization, J.B.J.H., I.v.d.K., L.A.J.N. and P.P.S.; methodology, J.B.J.H., I.v.d.K., H.J.K. and L.A.J.N.; validation, I.v.d.K, H.J.K. and J.B.J.H.; formal analysis, I.v.d.K., J.B.J.H. and H.J.K.; investigation, J.B.J.H., I.v.d.K. and P.P.S.; data curation, I.v.d.K., J.B.J.H.; writing—original draft reparation, J.B.J.H., I.v.d.K.; writing—review and editing, J.B.J.H., I.v.d.K., L.A.J.N. and P.P.S.; visualization, I.v.d.K.; supervision, J.B.J.H.; project administration, J.B.J.H.; funding acquisition, J.B.J.H.; All authors have read and agreed to the published version of the manuscript.

Funding: This research was financed by the Waddenfonds (01755849/WF-2019/200914).

Institutional Review Board Statement: This research was performed in line with the animal welfare protocol number VD7110020197584, approved by the Centrale Commissie Dierproeven (The Netherlands) on 5 July 2019.

Informed Consent Statement: Not applicable

Data Availability Statement: The data presented in this study are available on demand from the first author at: jeroen.huisman@hvhl.nl.

Acknowledgments: We thank all students that helped with the monitoring and the water authority Hunze en Aa's for assistance in the design and execution of field work. We would like to thank Joop van Eerbeek, Devika Narain, Tristan da Graca, and Jimmy van Rijn for their help with the VIE marking and field work.

Conflicts of Interest: The authors declare no conflict of interest

References

1. Dafforn, K.A.; Simpson, S.L.; Kelaher, B.P.; Clark, G.F.; Komyakova, V.; Wong, C.K.; Johnston, E.L. The challenge of choosing environmental indicators of anthropogenic impacts in estuaries. *Environ. Pollut.* **2012**, *163*, 207–217. [CrossRef] [PubMed]
2. Lotze, H.K. Radical changes in the Wadden Sea fauna and flora over the last 2000 years. *Helgol. Mar. Res.* **2005**, *59*, 71–83.
3. Van Ginneken, V.J.; Maes, G.E. The European eel (*Anguilla anguilla*, Linnaeus), its lifecycle, evolution and reproduction: A literature review. *Rev. Fish Biol. Fish.* **2005**, *15*, 367–398.
4. ICES. ICES: European eel (*Anguilla anguilla*) throughout its natural range. In *ICES Advice: Recurrent Advice*; Report; ICES: Copenhagen, Denmark, 2022. [CrossRef]
5. Feunteun, E. Management and restoration of European eel population (*Anguilla anguilla*): An impossible bargain. *Ecol. Eng.* **2002**, *18*, 575–591.
6. Harrison, A.J.; Walker, A.M.; Pinder, A.C.; Briand, C.; Aprahamian, M.W. A review of glass eel migratory behaviour, sampling techniques and abundance estimates in estuaries: Implications for assessing recruitment, local production and exploitation. *Rev. Fish Biol. Fish.* **2014**, *24*, 967–983.
7. Drouineau, H.; Briand, C.; Lambert, P.; Beaulaton, L. GEREM (Glass Eel Recruitment Estimation Model): A model to estimate glass eel recruitment at different spatial scales. *Fish Res.* **2016**, *174*, 68–80.
8. Cresci, A. A comprehensive hypothesis on the migration of European glass eels (*Anguilla anguilla*). *Biol. Rev.* **2020**, *95*, 1273–1286. [CrossRef]
9. Able, K.W.; Smith, J.M.; Caridad, J.F. American eel supply to an estuary and its tributaries: Spatial variation in Barnegat Bay, New Jersey. *Northeast. Nat.* **2015**, *22*, 53–68.

10. McCleave, J.D.; Kleckner, R.C. Selective tidal stream transport in the estuarine migration of glass eels of the American eel (*Anguilla rostrata*). *ICES J. Mar. Sci.* **1982**, *40*, 262–271. [CrossRef]
11. Trancart, T.; Lambert, P.; Rochard, E.; Daverat, F.; Coustillas, J.; Roqueplo, C. Alternative flood tide transport tactics in catadromous species: *Anguilla anguilla*, Liza ramada and *Platichthys flesus*. *Estuar. Coast. Shelf Sci.* **2012**, *99*, 191–198.
12. Jager, Z. *Processes of Tidal Transport and Accumulation of Larval Flounder (Platichthys flesus L.) in the Ems-Dollard Nursery*; Universiteit van Amsterdam: Amsterdam, The Netherlands, 1999.
13. Cresci, A.; Durif, C.M.; Paris, C.B.; Shema, S.D.; Skiftesvik, A.B.; Browman, H.I. Glass eels (*Anguilla anguilla*) imprint the magnetic direction of tidal currents from their juvenile estuaries. *Commun. Biol.* **2019**, *2*, 366.
14. Beaulaton, L.; Castelnaud, G. The efficiency of selective tidal stream transport in glass eel entering the Gironde (France). *Bull. Bulletin Français de la Pêche et de la Pisciculture* **2005**, *378–379*, 5–21. [CrossRef]
15. Stanev, E.V.; Jacob, B.; Pein, J. German Bight estuaries: An inter-comparison on the basis of numerical modeling. *Cont. Shelf Res.* **2019**, *174*, 48–65.
16. Schulz, K.; Burchard, H.; Mohrholz, V.; Holtermann, P.; Schuttelaars, H.M.; Becker, M.; Maushake, C.; Gerkema, T. Intratidal and spatial variability over a slope in the Ems estuary: Robust along-channel SPM transport versus episodic events. *Estuar. Coast. Shelf Sci.* **2020**, *243*, 106902.
17. Ysebaert, T.; van der Wal, J.T.; Tangelder, M.; de Groot, A.V.; Baptist, M.J. *Ecotopenkaart voor het Eems-Dollard Estuarium*; IMARES: Yerseke, The Netherlands, 2016.
18. Briand, C.; Sauvaget, B.; Girard, P.; Fatin, D.; Beaulaton, L. Push net fishing seems to be responsible for injuries and post fishing mortality in glass eel in the Vilaine estuary (France) in 2007. *Knowl. Manag. Aquat. Ecosyst.* **2012**, *404*, 2. [CrossRef]
19. Pavlov, D.; Nazarov, D.Y.; Zvezdin, A.; Kucheryavyi, A. Downstream migration of early larvae of the European river lamprey Lampetra fluviatilis. In *Doklady Biological Sciences*; Springer: Berlin/Heidelberg, Germany, 2014; p. 344.
20. Talke, S.A.; De Swart, H.E. *Hydrodynamics and Morphology in the Ems/Dollard Estuary: Review of Models, Measurements, Scientific Literature and the Effects of Changing Conditions*; Utrecht University Repository: Utrecht, The Netherlands, 2006.
21. Benndorf, J.; Wünsche, A.; Jürges, J.; Naulin, M. *Historic Developments of the Ems Estuary*; ECSA: Johannesburg, South Africa, 2021.
22. 92/43/EEC. Council Directive 92/43/EEC of 21 May 1992 on the conservation of natural habitats and of wild fauna and flora. *Off. J. L* **1992**, *206*, 7–50.
23. Verwilligen, J.; Vantorre, M.; Delefortrie, G.; Meinsma, J.; Van der Made, K. Manoeuvrability in proximity of nautical bottom in the harbour of Delfzijl. In Proceedings of the 33rd PIANC World Congress, San Francisco, CA, USA, 1–5 June 2014; pp. 1–18.
24. Josephson, D.C.; Robinson, J.M.; Weidel, B.C.; Kraft, C.E. Long-term retention and visibility of visible implant elastomer tags in brook trout. *N. Am. J. Fish. Manag.* **2008**, *28*, 1758–1761. [CrossRef]
25. Jungwirth, A.; Balzarini, V.; Zöttl, M.; Salzmann, A.; Taborsky, M.; Frommen, J.G. Long-term individual marking of small freshwater fish: The utility of Visual Implant Elastomer tags. *Behav. Ecol. Sociobiol.* **2019**, *73*, 49.
26. Imbert, H.; Beaulaton, L.; Rigaud, C.; Elie, P. Evaluation of visible implant elastomer as a method for tagging small European eels. *J. Fish Biol.* **2007**, *71*, 1546–1554. [CrossRef]
27. Walmsley, S.; Bremner, J.; Walker, A.; Barry, J.; Maxwell, D. Challenges to quantifying glass eel abundance from large and dynamic estuaries. *ICES J. Mar. Sci.* **2018**, *75*, 727–737. [CrossRef]
28. Piper, A.T.; Wright, R.M.; Kemp, P.S. The influence of attraction flow on upstream passage of European eel (*Anguilla anguilla*) at intertidal barriers. *Ecol. Eng.* **2012**, *44*, 329–336. [CrossRef]
29. Sheldon, M.; McCleave, J. Abundance of glass eels of the American eel, *Anguilla rostrata*, in mid-channel and near shore during estuarine migration. *Nat. Can.* **1985**, *112*, 425–430.
30. Bult, T.P.; Dekker, W. Experimental field study on the migratory behaviour of glass eels (*Anguilla anguilla*) at the interface of fresh and salt water. *ICES J. Mar. Sci.* **2007**, *64*, 1396–1401. [CrossRef]
31. FitzGerald, J.L.; Sheehan, T.F.; Kocik, J.F. Visibility of visual implant elastomer tags in Atlantic salmon reared for two years in marine net-pens. *N. Am. J. Fish. Manag.* **2004**, *24*, 222–227. [CrossRef]
32. Lazaridis, E. Lunar: Provides Functions to Calculate Lunar and Other Related Environmental Covariates. R-package 0.2-1. 2022. Available online: https://cran.r-project.org/web/packages/lunar/index.html (accessed on 30 May 2023).
33. Zuur, A.F. *A Beginner's Guide to Generalized Additive Models with R*; Highland Statistics Limited Newburgh: Newburgh, UK, 2012.
34. Burnham, K.; Anderson, D. *Model Selection and Inference: A Practical Information-Theoretic Approach*, 2nd ed.; Springer: New York, NY, USA, 2000.
35. Brooks, M.E.; Kristensen, K.; van Benthem, K.J.; Magnusson, A.; Berg, C.W.; Nielsen, A.; Skaug, H.J.; Maechler, M.; Bolker, B.M. glmmTMB Balances Speed and Flexibility among Packages for Zero-inflated Generalized Linear Mixed Modeling. *R J.* **2017**, *9*, 378–400. [CrossRef]
36. Lenth, R.V. Emmeans: Estimated Marginal Means. R-package 1.7.1-1. 2021. Available online: https://cran.r-project.org/web/packages/emmeans/index.html (accessed on 30 May 2023).
37. Hartig, F. DHARMa: Residual Diagnostics for Hierarchical (Multi-Level/Mixed) Regression Models. R Package Version 0.3. 2020. Available online: https://cran.r-project.org/web/packages/DHARMa/vignettes/DHARMa.html (accessed on 30 May 2023).
38. Aprahamian, M.; Wood, P. Estimation of glass eel (*Anguilla anguilla*) exploitation in the Severn Estuary, England. *Fish. Manag. Ecol.* **2021**, *28*, 65–75. [CrossRef]

39. Briand, C.; Fatin, D.; Feunteun, E.; Fontenelle, G. Estimating the stock of glass eels in an estuary by mark-recapture experiments using vital dyes. *Bull. Fr. Pêche Piscic.* **2005**, *378–379*, 23–46. [CrossRef]
40. Creutzberg, F. The role of tidal streams in the navigation of migrating elvers (*Anguilla vulgaris* Turt.). In *Orientierung der Tiere/Animal Orientation*; Springer: Berlin/Heidelberg, Germany, 1963; pp. 118–127.
41. Gascuel, D. Flow-carried and active swimming migration of the glass eel (*Anguilla anguilla*) in the tidal area of a small estuary on the French Atlantic coast. *Helgoländer Meeresunters* **1986**, *40*, 321–326. [CrossRef]
42. Pein, J.U.; Stanev, E.V.; Zhang, Y.J. The tidal asymmetries and residual flows in Ems Estuary. *Ocean Dyn.* **2014**, *64*, 1719–1741. [CrossRef]
43. Fitri, A.; Hashim, R.; Abolfathi, S.; Abdul Maulud, K.N. Dynamics of sediment transport and erosion-deposition patterns in the locality of a detached low-crested breakwater on a cohesive coast. *Water* **2019**, *11*, 1721. [CrossRef]
44. Kamphuis, J.; Meinsma, R. Succesfull approach to 'Keep the Sediment Navigable' in port of Delfzijl. In Proceedings of the World Dredging Congress and Exhibition, Brussels, Belgium, 3–7 June 2013; pp. 3–7.
45. Langdon, S.A.; Collins, A.L. Quantification of the maximal swimming performance of Australasian glass eels, Anguilla australis and Anguilla reinhardtii, using a hydraulic flume swimming chamber. *N. Z. J. Mar. Freshw. Res.* **2000**, *34*, 629–636. [CrossRef]
46. Garrelts, E.; Harten, H.; Hovers, G.; Lucht, F.; Oebius, H.; Ohlmeyer, F.; Rohde, H.; Sindern, J.; Vollmers, H.-J.; Wigand, V. Die Ausbauten der Mündungsstrecken der deutschen Tideströme und deren Einfluß auf die Sandbewegung. Deutsche Beiträge. 23. In Proceedings of the Internationaler Schifffahrtskongreß, Ottawa, ON, Canada, 12 July 1973; pp. 197–228.
47. Arribas, C.; Fernández-Delgado, C.; Oliva-Paterna, F.J.; Drake, P. Oceanic and local environmental conditions as forcing mechanisms of the glass eel recruitment to the southernmost European estuary. *Estuar. Coast. Shelf Sci.* **2012**, *107*, 46–57. [CrossRef]
48. Kroes, R.; Van Loon, E.; Goverse, E.; Schiphouwer, M.; Van der Geest, H. Attraction of migrating glass eel (*Anguilla anguilla*) by freshwater flows from water pumping stations in an urbanized delta system. *Sci. Total Environ.* **2020**, *714*, 136818. [CrossRef] [PubMed]
49. Larinier, M. *Dams and Fish Migration*; World Commission on Dams: Toulouse, France, 2000.
50. Wuenschel, M.; Able, K. Swimming ability of eels (*Anguilla rostrata, Conger oceanicus*) at estuarine ingress: Contrasting patterns of cross-shelf transport? *Mar. Biol.* **2008**, *154*, 775–786.

fishes

MDPI

Article

Detecting Japanese Eels (*Anguilla japonica*) and Revealing Their Distribution in Taiwanese Rivers by Environmental DNA Analysis

Hsiang-Yi Hsu †, Kai-Jen Wu † and Yu-San Han *

Institute of Fisheries Science, College of Life Science, National Taiwan University, 1, Sec. 4, Roosevelt Rd., Taipei 10617, Taiwan; hsiangyihsu@ntu.edu.tw (H.-Y.H.); r09b45008@ntu.edu.tw (K.-J.W.)
* Correspondence: yshan@ntu.edu.tw; Tel.: +886-2-33663726; Fax: +886-2-33669449
† These authors contributed equally to this work.

Abstract: The Japanese eel (*Anguilla japonica*) is the most prevalent freshwater eel species in Taiwan. However, its population has undergone a significant decline in recent decades due to factors such as overfishing, habitat destruction, and the effects of climate change. Urgent action is needed to conserve this species. Before implementing conservation measures, it is imperative to ascertain the distribution of Japanese eels in Taiwan's rivers. This study's primary objective was to assess the effectiveness of eDNA analysis as a method for detecting Japanese eels. To achieve this goal, we compared eDNA analysis data with results obtained from electrofishing, with the Fengshan and Shimen Rivers serving as our designated test sites. Additionally, we collected water samples from 34 other rivers across Taiwan to comprehensively assess the species' wider distribution using eDNA analysis. Our findings demonstrated eDNA analysis's viability for detecting Japanese eels. Of the 36 rivers tested, Japanese eel DNA was detected in samples from 21 rivers, scattered across northern, eastern, southern, and western Taiwan, with no specific concentration in any region. We also noted reduced detectability of Japanese eel DNA in highly polluted rivers, indicating that river pollution may have a potential impact on their population. In the future, expanding eDNA analysis to more rivers could identify additional rivers that Japanese eels inhabit. Subsequently, resource management and conservation efforts can be focused on these identified habitats. Furthermore, developing advanced eDNA-based methods for estimating the abundance or biomass of Japanese eels could enhance the flexibility of management and conservation measures.

Keywords: conservation; distribution; eDNA analysis; electrofishing; pollution

Key Contribution: Our findings confirm the effectiveness of eDNA analysis in detecting Japanese eels in Taiwan. Japanese eel DNA was found in samples from 21 of 36 tested rivers, spread across different regions but with reduced detectability in highly polluted rivers, suggesting a possible impact of pollution on their population.

Citation: Hsu, H.-Y.; Wu, K.-J.; Han, Y.-S. Detecting Japanese Eels (*Anguilla japonica*) and Revealing Their Distribution in Taiwanese Rivers by Environmental DNA Analysis. *Fishes* **2023**, *8*, 483. https://doi.org/10.3390/fishes8100483

Academic Editor: Jose Martin Pujolar

Received: 13 September 2023
Revised: 26 September 2023
Accepted: 26 September 2023
Published: 27 September 2023

1. Introduction

Freshwater eels, belonging to the genus *Anguilla*, exhibit a catadromous life cycle: growing in freshwater rivers and estuaries but spawning in the open ocean [1–4]. This complex life cycle encompasses five principal stages: leptocephalus, glass eel, elver, yellow eel, and silver eel [1,5]. While the leptocephalus and glass eel stages primarily inhabit oceans, the elver, yellow eel, and silver eel stages are mainly found in estuaries and freshwater environments. There are a total of 19 recognized species and subspecies within the freshwater eel genus [6–8].

Among these species, *Anguilla japonica*, commonly known as the Japanese eel, is prevalent in East Asia, encompassing Japan, Taiwan, China, South Korea, and North Korea. It also represents a relatively abundant freshwater eel species in Taiwan [9–11]. The recruitment of Japanese glass eels in Taiwan occurs during the winter months, spanning from October to February [5,9]. However, Japanese eel populations have undergone a drastic decline of more than 90% in recent decades, attributed to overfishing, habitat degradation, and the global impact of climate change [12,13]. River pollution is also believed to negatively affect freshwater eels [14,15]. Consequently, the International Union for Conservation of Nature included the Japanese eel on the Red List of Threatened Species in 2014 as an endangered species [16,17]. Urgent measures are now required to manage and conserve Japanese eel resources.

Investigating the distribution and habitat utilization of endangered or rare species is crucial for their resource management and conservation efforts [18]. Traditional capture-based methods, such as electrofishing, traps, and nets, have been employed to assess fish distribution [19,20], but these methods are labor-intensive and time-consuming. Moreover, the data collected through these means can often be sporadic and unevenly distributed [16].

Recently, environmental DNA (eDNA) extracted from soil or water has emerged as a valuable tool for detecting endangered fish species without the need for direct capture and for assessing biodiversity in specific aquatic environments [21–28]. All fish release DNA into their surrounding environment through slime, scales, epidermal cells, and feces. Compared to capture-based methods, eDNA analysis offers an efficient and non-invasive approach for tracing fish species in aquatic ecosystems [26,29,30].

Freshwater eels exhibit conspicuous hiding behaviors during the daytime [31] and migrate between saline and freshwater habitats during their growth stages [32–34], making them challenging to find using traditional capture-based methods. Consequently, eDNA analysis holds promise as a potential method for detecting freshwater eels in rivers [35–37]. In the past, there has been a lack of comprehensive national-scale surveys of the distribution of Japanese eels in Taiwan's rivers. Furthermore, there has been limited investigation into the precise impact of river pollution on Japanese eel distributions and populations.

In this study, we collected water samples from designated study sites along the Fengshan and Shimen Rivers for eDNA analysis. Subsequently, electrofishing was conducted at these sites to find Japanese eels. To gauge the effectiveness of eDNA analysis, we compared the data obtained from eDNA analysis with those from electrofishing. Additionally, we extended our water sampling efforts to encompass an additional 34 rivers across Taiwan, enabling us to provide an overall assessment of Japanese eel distributions and to better understand the influence of river pollution.

2. Materials and Methods

2.1. Study Area

According to previous studies, the sizes of the Fengshan and Shimen Rivers were found to be suitable for electrofishing [35,36]. Therefore, we initially selected these two rivers to conduct eDNA analysis and electrofishing to assess the effectiveness of eDNA analysis in detecting Japanese eels. In each river, a total of four study sites were selected, spanning from the lower reaches to the upper reaches (Figure 1). The specific coordinates of the study sites are listed in Tables S1 and S2. Water sampling and electrofishing were conducted at all study sites in the Fengshan and Shimen Rivers on sunny days. Because rainy weather prevents electrofishing and can dampen both equipment and survey personnel, increasing the risk of electric shock, we prefer clear and sunny days for our surveys in Taiwan. Rainfall can also raise river water levels, dilute eDNA concentrations, increase water turbidity, and complicate water sample filtration, adding complexity to our eDNA analysis. At each study site, water sampling and electrofishing were carried out simultaneously and repeated at three different time points between December 2021 and February 2022. Consequently, a total of 24 surveys, encompassing both water sampling and electrofishing, were carried out in the course of this research.

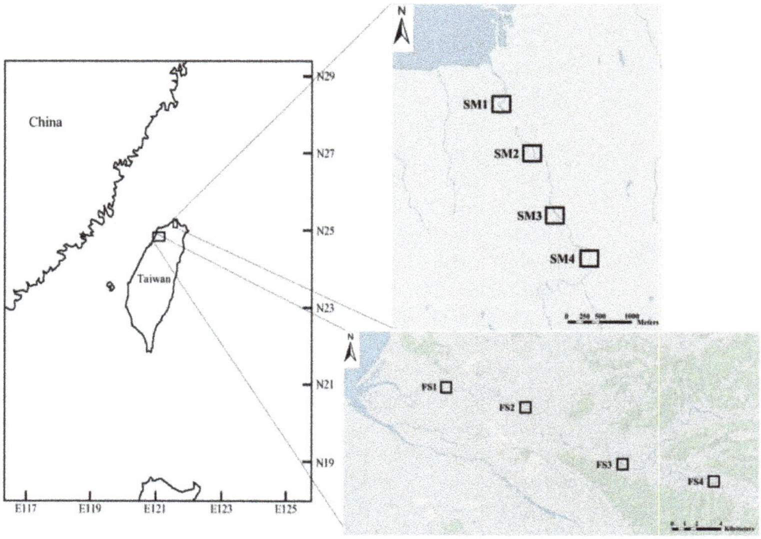

Figure 1. Study sites in the Fengshan and Shimen Rivers FS indicates the study site in the Fengshan River, and SM indicates the study site in the Shimen River. There are four study sites in the Fengshan and Shimen Rivers, respectively.

Furthermore, water samples were collected from an additional 34 rivers in Taiwan to perform eDNA analysis to detect Japanese eels. In each river, water samples were collected from a single site, all of which were located in the lower reaches of the rivers, offering a greater likelihood of detecting Japanese eel DNA. The precise coordinates of these sites can be found in Table S3. The fieldwork in these rivers, including water sampling and electrofishing, was approved by the Fisheries Agency of the Council of Agriculture, Executive Yuan, Taiwan (approval code: FA-1101254481, granted on 12 April 2021), in accordance with the objectives of scientific research.

2.2. Water Sampling and eDNA Extraction

The water sampling methodology in this study was adapted from a previous study [36]. In the Fengshan and Shimen Rivers, sampling was carried out at the downstream end of all study sites due to the electrofishing. In 34 other rivers, sampling was conducted at accessible locations in the lower reaches. A total of two replicates, each consisting of 1 L of water, were collected at the center of the river by submerging a Kartell PP bottle (Kartell, Noviglio, MI, Italy) to a depth of approximately 10 cm.

Upon collection, each water sample was immediately field filtered using a 47-mm mixed cellulose ester membrane filter (pore size: 0.8 μm; ADVANTEC®, Toyo Roshi Kaisha, Ltd., Tokyo, Japan) with a filtering device, encompassing a water filter funnel (MF 47 mm, 500 mL Magnetic Filter Funnel, Rocker Scientific, Kaohsiung, Taiwan) and a dry vacuum pump (Welch™ U2511C-02 Dry Vacuum Pump, Welch Allyn, New York, NY, USA). Following filtration, the membrane filter was placed into a 2-mL Eppendorf tube containing 600 μL of Longmire's solution to prevent eDNA degradation [38]. The Eppendorf tube, containing the membrane in Longmire's solution, was then stored at room temperature for two days to facilitate the dissolution of eDNA into Longmire's solution for subsequent eDNA extraction. To maintain aseptic conditions, the used bottles, filter funnels, and measuring cups were decontaminated using 0.1% sodium hypochlorite and were rinsed twice with double distilled water before their next use. Additionally, 1 L of double distilled water was vacuum-filtered through a membrane filter to serve as a negative control.

Total eDNA was extracted from each 2-mL Eppendorf tube containing the filter membrane and Longmire's solution using the EasyPure Genomic DNA Spin kit (Bioman, Taipei, Taiwan), following the manufacturer's instructions. The purified eDNA was stored at −20 °C until the real-time PCR analysis was conducted.

2.3. Electrofishing

Electrofishing was conducted for a duration of 1 to 1.5 h at each study site along the Fengshan and Shimen Rivers during each investigation. A battery-powered electroshocker, carried in a backpack, was utilized for electrofishing, inducing temporary incapacitation in eels. The electrofishing procedure at each study site followed a zigzag survey pattern within a 200 m-long section, per established protocols.

Skin coloration (marbled and plain) and fin difference ratio (dorsal and anal fins) can assist in distinguishing Japanese eels from other eel species in Taiwan [1,39,40]. The Japanese eel belongs to the plain eels, and its fin difference ratio is approximately 9. Moreover, the growth stages of the captured Japanese eels were determined based on morphometric characteristics, as outlined in previous studies [41,42]. Subsequently, the captured Japanese eels were released back into the same locations where they were initially caught.

2.4. Real-Time PCR

All the eDNA samples, including those extracted from river water samples and the negative controls, underwent real-time PCR analysis targeting the mitochondrial cytochrome b (*cytb*) gene, with three replicates conducted on a Bio-Rad MyIQ system (Bio-Rad, Hercules, CA, USA). The Japanese eel's *cytb* gene was amplified using a species-specific primer pair (Table 1). The primer design process involved selecting a nucleotide variation interval that exhibited relatively higher divergence and determining the sequences of the primer pair using the Primer-BLAST tool on NCBI. The highly variable nucleotide interval was identified through a comparative analysis of *cytb* sequences from four distinct Anguillid eel species available in Genbank: *A. japonica* (Accession Number: AB021772.1), *A. bicolor pacifica* (Accession Number: AB021774.1), *A. marmorata* (Accession Number: AB021778.1), and *A. luzonensis* (Accession Number: FJ170073.1) (Figure S1). Utilizing the highly variable nucleotide interval as a reference, we determined the sequences of the primer pairs using the Primer-BLAST tool. Parameters were set as follows: primer melting temperature ranged from 57 to 63 °C, and the desired PCR product size fell within the range of 50 to 200 base pairs. In addition, we conducted PCR (polymerase chain reaction) experiments by mixing the Japanese eel-specific primer pair with DNA from other eel species, including *A. bicolor pacifica*, *A. marmorata*, and *A. luzonensis*. Through gel electrophoresis results, it was confirmed that the Japanese eel-specific primer pair did not exhibit cross-reactivity with the DNA from these eel species, namely *A. bicolor pacifica*, *A. marmorata*, and *A. luzonensis* (Figure S2). Furthermore, we validated the sequence of PCR products derived from the combination of the Japanese eel-specific primer pair and Japanese eel DNA through Sanger sequencing.

Table 1. Japanese eel-specific primer pair of mitochondrial *cytb* for real-time PCR (F: forward; R: reverse).

Species	Primer Sequences	Amplicon Size (bp)	PCR Efficiency
A. japonica	AJ-F: 5′–ATGGATGATTCATCCGAAAT–3′ AJ-R: 5′–GTGTAGGTAGAGGCAGATAAAG–3′	68	98.5%

Each real-time PCR reaction was carried out in a 20-μL mixture, which included 10 μL of 2x SYBR green supermix (Bionova, Fremont, CA, USA), 0.5 μL each of the forward and reverse primers (resulting in a final concentration of 250 nM each), 2 μL of the eDNA sample, and 7 μL of double distilled water. The real-time PCR conditions consisted of the following steps: (1) pre-incubation at 95 °C for 3 min; and (2) amplification (40 cycles) at 95 °C for 30 s and 60 °C for 30 s.

All the real-time PCR data from the eDNA samples were reported as the mean Cq (quantification cycle) values ± standard deviations [43], calculated based on the Cq values obtained from six real-time PCR replicates across two sample replicates. It is important to note that the negative control did not yield a detectable Cq value.

2.5. River Pollution Index (RPI) in Taiwan

A comprehensive measure, known as the River Pollution Index (RPI), serves as an indicator of river water quality in Taiwan. The methodology for calculating the RPI and the corresponding RPI values for rivers in Taiwan are sourced from the website of the Environmental Protection Agency, Taiwan Executive Yuan (https://wq.epa.gov.tw/EWQP/zh/EnvWaterMonitoring/River.aspx, accessed on 25 September 2022). The RPI is determined based on four key parameters: dissolved oxygen (DO), biochemical oxygen demand (BOD), suspended solids (SS), and ammonia nitrogen (NH3-N). These parameters collectively assess the level of pollution in river water (see Table S4). The formula for calculating the RPI is as follows:

$$\text{River Pollution Index (RPI)} = \frac{1}{4}\sum_{i=1}^{4} Si$$

where Si represents the pollution score for each water parameter i, with scores ranging from 1 to 10. The degree of water pollution is categorized into four levels: non/mildly polluted (score range: $S \leq 2.0$), lightly polluted (score range: $2.0 < S \leq 3.0$), moderately polluted (score range: $3.1 \leq S \leq 6.0$), and severely polluted (score range: $S > 6.0$).

2.6. Data Analysis

To evaluate the efficacy of eDNA analysis in detecting Japanese eels in the Fengshan and Shimen Rivers, the occupied percentage of surveys in which Japanese eels were captured or not captured by electrofishing, within the surveys showing positive or negative Japanese eel eDNA detection, was calculated and visually represented using pie charts. Additionally, binary logistic regression analysis was employed to determine whether positive or negative Japanese eel eDNA detection (independent variable; *x*-axis) could predict the capture or non-capture of Japanese eels (dependent variable; *y*-axis) at the study sites. Furthermore, binary logistic regression analysis was also employed to investigate whether the River Pollution Index (RPI) values of the study sites across 21 rivers (independent variable; *x*-axis) could predict positive or negative Japanese eel eDNA detection (dependent variable; *y*-axis). The binary logistic regression analyses were conducted using SPSS statistical software (version 27).

3. Results

3.1. eDNA Analysis and Electrofishing at Fengshan and Shimen Rivers Study Sites

OF all 24 surveys, Japanese eel eDNA was detected in 11 of these surveys, while it was not detected in the remaining 13 surveys (Table 2). Remarkably, in the 11 surveys in which Japanese eel eDNA was detected, Japanese eels were successfully captured in 10 of them, representing a capture proportion of 90.9% (Figure 2A). Conversely, in the 13 surveys where Japanese eel eDNA was not detected, no Japanese eels were captured (Figure 2B). The binary logistic regression analysis affirmed a significant correlation between the presence or absence of Japanese eel eDNA detection and the capture or non-capture of Japanese eels ($p < 0.001$; *Exp (B)* = 1.615×10^{10}; 95% confidence interval for *Exp (B)*: upper = no value, lower = 0). The *Exp (B)* value of 1.615×10^{10} indicates that the probability of capturing Japanese eels at a site with positive eDNA detection is 1.615×10^{10} times that at a site with negative eDNA detection.

Table 2. The number of captured Japanese eels and the data of eDNA detection at the study sites in the Fengshan and Shimen Rivers.

Study Site	Date	Japanese Eel	
		eDNA (Cq)	Captures
FS1	December 2021	+, (30.3 ± 0.4)	2
FS2		−, (N/A)	0
FS3		−, (N/A)	0
FS4		−, (N/A)	0
FS1	January 2022	+, (30.5 ± 0.3)	3
FS2		−, (N/A)	0
FS3		+, (28.4 ± 0.4)	2
FS4		−, (N/A)	0
FS1	February 2022	+, (32.1 ± 0.3)	5
FS2		−, (N/A)	0
FS3		+, (31.8 ± 0.2)	3
FS4		−, (N/A)	0
SM1	December 2021	+, (32.7 ± 0.3)	3
SM2		+, (31.5 ± 0.4)	0
SM3		−, (N/A)	0
SM4		−, (N/A)	0
SM1	January 2022	+, (28.7 ± 0.4)	4
SM2		+, (30.2 ± 0.2)	5
SM3		−, (N/A)	0
SM4		−, (N/A)	0
SM1	February 2022	+, (32.7 ± 0.5)	13
SM2		+, (32.1 ± 0.4)	12
SM3		−, (N/A)	0
SM4		−, (N/A)	0

"+": positive detection; "−": negative detection; "N/A": no Cq value.

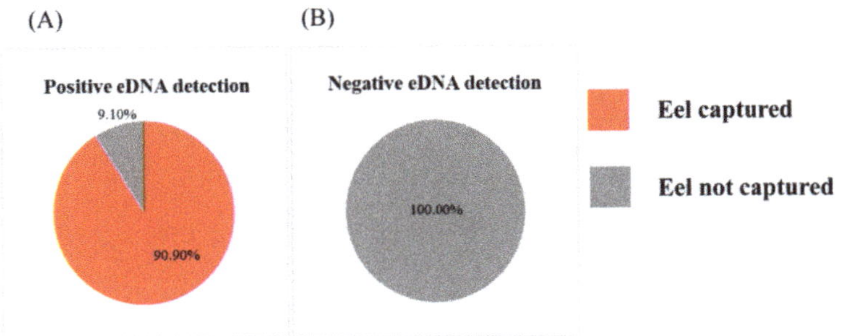

Figure 2. Pie chart for assessing the effectiveness of eDNA analysis in detecting Japanese eels (**A**). The occupied percentage of the surveys in which Japanese eels have been captured or not captured by electrofishing within the surveys having positive Japanese eel eDNA detection (n = 11). (**B**) The occupied percentage of the surveys in which Japanese eels have been captured or not captured by electrofishing within the surveys having negative Japanese eel eDNA detection (n = 13).

Among the eight study sites in the Fengshan and Shimen Rivers, Japanese eel eDNA was detected at four study sites, namely FS1, FS3, SM1, and SM2, where Japanese eels were also successfully captured. These four study sites are primarily situated in the lower and middle reaches, suggesting a preference for Japanese eels to inhabit these areas within the Fengshan and Shimen Rivers. Specifically, 15 Japanese eels were captured in the Fengshan River and 37 Japanese eels in the Shimen River (Table 2), all of which were in the yellow eel stage.

3.2. Survey of Japanese Eel Distribution across Taiwan's Rivers

To assess the distribution of Japanese eels across Taiwan's rivers, water samples were collected from the lower reaches of 34 rivers for eDNA analysis. Among these 34 rivers, Japanese eel eDNA was detected in 19 of them (Figure 3). This discovery indicates that Japanese eels are distributed widely across Taiwan, spanning the northern, eastern, western, and southern regions. Detailed information regarding the detected Cq values of Japanese eel eDNA in each river can be found in Table 3.

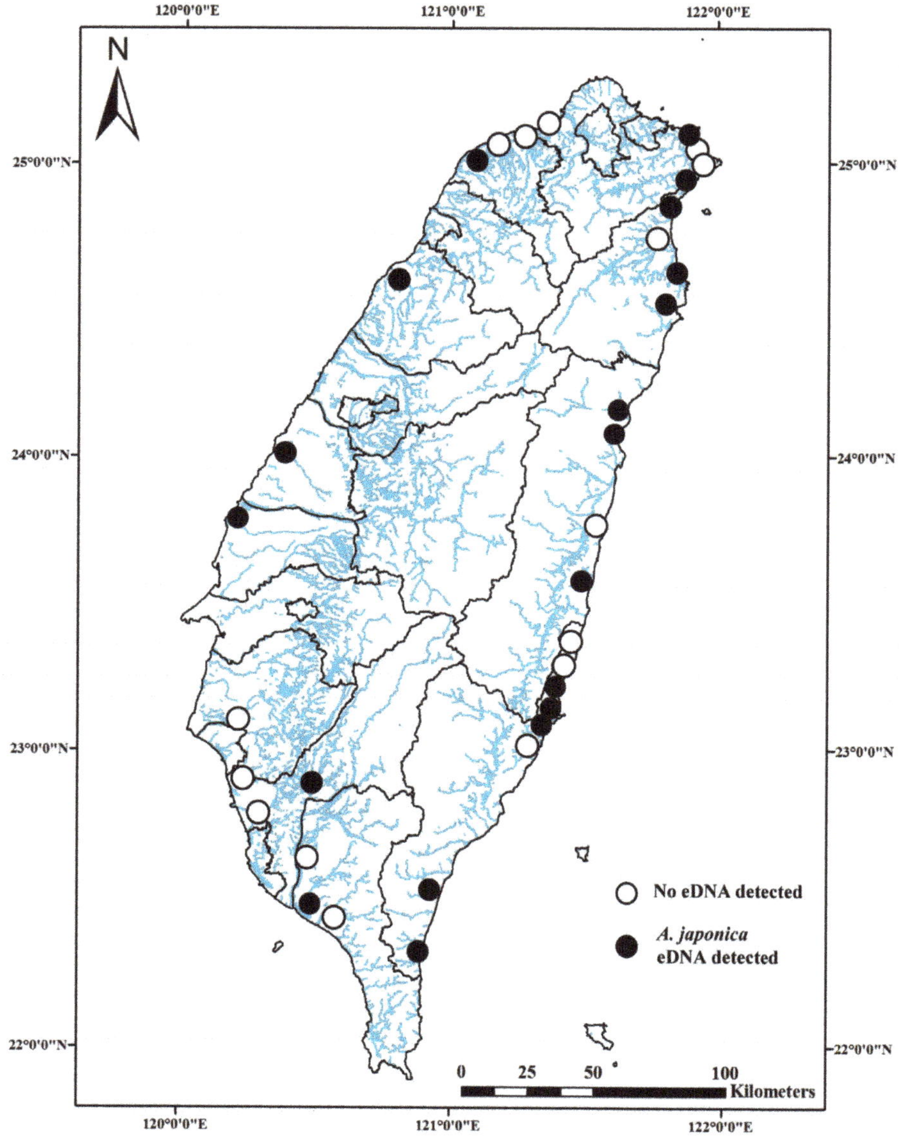

Figure 3. Overview of the Japanese eel eDNA detection at 34 river basins in Taiwan.

Table 3. Cq values of Japanese eel eDNA and RPI values of waters at the sampling sites of 36 target rivers in this study.

Name of River	eDNA Detection of Japanese Eel	RPI Value	Level of Pollution
Beishikeng River	+, 35.0 ± 0.4	N	N
Hemei River	–	N	N
Jingshawan River	–	N	N
Daxi River	+, 27.2 ± 0.3	1	non/mildly polluted
Gengfang River	+, 29.8 ± 0.3	N	N
Dezikou River	–	3.75	moderately polluted
Xinchen River	+, 28.9 ± 0.2	1	non/mildly polluted
Dongao North River	+, 27.8 ± 0.3	1	non/mildly polluted
Liwu River	+, 34.6 ± 0.4	1	non/mildly polluted
Sanzhan River	+, 31.5 ± 0.4	1	non/mildly polluted
Shuliao River	–	N	N
Xiuguluan River	+, 38.1 ± 0.2	3.25	moderately polluted
Ningpu River	–	N	N
Dabin River	–	N	N
Duwei River	+, 32.3 ± 0.2	N	N
Fujia River	+, 31.3 ± 0.3	1.25	non/mildly polluted
Hsinkang River	+, 33.3 ± 0.2	N	N
Mawu River	–	1.25	non/mildly polluted
Taimali River	+, 30.2 ± 0.3	N	N
Anshuo River	+, 35.9 ± 0.4	N	N
Nanwan River	–	N	N
Nankan River	–	4	moderately polluted
Shuangxikou River	–	N	N
Xinwu River	+, 36.0 ± 0.3	2.25	lightly polluted
Zhonggang River	+, 32.5 ± 0.3	2	non/mildly polluted
Oldzhuoshui River	+, 34.5 ± 0.3	2.25	lightly polluted
Newhuwei River	+, 31.4 ± 0.2	4.25	moderately polluted
Yanshui River	–	6.25	severely polluted
Erren River	–	4.75	moderately polluted
Tianliao River	+, 31.1 ± 0.2	N	N
Dianbao River	–	7.25	severely polluted
Gaoping River	–	3.25	moderately polluted
Donggang River	+, 37.9 ± 0.3	5.5	moderately polluted
Linbian River	–	1.5	non/mildly polluted
Fengshan River	+, 30.5 ± 0.3	2.25	lightly polluted
Shimen River	+, 28.7 ± 0.3	N	N

"+": positive detection; "–": negative detection; "N": no RPI value or unknown pollution level.

3.3. RPI and Cq Values for Taiwan's Lower Reaches of Thirty-Six Study Rivers

The RPI values for the lower reaches of the study river basins were sourced from the website of the Environmental Protection Agency, Taiwan Executive Yuan. Among the 36 study rivers, RPI values were available for 21 of them (Table 3). These 21 river basins were categorized into four levels of water pollution based on their pollution point: non/mildly polluted (nine river basins), lightly polluted (three river basins), moderately polluted (seven river basins), and severely polluted (two river basins).

Within the nine non/mildly polluted rivers, Japanese eel eDNA was detected in seven of them. In the three lightly polluted rivers, Japanese eel eDNA was detected in all cases. Among the seven moderately polluted rivers, Japanese eel eDNA was detected in three. However, in the two severely polluted rivers, Japanese eel eDNA was not detected in either of them. Binary logistic regression analysis revealed a significant correlation between the RPI values of the study sites and the presence or absence of Japanese eel eDNA detection at those sites ($p < 0.05$; *Exp* (*B*) = 0.539; 95% confidence interval for *Exp* (*B*): upper = 0.990, lower = 0.294). The *Exp* (*B*) value of 0.539 indicates that, for every one-unit increase in RPI, the probability of detecting Japanese eel eDNA as positive was 0.539 times that of detecting it as negative.

4. Discussion

Environmental DNA (eDNA) analysis is a non-invasive method widely utilized for tracing aquatic species, and its effectiveness has been substantiated in numerous studies [24,36,37,44–46]. However, the accuracy and detection probability of eDNA analysis can be influenced, and conducting multiple sample repeats can enhance its effectiveness [47]. Additionally, the precision of eDNA analysis may be susceptible to false negatives or false positives, typically arising during sample collection [48]. Nevertheless, our eDNA survey results in the Fengshan and Shimen Rivers appear to be free from false negatives or positives. Among our surveys, Japanese eel eDNA was detected in only one instance, specifically at SM2, without any accompanying captures of Japanese eels. This occurrence can likely be attributed to the low population density of Japanese eels, their migratory behavior, their tendency to hide, or their reduced activity during daytime, diminishing the likelihood of their detection via electrofishing [49]. Consequently, this fact could explain why Japanese eel eDNA was detected in only one survey at SM2, but no captures were made. Based on our survey findings in the Fengshan and Shimen Rivers, eDNA analysis may exhibit superior sensitivity in detecting the presence and distribution of Japanese eels compared to capture-based methods. This finding suggests that eDNA analysis could potentially offer a more reliable means of assessment.

Following our field survey of the Fengshan and Shimen Rivers, it became evident that Japanese eels exhibit a preference for inhabiting the lower and middle reaches of these rivers. This finding aligns with numerous studies that have reported a decrease in the abundance of Japanese eels with increasing distance from the river mouth [10,36,50,51]. Furthermore, several studies have indicated a habitat partition between Japanese eels and giant-mottled eels (*Anguilla marmorata*) within river basins, with the latter primarily occupying the middle and upper reaches [10,36]. Subsequently, this habitat partition can be substantiated through eDNA analysis, and understanding the habitat distribution of Japanese and giant-mottled eels holds significant implications for their conservation efforts.

Evaluating the abundance of eels in rivers is a critical aspect of eel resource conservation. If eDNA analysis proves capable of estimating eel abundance in rivers, it could offer a highly convenient and practical approach for assessing eel resources. Several prior studies have indicated a significant, positive correlation between eDNA concentrations and the abundance of anguillid eels in rivers [35,36]. Additionally, some research has revealed a positive association between eDNA concentration and fish biomass in experimental ponds [52,53]. However, it is important to note that eDNA concentrations in river water can be influenced by numerous factors, including fish species shedding rates, population size, population structure, water temperature, food availability, water velocity, and chemical constituents [54]. The inherent complexities of these variables pose challenges in quantitatively assessing fish species abundance using the current eDNA analytical framework. Therefore, it should be acknowledged that eDNA analysis may not entirely replace the need for catch per unit effort (CPUE) data [55]. In our forthcoming research, we will continue to investigate whether eel eDNA concentration in rivers accurately reflects eel abundance or biomass in those rivers. We aim to develop a method for estimating eel resources in freshwater habitats.

Numerous studies have consistently reported a greater abundance of eels in the lower reaches of river basins [36,56–58]. Additionally, it has been observed that higher salinity levels, increased ionic content, a stable pH, and temperature conditions can enhance the preservation of eDNA [59]. This finding aligns with our findings from the Fengshan and Shimen Rivers survey, which demonstrated a greater likelihood of detecting Japanese eel eDNA in the lower reaches. Consequently, for the nationwide eDNA analysis, we collected water samples exclusively from the lower reaches of 34 river basins. The results of our national survey revealed that Japanese eel eDNA was not uniformly detected in all rivers, but it was present in rivers spanning the north, east, south, and west regions of Taiwan. In our previous research, we explored the transport and distribution mechanisms of Japanese glass eels through ocean currents [9,60]. The outcomes of these studies indicated that Japanese glass eels have the potential to disperse throughout the entirety of Taiwan.

Water pollution in rivers can significantly impact the survival and abundance of aquatic organisms, including anguillid eels [14,56,61]. In the present investigation, we employed the River Pollution Index (RPI) as the metric for assessing the pollution levels of the rivers. However, among the 36 study rivers, only 21 provided measurable RPI values, categorizing them as non/mildly polluted (nine rivers), lightly polluted (three rivers), moderately polluted (seven rivers), and severely polluted (two rivers). The results of binary logistic regression indicated that Japanese eel eDNA was less detectable in more polluted rivers. This phenomenon may be attributed to the smaller population sizes of Japanese eels in highly polluted rivers. Furthermore, previous research has suggested that pollutants can negatively influence eDNA analysis [26,27,62]. Organic substances in the water can inhibit the qPCR reaction, and these substances may indirectly lower the water's pH, accelerating the degradation of eDNA [63]. Consequently, the probability of detecting Japanese eel eDNA in polluted water may be significantly reduced. It is worth noting that Taiwan still has many rivers that are severely impacted by industrialization and pollution. Therefore, further studies are necessary to comprehensively examine the effects of water pollution on Japanese eel resources and eDNA detection.

5. Conclusions

The findings from the surveys conducted in the Fengshan and Shimen Rivers indicate that eDNA analysis is a viable method for detecting Japanese eels. Notably, this study marks the inaugural nationwide investigation into the distribution of Japanese eels in Taiwan's rivers utilizing eDNA analysis. The eDNA analysis outcomes unveiled a potential presence of Japanese eels across the river basins spanning the entirety of Taiwan. However, Japanese eel eDNA was less discernible in severely polluted rivers, hinting at potential correlations between water pollution and both the population sizes of Japanese eels and the detectability of their eDNA. In the broader context of East Asia, the management and conservation of Japanese eel resources hold paramount and urgent significance. Developing effective methods to detect Japanese eels and accurately assess their abundance or biomass show great potential in supporting the sustainable management and conservation efforts for this species.

Supplementary Materials: The following supporting information can be downloaded at: https://www.mdpi.com/article/10.3390/fishes8100483/s1, Figure S1: Comparison of *cytb* sequences among four different Anguillid eel species found in Taiwan and the location of *Anguilla japonica*-specific primers for real-time PCR of *cytb*; Figure S2: Gel electrophoresis following polymerase chain reaction (PCR) with the *A. japonica*-specific *cytb* primer pair; Table S1: Detail coordinates of the study sites at Fengshan River; Table S2: Detail coordinates of the study sites at Shimen River; Table S3: Detail coordinates of the water sampling sites of 34 river basins in a national scale survey; Table S4: Calculation and comparison benchmark of River Pollution Index.

Author Contributions: Conceptualization, Y.-S.H.; data curation, K.-J.W.; formal analysis, H.-Y.H. and K.-J.W.; funding acquisition, Y.-S.H.; investigation, H.-Y.H. and K.-J.W.; methodology, H.-Y.H. and K.-J.W.; project administration, Y.-S.H.; supervision, Y.-S.H.; validation, H.-Y.H.; writing—original draft, H.-Y.H.; writing—review and editing, H.-Y.H., K.-J.W. and Y.-S.H. All authors have read and agreed to the published version of the manuscript.

Funding: This research was funded by the National Science and Technology Council, Executive Yuan, Taiwan (MOST 109-2313-B-002-001-MY2 and MOST 111-2313-B-002-016-MY3).

Institutional Review Board Statement: The fieldwork in rivers, including water sampling and electrofishing, was approved by the Fisheries Agency of the Council of Agriculture, Executive Yuan, Taiwan (approval code: FA-1101254481, granted on 12 April 2021), in accordance with the objectives of scientific research.

Data Availability Statement: All the data are provided within this manuscript.

Conflicts of Interest: The authors declare no conflict of interest.

References

1. Tesch, F.W. *The Eel*; Blackwell Science: Oxford, UK, 2003.
2. Aoyama, J.; Wouthuyzen, S.; Miller, M.J.; Sugeha, H.Y.; Kuroki, M.; Watanabe, S.; Syahailatua, A.; Tantu, F.Y.; Hagihara, S.; Triyanto; et al. Reproductive Ecology and Biodiversity of Freshwater Eels around Sulawesi Island Indonesia. *Zool. Stud.* **2018**, *57*, e30. [PubMed]
3. Chen, S.C.; Chang, C.R.; Han, Y.S. Seaward Migration Routes of Indigenous Eels, *Anguilla japonica*, *A. marmorata*, and *A. bicolor pacifica*, via Satellite Tags. *Zool. Stud.* **2018**, *57*, e21.
4. Higuchi, T.; Watanabe, S.; Manabe, R.; Kaku, T.; Okamura, A.; Yamada, Y.; Miller, M.J.; Tsukamoto, K. Tracking *Anguilla japonica* Silver Eels Along the West Marina Ridge Using Pop-up Archival Transmitting Tags. *Zool. Stud.* **2018**, *57*, e24.
5. Han, Y.S.; Wu, C.R.; Iizuka, Y. Batch-like Arrival Waves of Glass Eels of *Anguilla japonica* in Offshore Waters of Taiwan. *Zool. Stud.* **2016**, *55*, e36. [PubMed]
6. Arai, T. Taxonomy and Distribution. In *Biology and Ecology of Anguillid Eels*; Arai, T., Ed.; CRC Press: Boca Raton, FL, USA, 2016; pp. 1–20.
7. Kaifu, K.; Stein, F.; Dekker, W.; Walker, N.; Dolloff, C.A.; Steele, K.; Aguirre, A.A.; Nijman, V.; Siriwat, P.; Sasal, P. Global exploitation of freshwater eels (genus *Anguilla*): Fisheries, stock status and illegal trade. In Proceedings of the First International Eel Science Symposium, Zoological Society of London, London, UK, 13–15 June 2017; 5m Books Ltd.: Essex, UK, 2017; pp. 377–422.
8. Watanabe, S.; Aoyama, J.; Tsukamoto, K. A new species of freshwater eel *Anguilla luzonensis* (Teleostei: Anguillidae) from Luzon Island of the Philippines. *Fish. Sci.* **2009**, *75*, 387–392. [CrossRef]
9. Han, Y.S.; Yambot, A.V.; Zhang, H.; Hung, C.L. Sympatric Spawning but Allopatric Distribution of *Anguilla japonica* and *Anguilla marmorata*: Temperature- and Oceanic Current-Dependent Sieving. *PLoS ONE* **2012**, *7*, e37484. [CrossRef]
10. Hsu, H.Y.; Chen, H.W.; Han, Y.S. Habitat Partitioning and its Possible Genetic Background Between Two Sympatrically Distributed Eel Species in Taiwan. *Zool. Stud.* **2019**, *58*, e27. [PubMed]
11. Shiao, J.C.; Iizuka, Y.; Chang, C.W.; Tzeng, W.N. Disparities in habitat use and migratory behavior between tropical eel *Anguilla marmorata* and temperate eel *A. japonica* in four Taiwanese rivers. *Mar. Ecol. Prog. Ser.* **2003**, *261*, 233–242. [CrossRef]
12. Chen, J.Z.; Huang, S.L.; Han, Y.S. Impact of long-term habitat loss on the Japanese eel *Anguilla japonica*. *Estuar. Coast. Shelf Sci.* **2014**, *151*, 361–369. [CrossRef]
13. Tsukamoto, K.; Aoyama, J.; Miller, M.J. Present status of the Japanese eel: Resources and recent research. *Am. Fish. Soc. Symp.* **2009**, *58*, 21–35.
14. Yamaguchi, N.; Gazzard, D.; Scholey, G.; Macdonald, D.W. Concentrations and hazard assessment of PCBs, organochlorine pesticides and mercury in fish species from the Upper Thames: River pollution and its potential effects on top predators. *Chemosphere* **2003**, *50*, 265–273. [CrossRef]
15. Agradi, E.; Baga, R.; Cillo, F.; Ceradini, S.; Heltai, D. Environmental contaminants and biochemical response in eel exposed to Po river water. *Chemosphere* **2000**, *41*, 1555–1562. [CrossRef]
16. Jacoby, D.M.P.; Casselman, J.M.; Crook, V.; DeLucia, M.B.; Ahn, H.; Kaifu, K.; Kurwie, T.; Sasal, P.; Silfvergrip, A.M.C.; Smith, K.G.; et al. Synergistic patterns of threat and the challenges facing global anguillid eel conservation. *Glob. Ecol. Conserv.* **2015**, *4*, 321–333. [CrossRef]
17. IUCN. *The IUCN Red List of Threatened Species*; IUCN: Gland, Switzerland, 2014.
18. Jackson, D.A.; Peres-Neto, P.R.; Olden, J.D. What controls who is where in freshwater fish communities—The roles of biotic, abiotic, and spatial factors. *Can. J. Fish. Aquat. Sci.* **2001**, *58*, 157–170.
19. Murphy, B.R.; Willis, D.W. (Eds.) *Fisheries Techniques*, 2nd ed.; American Fisheries Society: Bethesda, MD, USA, 1996; p. 732.
20. Snyder, D.E. Invited overview: Conclusions from a review of electrofishing and its harmful effects on fish. *Rev. Fish. Biol. Fisher* **2003**, *13*, 445–453. [CrossRef]
21. Janosik, A.M.; Johnston, C.E. Environmental DNA as an effective tool for detection of imperiled fishes. *Environ. Biol. Fish.* **2015**, *98*, 1889–1893. [CrossRef]
22. Bergman, P.S.; Schumer, G.; Blankenship, S.; Campbell, E. Detection of Adult Green Sturgeon Using Environmental DNA Analysis. *PLoS ONE* **2016**, *11*, e0153500. [CrossRef] [PubMed]
23. Thomsen, P.F.; Kielgast, J.; Iversen, L.L.; Wiuf, C.; Rasmussen, M.; Gilbert, M.T.P.; Orlando, L.; Willerslev, E. Monitoring endangered freshwater biodiversity using environmental DNA. *Mol. Ecol.* **2012**, *21*, 2565–2573. [CrossRef] [PubMed]
24. Miya, M.; Gotoh, R.O.; Sado, T. MiFish metabarcoding: A high-throughput approach for simultaneous detection of multiple fish species from environmental DNA and other samples. *Fish. Sci.* **2020**, *86*, 939–970. [CrossRef]
25. Darling, J.A.; Mahon, A.R. From molecules to management: Adopting DNA-based methods for monitoring biological invasions in aquatic environments. *Environ. Res.* **2011**, *111*, 978–988. [CrossRef]
26. Rees, H.C.; Maddison, B.C.; Middleditch, D.J.; Patmore, J.R.M.; Gough, K.C. The detection of aquatic animal species using environmental DNA—A review of eDNA as a survey tool in ecology. *J. Appl. Ecol.* **2014**, *51*, 1450–1459. [CrossRef]

27. Dougherty, M.M.; Larson, E.R.; Renshaw, M.A.; Gantz, C.A.; Egan, S.P.; Erickson, D.M.; Lodge, D.M. Environmental DNA (eDNA) detects the invasive rusty crayfish *Orconectes rusticus* at low abundances. *J. Appl. Ecol.* **2016**, *53*, 722–732. [CrossRef] [PubMed]
28. Yamamoto, S.; Masuda, R.; Sato, Y.; Sado, T.; Araki, H.; Kondoh, M.; Minamoto, T.; Miya, M. Environmental DNA metabarcoding reveals local fish communities in a species-rich coastal sea. *Sci. Rep.* **2017**, *7*, 40368. [CrossRef] [PubMed]
29. Lodge, D.M.; Turner, C.R.; Jerde, C.L.; Barnes, M.A.; Chadderton, L.; Egan, S.P.; Feder, J.L.; Mahon, A.R.; Pfrender, M.E. Conservation in a cup of water: Estimating biodiversity and population abundance from environmental DNA. *Mol. Ecol.* **2012**, *21*, 2555–2558. [CrossRef]
30. Bohmann, K.; Evans, A.; Gilbert, M.T.P.; Carvalho, G.R.; Creer, S.; Knapp, M.; Yu, D.W.; de Bruyn, M. Environmental DNA for wildlife biology and biodiversity monitoring. *Trends Ecol. Evol.* **2014**, *29*, 358–367. [CrossRef] [PubMed]
31. Aoyama, J.; Watanabe, S.; Ishikawa, S.; Nishida, M.; Tsukamoto, K. Are morphological characters distinctive enough to discriminate between two species of freshwater eels, *Anguilla celebesensis* and *A. interioris*? *Ichthyol. Res.* **2000**, *47*, 157–161. [CrossRef]
32. Moriarty, C. The Yellow Eel. In *Eel Biology*; Aida, K., Tsukamoto, K., Yamauchi, K., Eds.; Springer: Tokyo, Japan, 2003. [CrossRef]
33. Tzeng, W.N.; Shiao, J.C.; Iizuka, Y. Use of otolith Sr: Ca ratios to study the riverine migratory behaviors of Japanese eel *Anguilla japonica*. *Mar. Ecol. Prog. Ser.* **2002**, *245*, 213–221. [CrossRef]
34. Briones, A.A.; Yambot, A.V.; Shiao, J.C.; Iizuka, Y.; Tzeng, W.N. Migratory pattern and habitat use of tropical eels *Anguilla* spp. (Teleostei: Anguilliformes: Anguillidae) in the Philippines, as revealed by otolith microchemistry. *Raffles B Zool.* **2007**, *14*, 141–149.
35. Itakura, H.; Wakiya, R.; Yamamoto, S.; Kaifu, K.; Sato, T.; Minamoto, T. Environmental DNA analysis reveals the spatial distribution, abundance, and biomass of Japanese eels at the river-basin scale. *Aquat. Conserv.* **2019**, *29*, 361–373. [CrossRef]
36. Itakura, H.; Wakiya, R.; Sakata, M.K.; Hsu, H.Y.; Chen, S.C.; Yang, C.C.; Huang, Y.C.; Han, Y.S.; Yamamoto, S.; Minamoto, T. Estimations of Riverine Distribution, Abundance, and Biomass of Anguillid Eels in Japan and Taiwan Using Environmental DNA Analysis. *Zool. Stud.* **2020**, *59*, e17.
37. Kasai, A.; Yamazaki, A.; Ahn, H.; Yamanaka, H.; Kameyama, S.; Masuda, R.; Azuma, N.; Kimura, S.; Karaki, T.; Kurokawa, Y.; et al. Distribution of Japanese Eel *Anguilla japonica* Revealed by Environmental DNA. *Front. Ecol. Evol.* **2021**, *9*, 621461. [CrossRef]
38. Kelly, E.W.; Kathryn, P.H.; Antoinette, J.P. No filters, no fridges: A method for preservation of water samples for eDNA analysis. *BMC Res. Notes* **2016**, *9*, 298. [CrossRef]
39. Hsu, H.Y.; Lin, Y.T.; Huang, Y.C.; Han, Y.S. Skin coloration and habitat preference of the freshwater *Anguilla* eels. *Int. J. Aquac. Fish. Sci.* **2020**, *6*, 96–101. [CrossRef]
40. Aoyama, J.; Yoshinaga, T.; Shinoda, A.; Shirotori, F.; Yambot, A.V.; Han, Y.S. Seasonal Changes in Species Composition of Glass Eels of the Genus *Anguilla* (Teleostei: Anguillidae) Recruiting to the Cagayan River, Luzon Island, the Philippines. *Pac. Sci.* **2015**, *69*, 263–270. [CrossRef]
41. Han, Y.S.; Liao, I.C.; Huang, Y.S.; He, J.T.; Chang, C.W.; Tzeng, W.N. Synchronous changes of morphology and gonadal development of silvering Japanese eel Anguilla japonica. *Aquaculture* **2003**, *219*, 783–796. [CrossRef]
42. Lin, Y.S.; Tzeng, C.S.; Hwang, J.K. Reassessment of morphological characteristics in freshwater eels (genus *Anguilla*, Anguillidae) shows congruence with molecular phylogeny estimates. *Zool. Scr.* **2005**, *34*, 225–234. [CrossRef]
43. Bustin, S.A.; Benes, V.; Garson, J.A.; Hellemans, J.; Huggett, J.; Kubista, M.; Mueller, R.; Nolan, T.; Pfaffl, M.W.; Shipley, G.L.; et al. The MIQE guidelines: Minimum information for publication of quantitative real-time PCR experiments. *Clin. Chem.* **2009**, *55*, 611–622. [CrossRef]
44. Sakata, M.K.; Maki, N.; Sugiyama, H.; Minamoto, T. Identifying a breeding habitat of a critically endangered fish, *Acheilognathus typus*, in a natural river in Japan. *Sci. Nat-Heidelb.* **2017**, *104*, 100. [CrossRef]
45. Jerde, C.L. Can we manage fisheries with the inherent uncertainty from eDNA? *J. Fish. Biol.* **2021**, *98*, 341–353. [CrossRef]
46. Wilcox, T.M.; McKelvey, K.S.; Young, M.K.; Sepulveda, A.J.; Shepard, B.B.; Jane, S.F.; Whiteley, A.R.; Lowe, W.H.; Schwartz, M.K. Understanding environmental DNA detection probabilities: A case study using a stream-dwelling char *Salvelinus fontinalis*. *Biol. Conserv.* **2016**, *194*, 209–216. [CrossRef]
47. Mauvisseau, Q.; Burian, A.; Gibson, C.; Brys, R.; Ramsey, A.; Sweet, M. Influence of accuracy, repeatability and detection probability in the reliability of species-specific eDNA based approaches. *Sci. Rep.* **2019**, *9*, 580. [CrossRef] [PubMed]
48. Buxton, A.; Matechou, E.; Griffin, J.; Diana, A.; Griffiths, R.A. Optimising sampling and analysis protocols in environmental DNA studies. *Sci. Rep.* **2021**, *11*, 11637. [CrossRef]
49. Sint, D.; Kolp, B.; Rennstam Rubbmark, O.; Füreder, L.; Traugott, M. The amount of environmental DNA increases with freshwater crayfish density and over time. *Environ. DNA* **2022**, *4*, 417–424. [CrossRef]
50. Yokouchi, K.; Sudo, R.; Kaifu, K.; Aoyama, J.; Tsukamoto, K. Biological characteristics of silver-phase Japanese eels, *Anguilla japonica*, collected from Hamana Lake, Japan. *Coast. Mar. Sci.* **2009**, *33*, 1–10.
51. Kaifu, K.; Tamura, M.; Aoyama, J.; Tsukamoto, K. Dispersal of yellow phase Japanese eels *Anguilla japonica* after recruitment in the Kojima Bay-Asahi River system, Japan. *Environ. Biol. Fish.* **2010**, *88*, 273–282. [CrossRef]
52. Doi, H.; Takahara, T.; Minamoto, T.; Matsuhashi, S.; Uchii, K.; Yamanaka, H. Droplet Digital Polymerase Chain Reaction (PCR) Outperforms Real-Time PCR in the Detection of Environmental DNA from an Invasive Fish Species. *Environ. Sci. Technol.* **2015**, *49*, 5601–5608. [CrossRef]

53. Klymus, K.E.; Richter, C.A.; Chapman, D.C.; Paukert, C. Quantification of eDNA shedding rates from invasive bighead carp *Hypophthalmichthys nobilis* and silver carp *Hypophthalmichthys molitrix*. *Biol. Conserv.* **2015**, *183*, 77–84. [CrossRef]
54. Iversen, L.L.; Kielgast, J.; Sand-Jensen, K. Monitoring of animal abundance by environmental DNA—An increasingly obscure perspective: A reply to Klymus et al., 2015. *Biol. Conserv.* **2015**, *192*, 479–480. [CrossRef]
55. Johnsen, S.I.; Strand, D.A.; Rusch, J.C.; Vralstad, T. Environmental DNA (eDNA) Monitoring of Noble Crayfish *Astacus astacus* in Lentic Environments Offers Reliable Presence-Absence Surveillance—But Fails to Predict Population Density. *Front. Environ. Sci.* **2020**, *8*, 612253. [CrossRef]
56. Itakura, H.; Kaino, T.; Miyake, Y.; Kitagawa, T.; Kimura, S. Feeding, condition, and abundance of Japanese eels from natural and revetment habitats in the Tone River, Japan. *Environ. Biol. Fish.* **2015**, *98*, 1871–1888. [CrossRef]
57. Itakura, H.; Miyake, Y.; Kitagawa, T.; Kimura, S. Site fidelity, diel and seasonal activities of yellow-phase Japanese eels (*Anguilla japonica*) in a freshwater habitat as inferred from acoustic telemetry. *Ecol. Freshw. Fish.* **2018**, *27*, 737–751. [CrossRef]
58. Wakiya, R.; Kaifu, K.; Azechi, K.; Tsukamoto, K.; Mochioka, N. Evaluation of downward movements of Japanese eel *Anguilla japonica* inhabiting brackish water areas. *J. Fish. Biol.* **2020**, *96*, 516–526. [CrossRef] [PubMed]
59. Collins, R.A.; Wangensteen, O.S.; O'Gorman, E.J.; Mariani, S.; Sims, D.W.; Genner, M.J. Persistence of environmental DNA in marine systems. *Commun. Biol.* **2018**, *1*, 185. [CrossRef] [PubMed]
60. Hsiung, K.M.; Lin, Y.T.; Han, Y.S. Current Dependent Dispersal Characteristics of Japanese Glass Eel around Taiwan. *J. Mar. Sci. Eng.* **2022**, *10*, 98. [CrossRef]
61. Dudgeon, D.; Arthington, A.H.; Gessner, M.O.; Kawabata, Z.I.; Knowler, D.J.; Leveque, C.; Naiman, R.J.; Prieur-Richard, A.H.; Soto, D.; Stiassny, M.L.J.; et al. Freshwater biodiversity: Importance, threats, status and conservation challenges. *Biol. Rev.* **2006**, *81*, 163–182. [CrossRef] [PubMed]
62. Jane, S.F.; Wilcox, T.M.; McKelvey, K.S.; Young, M.K.; Schwartz, M.K.; Lowe, W.H.; Letcher, B.H.; Whiteley, A.R. Distance, flow and PCR inhibition: eDNA dynamics in two headwater streams. *Mol. Ecol. Resour.* **2015**, *15*, 216–227. [CrossRef] [PubMed]
63. Seymour, M.; Durance, I.; Cosby, B.J.; Ransom-Jones, E.; Deiner, K.; Ormerod, S.J.; Colbourne, J.K.; Wilgar, G.; Carvalho, G.R.; de Bruyn, M.; et al. Acidity promotes degradation of multi-species environmental DNA in lotic mesocosms. *Commun. Biol.* **2018**, *1*, 4. [CrossRef]

fishes

MDPI

Article

The Eel Ascending: The Influence of Lateral Slope, Climbing Substrate and Flow Rate on Eel Pass Performance

Adam T. Piper [1,*], Paula J. Rosewarne [2], Charlotte Pike [3] and Rosalind M. Wright [4]

1 Institute of Zoology, Zoological Society of London, Regent's Park, London NW1 4RY, UK
2 Independent Researcher, Purleigh CM3 6PZ, UK
3 Conservation and Policy, Zoological Society of London, Regent's Park, London NW1 4RY, UK
4 Environment Agency, Rivers House, Threshelfords Business Park, Inworth Road, Feering CO5 9SE, UK
* Correspondence: adam.piper@ioz.ac.uk

Abstract: Optimising the design of passage facilities to restore fluvial connectivity for juvenile European eel (*Anguilla anguilla*) is a key priority within conservation efforts for the species, across the majority of its freshwater range. Employing an experimental setup that simulated gravity-fed upstream eel passes, this study demonstrated that novel V-profile passes, which incorporate two lateral slopes (15°), performed better than laterally flat passes over the flow rates tested (0.2–0.6 L s^{-1}). For the small eel size used (60–80 mm length), the bristle substrate consistently outperformed studs, but the lateral slope had a greater effect on passage metrics than the substrate choice. Our findings strongly support the use of V-shaped channels for upstream migrating eel at fish passage facilities, particularly in scenarios where flow rates may be elevated and/or fluctuating, such as for gravity-fed passes.

Keywords: European eel; *Anguilla anguilla*; fish passage facilities; fish behaviour; eel fishway; climbing substrate

Key Contribution: The incorporation of a lateral slope, such as within a V-shaped channel, is recommended for fish passes targeting small upstream migrating eel, particularly in scenarios where flow rates may be elevated and/or fluctuating, such as for gravity-fed passes.

Citation: Piper, A.T.; Rosewarne, P.J.; Pike, C.; Wright, R.M. The Eel Ascending: The Influence of Lateral Slope, Climbing Substrate and Flow Rate on Eel Pass Performance. *Fishes* **2023**, *8*, 612. https://doi.org/10.3390/fishes8120612

Academic Editor: Jose Martin Pujolar

Received: 11 November 2023
Revised: 14 December 2023
Accepted: 14 December 2023
Published: 18 December 2023

1. Introduction

The population status of the European eel (*Anguilla anguilla*) is of significant concern, with continuing International Council for the Exploration of the Sea (ICES) advice to reduce human impacts to as close to zero as possible, and the species listed as Critically Endangered on the International Union for Conservation of Nature's Red List [1,2]. As a facultative catadromite, juveniles hatched in the sea migrate to continental waters, and a significant proportion enter freshwater where they grow for ca. 3–15 years, then ultimately undertake an oceanic spawning migration as maturing adults [3–5]. Artificial structures such as weirs and dams, which are numerous in our waterways (e.g., an estimated 1.2 million across Europe [6] and 16,000 in the UK alone [7]), can impede upstream migration and prevent eel from accessing freshwater habitats [8,9]. Accordingly, restoring connectivity at barriers is a conservation priority for the European eel, legislated for under the European Union eel recovery plan (EU Council Regulation No 1100/2007) [10]).

The installation of fish passage facilities is a widespread and proven approach to mitigate the negative effects of barriers on fluvial connectivity for fish [11,12]. Eel have poor burst and sustained swimming capabilities compared to other anadromous and potadromous fishes, and multispecies technical passes often function poorly for them [8]. Eel-specific upstream passes are, therefore, designed to create lower velocity routes and usually seek to exploit the anguilliform climbing behaviour typical of the juvenile life stages [13]. There are a wide variety of designs in use, ranging from the simple addition of artificial climbing substrate to the barrier face to technical up-and-over passes which provide a migration route that completely circumvents the structure [14]. Various climbing materials have been developed specifically, while others are adapted from use in other industries (e.g., drainage, horticulture). The most commonly used substrates in England are bristle boards and eel tiles (studs) [15]. Bristle boards, which comprise tufts of stiff synthetic fibres on a backing board, are deployed both horizontally, usually lining the channel of up-and-over type passes, and vertically, mounted parallel to the wing wall of a weir with the bristles protruding towards the wall. Both installation types have been shown effective at passing pigmented glass eel through to yellow eel life stages [16–18]. Narrow spacing of the bristle clusters (15–20 mm) is most suited to small eel because they require substantial support during climbing [19,20]. Studded substrates are considered a more robust alternative to bristles in high-velocity environments and are generally less prone to clogging by debris [15]. Studs arranged in staggered rows (quincunx) are the most effective [21], with higher-density arrangements most suited to small size classes [22]. In trials on a model crump weir, the addition of studded substrate oriented horizontally achieved a 46% improvement in passage efficiency for yellow eel (424 +/− 76 mm total length) [23] and 67% for pigmented glass eel [22].

The positive response of juvenile eel to rheotactic cues is well documented and fundamental to the pass design [24,25]. The optimum flow rate through the pass represents a balance between sufficient flow to attract migrating eel and stimulate them to ascend, while maintaining velocities comfortably within swimming capabilities [12,13]. In up-and-over type passes, water is pumped to the crest where it splits to either flow down the pass length as conveyance flow or down the opposite side of the pass crest to wash ascended eel back into the watercourse upstream of the structure [14]. The relatively high-cost inputs for such passes, particularly in the set-up and maintenance of electrical pumps, can be reduced by using river flow, as within gravity-fed passes. These are placed within the watercourse, often attached to the sides of weirs, with the upstream exit submerged to allow water to naturally flow through them [15]. Whereas pumped passes offer the advantage of high consistency in the conveyance flow rate, it is necessary for gravity-fed passes to operate over a broad range of ambient water levels and resultant flow rates. In a wide-reaching review of facilities in Europe and North America, passes typically operated at flow levels equivalent to 0.14–1.1 L s^{-1} per m width of pass [26]. However, such measurements of total discharge only partially describe the hydrodynamics within the pass, with factors such as longitudinal slope and substrate properties exerting a strong influence on localised flow velocities and turbulence [27,28]. Despite the key role of flow in the functioning of eel passes, few studies have tested pass designs under different flow scenarios and/or conducted flow velocity measurements (but see [16,23]).

The design of a fishway should aim to facilitate passage without inducing delay, stress, disease or injury and without demanding additional energy expenditure [29]. Steeper longitudinal gradients require a higher energy expenditure per metre of the pass, but there is an obvious trade-off with pass length; a steeper pass provides opportunity to employ a shorter pass to surmount an equivalent elevation [30]. Several studies have compared eel passage efficiency on substrates with various longitudinal slopes (25–70°), and in general, the shallower slopes were associated with the highest passage rates [18,19,21]. Far less attention has been given to the lateral slope. The majority of constructed passes currently operating in England are rectangular in their cross-section and aligned flat on the horizontal plane, i.e., with no lateral slope [15]. Variations in lateral slope, created by tilting the pass channel or substrate bed in a flat-bottomed pass, or using a V- or U-shaped cross-section, can achieve a greater diversity of flow velocities and enhance passage [21]. It is hypothesised that the hydrodynamic heterogeneity created will provide suitable ascent conditions for a broad range of eel size classes within the same facility [13]. Although not widely tested for eel, a 15° lateral slope is currently recommended within New Zealand's fish passage guidance for small structures with the aim of benefiting a range of native species including *Galaxias* spp., red fin bullies (*Gobiomorphus huttoni*) and long-finned and short-finned eel

(*Anguilla dieffenbachia* and *Anguilla australis*), all of which are poor swimmers and employ climbing when ascending structures [31,32]. Within the context of gravity-fed passes, the integration of a lateral slope may facilitate their effective operation over a wider range of flow rates.

The aim of this study was to quantify the effect of substrate, lateral slope and flow rate on the efficacy of passage facilities for enhancing the upstream migration of juvenile European eel. A novel pass design, which incorporates two lateral slopes to create a symmetrical V-shaped channel, was trialled alongside the traditional flat-channelled equivalent. The passes were furnished with either bristles or studs and tested under a range of flow rates. The experimental set-up emulated gravity-fed passes, for which there are currently large knowledge gaps regarding their optimum operating criteria and design.

2. Materials and Methods

2.1. Experimental Setup

Trials were conducted from May to August 2020 at an outdoor research facility in Essex, UK. Four test passes were constructed within two filming tents, to exclude ambient light, and each pass comprised a high-density polyethylene ramp (2 m length), a header tank (0.7 m $\times$ 0.4 m $\times$ 0.35 m) and a footer tank (0.6 $\times$ 1 m) (Figure 1). Two ramps were furnished with bristle climbing substrate (nylon bristles, 22 mm spacing in staggered rows, manufactured by Aquatic Control Engineering, Rampton, UK) and two with stud substrate (17 mm spacing in staggered rows, manufactured by Berry Escott Engineering, Bridgwater, UK). Within each substrate type, one ramp had a flat lateral slope (0.2 m width), and one was V-shaped in cross-section (0.24 m horizontal width, 15° lateral slopes). For each pass, water was recirculated from the footer tank to the header tank using a separate submersible pump (120 L min^{-1}) and inline valve, and from there it flowed over the narrow crest and down the ramp, simulating a gravity-fed design. The pump intake was located underneath the pass, approximately 0.8 m away from the downstream entrance to minimise its influence on flow patterns at the entrance. The flow volume delivered down the pass was regulated using a ball valve fitted inline to the pipework (25 mm diameter) between the pump and the header tank to create five flow levels: 0.2, 0.3, 0.4, 0.5 and 0.6 L s^{-1}. The flow was measured at the start and end of each trial using an ultrasonic clamp-on pipe flow meter (DMTFP, Dynaflox, Shanghai, China) and the actual flow rates differed from the desired flow rates by a maximum of 0.02 L s^{-1}. No additional attraction flow was provided to the downstream entrance of the ramps.

All test ramps were set with a longitudinal slope of 30°. The footer tanks were filled to a water depth of 0.35 m, and the ramps were positioned so that the lower 0.25 m length of the ramp was submerged. A modified keep net (0.5 m length, 0.4 m width, 0.4 height) was placed over the lower 0.25 m of each ramp and secured with a shock cord to retain test subjects in the vicinity of the pass entrance.

The water temperature was maintained using a water heater and aquarium chiller, both automatically controlled (D-D Dual Temp Controller, D-D, Ilford, UK). The actual temperature was logged every 15 min throughout the experimental period (Tiny Tag, Gemini dataloggers, Chichester, UK) (mean = 19.82 $\pm$ 0.59 °C, $\pm$ SD). The water in the footer tanks was continuously aerated during the trials and replaced by 50% daily with dechlorinated water.

Upstream eel migration occurs predominantly during darkness [3]. To enable observation of eel behaviour during the night-time trials, test subjects were marked with fluorescent visible implant elastomer (VIE) (Northwest Marine Technology Inc., Anacortes, WA, USA), which has been shown to cause no adverse effect on juvenile eel behaviour (59–158 mm total length) [33]. The test ramps were lit from above with blue-black light (peak wavelength = 365 nm), which is outside the spectral sensitivity of European eel (kmax = 482 nm) [34,35]. Two cameras (HDCC500, Abus, Wetter, Germany and SDN-550, Samsung, Yeongtong-gu, South Korea) were mounted above each pass, parallel to the ramps, to record eel activity in two marked sections: (1) from 0.25 to 0.5 m upstream from the pass entrance and (2) the crest (Figure 1).

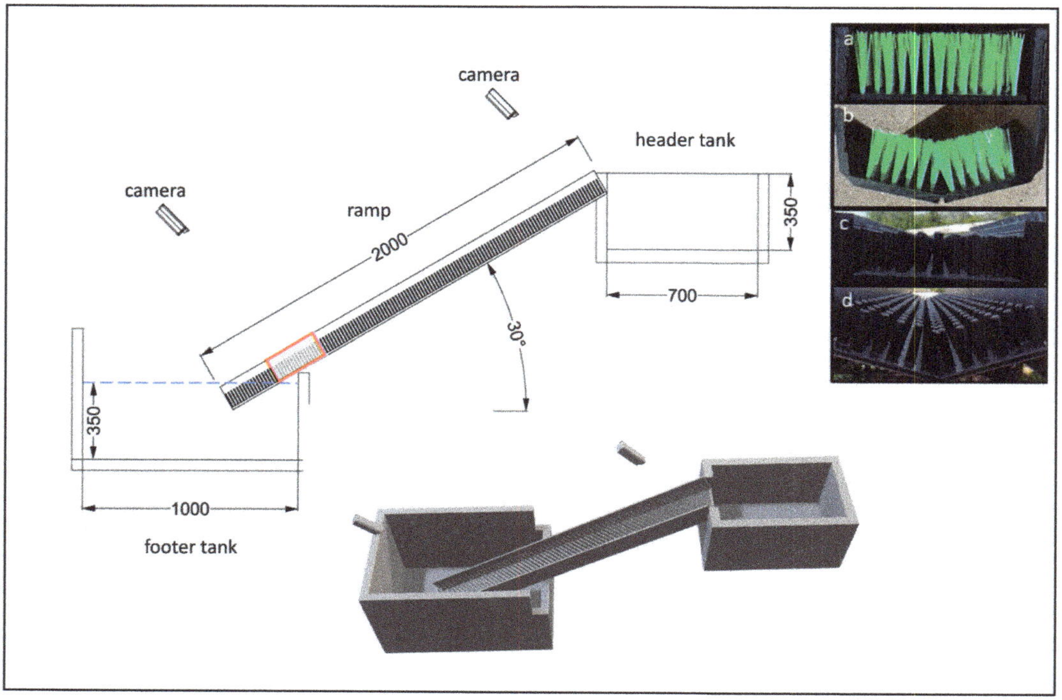

Figure 1. Simplified schematic of the experimental set-up with the filming zone in which eel were tracked indicated by a red outline and the water level in the footer tank by a blue dashed line. Inset photographs show the four substrate and lateral slope combinations tested: (**a**) laterally flat bristle; (**b**) V-profile bristle; (**c**) laterally flat studs; and (**d**) V-profile studs.

2.2. Fish Capture and Marking

Actively migrating juvenile eel were sourced by the Environment Agency close to the tidal limit of the Chelmer and Blackwater Navigation at Beeleigh, Essex, UK (51.743° N, 0.662° E) using a small pumped pass and trap. Captured eel were transported to the research facility in aerated river water and held outside in enclosed aerated tanks (500 L) of dechlorinated tap water maintained at 17 ± 1 °C using a heater, cooler and temperature control system, as above. A transparent panel (0.15 m diameter) in the lid of each tank enabled natural light to enter. The eel were acclimated to holding conditions for a minimum of 48 h and fed defrosted frozen bloodworm, artemia and daphnia every other day, followed a few hours later by a 70% water change. Eel were not fed on the day they undertook trials.

On the morning of a trial, test subjects were collected from the holding tank by random sweeps of a hand net at all heights in the water column. Prior to marking, eel were anaesthetised using an alcoholic solution of eugenol (10%) administered by an anaesthetic bath (3.5 mL L^{-1}) and measured (total length, mm). Only individuals of length 60–80 mm were retained for marking; all others were placed in a recovery tank (500 L) and subsequently released close to the site of capture. VIE of either fluorescent pink or orange was injected subcutaneously anterior to the dorsal fin to create a mark of approximately 3 mm in length (Figure 2). After recovery, eel were transferred to perforated holding tubes (200 mm length, 100 mm diameter), 20 individuals per tube, and placed in the appropriate footer tank. Procedures were subject to ethical approval by the Zoological Society of London Ethics Committee and conducted under Home Office licence (PPL 7008909). No eel died during anaesthesia or recovery. The mean length of test subjects was 73 mm ± 4.94 (S.D.).

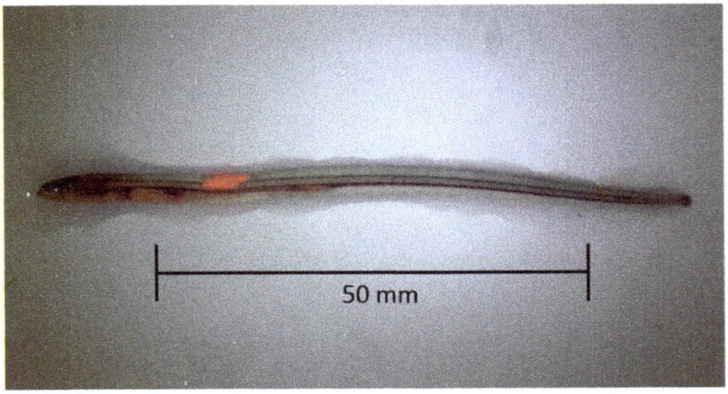

Figure 2. A study subject with visible implant elastomer (VIE) injected subcutaneously anterior to the dorsal fin.

2.3. Experimental Protocol

Trials were conducted in darkness (20:00–02:00), and each lasted 2 h with up to two trials per ramp per night. At the start of each trial, 20 eel were transferred in their holding tube and emptied into the net cage at the base of the test ramp. Video footage was recorded continuously at 25 fps (HD Analogue Recorder HDCC9001, Abus, Wetter, Germany).

The four substrate/slope treatments were assigned to the ramps and retained throughout the study. The five flow rates were randomly assigned on a daily basis but remained consistent throughout one night's trials. A total of 100 trials were conducted, with five replicates for each substrate/slope type and flow rate combination. At the end of each trial, the recirculatory pumps were stopped and eel present on the ramps were manually washed into the holding nets. The header tank and holding net on each ramp were then emptied of eel. Individuals were used only once and subsequently released to the wild close to the site of capture.

2.4. Data Processing and Analysis

Video footage of the lower pass section (from 0.25 to 0.5 m) was analysed using Tracker 6.1.1 (https://physlets.org/tracker/index.html, accessed on 14 January 2022), which enabled the manual marking of individual eel and the extraction of the XY cartesian coordinates of their fluorescent VIE mark at a time (t). The large time input required for footage processing did not permit analysis of the entire ascent ramp. Occasionally, individuals climbed onto the vertical walls of the pass and ascended out of the flow; these attempts were excluded. The calibration scale was set using fluorescent distance marks on the passes. The footage was processed at a step rate of 5 frames, which, based on preliminary explorations, represented the best balance of processing time versus the risk of missing the eel. Trials affected by occasional power failures, technical issues with the filming equipment and pump failures were excluded from analysis, yielding a total of 71 trials with a minimum of 3 replicates per substrate/slope type and flow rate combination (Table S1).

The metrics derived from the eel tracks within the 0.25 m–0.5 m pass section were as follows: (1) *No. of attempts per trial*, where each attempt commenced when an eel crossed the 0.25 m mark and ended either when it ascended beyond the 0.5 m mark or descended below the 0.25 m mark. Eel were not individually recognisable in the footage, so individual attempt data were not obtainable; (2) *Success of attempt*—an attempt was deemed successful when the eel reached the 0.5 m distance mark, even if it was visibly washed down or volitionally descended after reaching this point; (3) *Track duration (s)*—calculated for each trajectory; and (4) *Distance ascended/time*—calculated for each marked step of the trajectory, where the direction of movement was up the pass, i.e., this excluded steps where the eel remained stationary or descended. Descent movements could not be reliably tracked because they often occurred very quickly, for example, when an individual was washed down in the flow. For the pass crest, the footage was watched back at 3× speed to extract the time at which the first eel in each trial successfully passed over the crest into the header tank. Crest ascent times were compared between substrate types using a Wilcoxon rank sum test. Two analysts processed all the footage in a random order with two trials repeated by both to gauge the consistency of the approach.

Generalised linear models (GLMs) with a binomial error distribution were used to investigate the influence of the substrate type (bristles or studs), lateral slope (flat or V-profile) and flow rate (numerical) on the success of ascent attempts (1 or 0 binary response). GLMs with a gamma error distribution and log link function were used to investigate the influence of the substrate type, lateral slope and flow rate on the distance ascended/time for each step (numerical response). The maximal models with third-order interaction terms were fitted, and a stepwise deletion with likelihood ratio tests was performed to arrive at the most parsimonious model with the lowest AIC value [36]. The suitability of the error structures was evaluated using plots of standardised residuals against plots of the fitted values. All analyses were conducted in R version 4.3.0. The number of degrees of freedom is denoted by DF and the standard deviation by SD throughout.

3. Results

A total of 4032 eel ascent attempts were tracked across the 71 trials analysed. Of these, 615 were successful for the 0.25–0.5 m pass section, with the overall percentage of successful attempts ranging from 0 to 89% across the trials (Figure 3, Table S1). In the best fitting model, lateral slope was the strongest predictor of an ascent attempt being successful (deviance = 18.6%, $p < 0.01$), followed by the substrate type (deviance = 8.5%, $p < 0.05$) and the interaction between them (deviance = 3.0%, $p < 0.01$). The flow rate and its interactions with lateral slope and substrate type were weak significant predictors and collectively explained 0.9% of the deviance (Table S2). The laterally sloped 'V' passes, particularly the V bristle pass, were associated with the highest percentages of success (Figure 3).

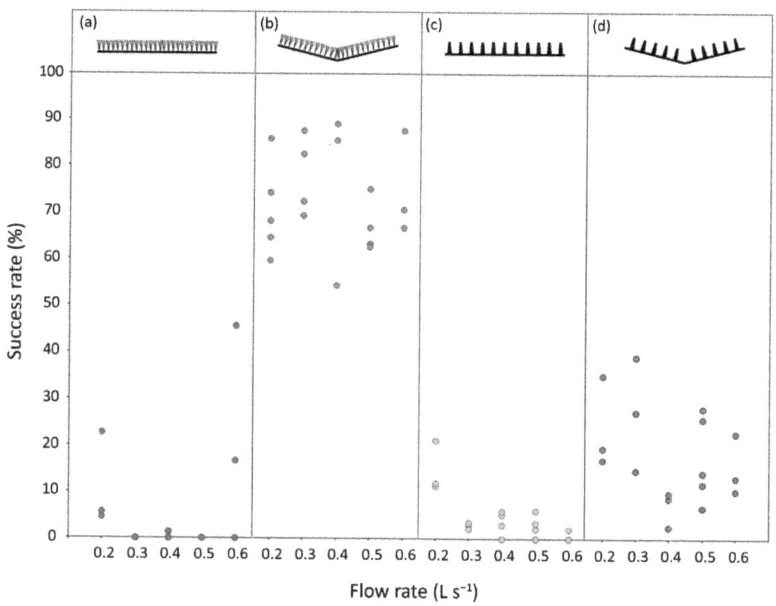

Figure 3. A strip plot showing the percentage of ascent attempts by eel in the 0.25 m–0.5 m pass section, which were successful, per trial. Trials were conducted with four different substrate/lateral slope combinations: (**a**) laterally flat bristle (red circles); (**b**) V-profile bristle (grey circles); (**c**) laterally flat studs (yellow circles); and (**d**) V-profile studs (blue circles). Trials were conducted under five different flow levels (0.2, 0.3, 0.4, 0.5 and 0.6 L s^{-1}).

The duration of ascent attempts in the 0.25–0.5 m pass section ranged from less than one second to 108 min, although 86% of attempts lasted less than 1 min (median = 2 s). The rate at which eel advanced up the passes ranged from 3.4×10^{-6} to 0.7 m s^{-1} (median = 0.008 m s^{-1}). In the best-fitting model, the lateral slope was the strongest predictor of the ascent step rate (deviance = 30.7%, $p < 0.01$), followed by the substrate type (deviance = 2.3%, $p < 0.01$) and the flow rate (deviance = 1.7%, $p < 0.01$). Interaction terms collectively explained 1% of the deviance (Table S2). The non-laterally sloped (i.e., flat) passes were associated with the fastest ascent rates (Figure 4).

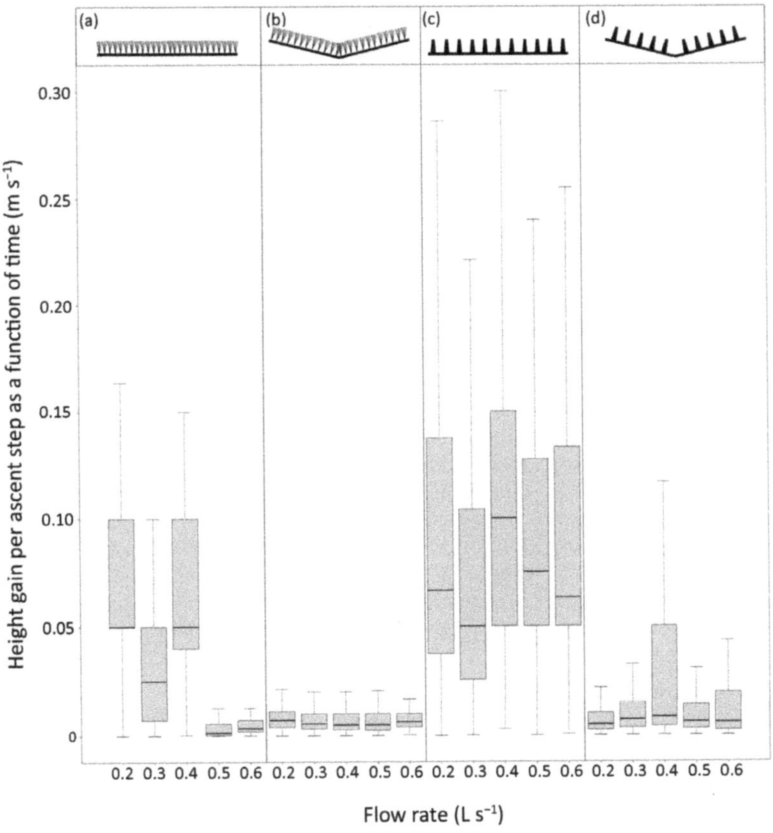

Figure 4. Boxplot of the height advanced up the pass per second, calculated for each step of eel tracks in the 0.25–0.5 m section of eel passes with four different substrate types and five different flow rates. Only steps where a gain in height was achieved are included. Thick line, median; boxes, interquartile ranges (Q1–Q3); whiskers, non-outlier range. (**a**) laterally flat bristle; (**b**) V-profile bristle; (**c**) laterally flat studs; and (**d**) V-profile studs.

Successful ascent of the entire 2 m ramp was observed in just 25 of the 71 trials, all within the laterally slopped passes; 72% in the V-shaped bristle pass and 28% in the V-shaped stud pass. The time between eel release and the first successful ascent of the crest ranged from 17 min to 1 h 58 min (median = 44 min) and did not vary between the two V-profile pass types (W = 37, $p = 0.12$).

4. Discussion

Eel passes are a widely used management tool to help restore fluvial connectivity for migrating eel and allow access to important freshwater habitats upstream of barriers [25,37]. Gravity-fed as opposed to pumped passes are a valuable, generally lower-cost option in scenarios where there is no electrical supply and/or high-frequency maintenance is not feasible. However, there is a current dearth of information on the optimum design and operating flows for such passes. Under the range of flows used in the current study (0.2–0.6 L s^{-1}), the V-shaped bristle pass was consistently the most effective for passing juvenile eel (60–80 mm, total length), while the overall effect of flow rate was minimal and non-linear.

The lateral slope followed by substrate type had the most influence on all the metrics of eel pass performance measured in the current study. Ascent of the full 2 m long ramp by at least one individual within the 2 h trial period was only observed for the V-shaped passes, with 2.5 times more successes for the bristle as opposed to the stud substrate. Despite numerous tracked attempts and successful ascents of the lower section of the ramp for all the substrate/slope combinations, the proportion of successful attempts was consistently highest for the V-shaped bristle pass followed by the V-shaped stud pass. The life stage of eel used was relatively small, a size class that typically dominates in areas close to the tidal limit of rivers [38,39]. Bristles are known to be effective for small eel, which rely on a combination of climbing and swimming to ascend structures [21,40,41]. The stud substrate tested, with narrow spacing between studs, has been previously shown suitable for similarly sized small eel (mean length 72 $\pm$ 3.9 mm, $\pm$ S.D.) [22], although a comparison of studded versus bristle substrates in eel 60–110 mm indicated that studs were both more attractive and yielded a higher successful passage rate once a climb was initiated [18]. However, direct comparison of the current results with previous studies is limited by differences in the specifications of materials used, flow rates and the development phase of the study subjects. Very little focus has been directed towards pass optimisation for the early unpigmented glass eel life phase, in no small part due to the logistical difficulties in tracking such small and transparent individuals. Here we demonstrated how the use of VIE marking and appropriate lighting and filming techniques can be applied to help resolve this bias. At elevated flow rates, as may be associated with gravity-fed passes, incorporating a lateral slope is a far more important design criterion than the selection of bristles or studs. In comparison, both of the laterally flat passes performed poorly, irrespective of the substrate.

The flow rates tested in the current study were selected to recreate conditions in gravity-fed passes, which, due to changes in ambient water level, are subject to a far greater range of operating flows than pumped passes. For small eel (60–110 mm, total length), it has been shown that effective passage in bristle and stud passes can be achieved with flows as low as 0.07 L s^{-1} [18], while high velocities result in individuals being repeatedly washed down the ramp [26]. The inability of any eel in the current study to ascend the full 2 m ramp in either of the laterally flat passes, even in a flow of 0.2 L s^{-1}, supports the idea that the presence of a low-velocity wetted margin is a crucial design feature for eel passes targeting small-size classes [13,42]. Although not tested in the current study, V-shaped passes would be expected to offer an advantage over single-sided laterally sloped passes because two wetted margins are created, one on either side of the central flow plume. Observations of ascending eel on the V-shaped passes in the video footage confirmed that successful ascents were generally in this zone. Further, the most common cause of failure during ascents was when an individual ascending within the wetted perimeter moved towards and entered the central flow plume and was subsequently washed down.

The absence of a strong relationship between the flow rate and the measured passage metrics suggested that, first, the minimum flow rate of 0.2 L s^{-1} was too high for the effective operation of the laterally flat passes for this early life stage of the eel, and second, the presence of the wetted perimeter in the V-shaped passes effectively decoupled the influence of the flow. Based on these findings, V-shaped passes would be expected to provide effective passage at even higher flows than those tested, so long as the large wetted perimeter is maintained. Under the tested flow rates of $\leq$0.6 L s^{-1}, the central flow plume did not extend to the edges of the pass, which also incidentally rendered the overall pass width to some extent irrelevant. The positive effect of the V-shape on passage would presumably be lost at the point where the increased flow caused the margins of the central flow plume to extend to the vertical walls of the pass, increasing water depth across the pass to a level that small individuals would be forced to swim rather than climb. Juvenile eel of the size studied can burst swim at 0.35–0.5 m s^{-1} for a few seconds, but the modelled sustained (30–40 min) swimming speeds are in the order of 0.2–0.3 m s^{-1} [43,44]. The rapidity of the ascent steps was highest for the laterally flat passes, and an examination of the footage revealed this was attributable to the high incidence of burst swimming behaviour, which was usually immediately followed by washdown. One outlier group in the laterally flat bristle pass under the 0.6 L s^{-1} flow rate exhibited an unexpectedly high success rate in the tracked section, although all eel were subsequently washed down. The reasons for this are unclear but may be due to exceptionally motivated individuals in this group despite efforts to avoid sampling bias among the wild-caught fish. Occasionally, eel on the laterally flat passes would rest within the flow by wrapping around a bristle cluster or stud before making another burst swim attempt. In the V-shaped passes, the ascent steps were generally slower because eel tended to climb rather than swim. Eel were also frequently observed resting in these passes, usually within the wetted perimeter or moving just outside of it towards the vertical walls of the ramp.

5. Conclusions

Our findings provide strong evidence for recommending the incorporation of a lateral slope for eel passes in scenarios where flow rates are likely to be elevated ($\geq$0.2 L s^{-1}) and/or fluctuate. The novel V-pass profiles consistently outperformed their flat counterparts, while selection of the most appropriate climbing substrate will be dependent on the life stage being targeted. For locations near the tidal limit where small eel dominate, our results suggest bristles will outperform studs, although there will be other site-based factors to consider such as the risk of clogging with debris. Further, we acknowledge that attraction to the pass is fundamental to effective operation, and this was not quantified in the current study. The architecture of the substrates tested renders measurement of localised hydrodynamic features within the pass (e.g., velocity and turbulence) unfeasible using conventional empirical methods such as electromagnetic flow meters or acoustic Doppler velocimeters. However, computational fluid dynamics modelling techniques provide future potential to explore, at a finer scale, the relationships between these hydrodynamic features and the tracked eel trajectories, and could provide a valuable next step to build on the current findings.

Supplementary Materials: The following supporting information can be downloaded at: https://www.mdpi.com/article/10.3390/fishes8120612/s1, Table S1: Summary of trials analysed and average attempt and success rates for eel tracked in the lower pass section (from 0.25 to 0.5 m), for 4 different substrate and lateral slope combinations, and 5 discharge levels.; Table S2: Summary of generalised linear model (GLM) comparisons using Akaike's information criterion (AIC) relative to the null model for the two modelled response variables.

Author Contributions: Conceptualization, A.T.P., P.J.R. and R.M.W.; methodology, A.T.P. and P.J.R.; formal analysis, P.J.R. and C.P.; investigation, A.T.P., P.J.R. and C.P.; resources, A.T.P. and R.M.W.; data curation, P.J.R.; writing—original draft preparation, A.T.P. and P.J.R.; writing—review and editing, A.T.P., P.J.R. and C.P.; visualization, A.T.P. and P.J.R.; supervision, A.T.P.; project administration, A.T.P.; funding acquisition, R.M.W. All authors have read and agreed to the published version of the manuscript.

Funding: This research was principally funded by the Environment Agency. ATP was part-funded by Research England.

Institutional Review Board Statement: The animal study protocol was approved by the Ethics Committee of The Zoological Society of London (ref: BME716, approved May 2020).

Data Availability Statement: Data are available upon reasonable request to the corresponding author.

Acknowledgments: We thank Dan Hayter and Ben Norrington for assistance with the collection of study animals and Chris Grzesiok for assistance with the design and production of Figure 1.

Conflicts of Interest: The funders had no role in the design of the study; in the collection, analyses, or interpretation of data; in the writing of the manuscript; or in the decision to publish the results.

References

1. ICES. *European Eel (Anguilla anguilla) throughout Its Natural Range*; ICES: Copenhagen, Denmark, 2022. [CrossRef]
2. Pike, C.; Crook, V.; Gollock, M. *Anguilla anguilla*. In *The IUCN Red List of Threatened Species 2020: E.T60344A152845178*; IUCN: Gland, Switzerland, 2020. [CrossRef]
3. Tesch, F.W. *The Eel*, 5th ed.; Blackwell Science Ltd.: Oxford, UK, 2003.
4. Aprahamian, M.W. Age Structure of Eel, *Anguilla anguilla* (L.), Populations in the River Severn, England, and the River Dee, Wales. *Aquac. Res.* **1988**, *19*, 365–376. [CrossRef]
5. Righton, D.; Piper, A.T.; Aarestrup, K.; Amilhat, E.; Belpaire, C.; Casselman, J.; Castonguay, M.; Díaz, E.; Dörner, H.; Faliex, E.; et al. Important Questions to Progress Science and Sustainable Management of Anguillid Eels. *Fish Fish.* **2021**, *22*, 762–788. [CrossRef]
6. Belletti, B.; Garcia de Leaniz, C.; Jones, J.; Bizzi, S.; Börger, L.; Segura, G.; Castelletti, A.; van de Bund, W.; Aarestrup, K.; Barry, J.; et al. More than One Million Barriers Fragment Europe's Rivers. *Nature* **2020**, *588*, 436–441. [CrossRef] [PubMed]
7. Environment Agency. *The Eel Manual—Eel and Elver Passes: A Guide to the Design and Implementation of Passage Solutions at Weirs, Tidal Gates and Sluices*; Environment Agency: Bristol, UK, 2011.
8. Feunteun, E. Management and Restoration of European Eel Population (*Anguilla anguilla*): An Impossible Bargain. *Ecol. Eng.* **2002**, *18*, 575–591. [CrossRef]
9. Jacoby, D.M.P.; Casselman, J.M.; Crook, V.; DeLucia, M.B.; Ahn, H.; Kaifu, K.; Kurwie, T.; Sasal, P.; Silfvergrip, A.M.C.; Smith, K.G.; et al. Synergistic Patterns of Threat and the Challenges Facing Global Anguillid Eel Conservation. *Glob. Ecol. Conserv.* **2015**, *4*, 321–333. [CrossRef]
10. *Council Regulation (EC) No. 1100/2007 of 18 September 2007 Establishing Measures for the Recovery of the Stock of European Eel*; European Union: Brussels, Belgium, 2007.
11. Roscoe, D.W.; Hinch, S.G. Effectiveness Monitoring of Fish Passage Facilities: Historical Trends, Geographic Patterns and Future Directions. *Fish. Fish.* **2010**, *11*, 12–33. [CrossRef]
12. Clay, C.H. *Design of Fishways and Other Fish Facilities*, 2nd ed.; Lewis Publishers: Boca Raton, FL, USA, 1995.
13. Porcher, J.P. Fishways for Eels. *Bull. Français Pêche Piscic.* **2002**, *364*, 147–155. [CrossRef]
14. Solomon, D.J.; Beach, M.H. *Manual for Provision of Upstream Migration Facilities for Eel and Elver*; Science Report SC020075/SR2; Environment Agency: Bristol, UK, 2007.
15. Rosewarne, P.J.; Wright, R.M. *Eel Passage—Design and Performance SC150001*; Environment Agency: Bristol, UK, 2022.
16. Kerr, J.R.; Karageorgopoulos, P.; Kemp, P.S. Efficacy of a Side-Mounted Vertically Oriented Bristle Pass for Improving Upstream Passage of European Eel (*Anguilla anguilla*) and River Lamprey (*Lampetra fluviatilis*) at an Experimental Crump Weir. *Ecol. Eng.* **2015**, *85*, 121–131. [CrossRef]
17. Briand, C.; Fatin, D.; Fontenelle, G.; Feunteun, E. Effect of Re-Opening of a Migratory Pathway for Eel (*Anguilla anguilla*, L.) at a Watershed Scale. *Bull. Français Pêche Piscic.* **2005**, *378–379*, 67–86. [CrossRef]
18. Watz, J.; Nilsson, P.A.; Degerman, E.; Tamario, C.; Calles, O. Climbing the Ladder: An Evaluation of Three Different Anguillid Eel Climbing Substrata and Placement of Upstream Passage Solutions at Migration Barriers. *Anim. Conserv.* **2019**, *22*, 452–462. [CrossRef]
19. Jellyman, P.; Bauld, J.; Crow, S.K. The Effect of Ramp Slope and Surface Type on the Climbing Success of Shortfin Eel (*Anguilla australis*) Elvers. *Mar. Freshw. Res.* **2017**, *68*, 1317–1324. [CrossRef]
20. Podgorniak, T.; Angelini, M.; De Oliveira, E.; Daverat, F.; Pierron, F. Selective Pressure of Fishways upon Morphological and Muscle Enzymatic Traits of Migrating Glass Eels. *Can. J. Fish. Aquat. Sci.* **2017**, *74*, 445–451. [CrossRef]
21. Voegtle, B.; Larinier, M. *Etude Sur Les Capacités de Franchissement Des Civelles et Anguillettes: Site Hydroélectrique de Tuilières Sur La Dordogne (24) Barrage Estuarien d'Arzal Sur La Vilaine (56)*; Hindustan Aeronautics Limited: Bangalore, India, 2000.
22. Vowles, A.S.; Don, A.M.; Karageorgopoulos, P.; Worthington, T.A.; Kemp, P.S. Efficiency of a Dual Density Studded Fish Pass Designed to Mitigate for Impeded Upstream Passage of Juvenile European Eels (*Anguilla anguilla*) at a Model Crump Weir. *Fish Manag. Ecol.* **2015**, *22*, 307–316. [CrossRef]
23. Vowles, A.S.; Don, A.M.; Karageorgopoulos, P.; Kemp, P.S. Passage of European Eel and River Lamprey at a Model Weir Provisioned with Studded Tiles. *J. Ecohydraulics* **2017**, *2*, 88–98. [CrossRef]
24. Feunteun, E.; Laffaille, P.; Robinet, T.; Briand, C.; Baisez, A.; Olivier, J.M.; Acou, A. A Review of Upstream Migration and Movements in Inland Waters by Anguillid Eels: Toward a General Theory. In *Eel Biology*; Aida, K., Tsukamoto, K., Yamauchi, K., Eds.; Springer: Tokyo, Japan, 2003; pp. 191–213.

25. Knights, B.; White, E.M. Enhancing Immigration and Recruitment of Eels: The Use of Passes and Associated Trapping Systems. *Fish Manag. Ecol.* **1998**, *5*, 459–471. [CrossRef]
26. Solomon, D.; Beach, M. *Fish Pass Design for Eel and Elver (Anguilla anguilla)*; Environment Agency: Bristol, UK, 2004.
27. Ibnu Syihab, A.B.M.; Verdin, P.G.; Wright, R.M.; Piper, A.T.; Rivas Casado, M. Computational Fluid Dynamics Simulations of Water Flow on a Studded Upstream Eel Pass. *River Res. Appl.* **2021**, *37*, 1279–1293. [CrossRef]
28. Padgett, T.E.; Thomas, R.E.; Borman, D.J.; Mould, D.C. Individual-Based Model of Juvenile Eel Movement Parametrized with Computational Fluid Dynamics-Derived Flow Fields Informs Improved Fish Pass Design. *R. Soc. Open Sci.* **2020**, *7*, 191505. [CrossRef]
29. Castro-Santos, T.; Cotel, A.; Webb, P. Fishway Evaluations for Better Bioengineering—An Integrative Approach. In *Challenges for Diadromous Fishes in a Dynamic Global Environment*; Haro, A., Moffit, C., Dadswell, M., Eds.; American Fisheries Society Symposium: Bethesda, MD, USA, 2009; Volume 69, pp. 557–575.
30. Noonan, M.J.; Grant, J.W.A.; Jackson, C.D. A Quantitative Assessment of Fish Passage Efficiency. *Fish Fish.* **2012**, *13*, 450–464. [CrossRef]
31. Baker, C.F.; Boubee, J.A.T. Upstream Passage of Inanga *Galaxias Maculatus* and Redfin Bullies *Gobiomorphus Huttoni* over Artificial Ramps. *J. Fish Biol.* **2006**, *69*, 668–681. [CrossRef]
32. Franklin, P.; Gee, E.; Baker, C.; Bowie, S. *New Zealand Fish Passage Guidelines for Structures Up to 4 Metres*; NIWA Taihoro Nukurangi: Auckland, New Zealand, 2018.
33. Imbert, H.; Beaulaton, L.; Rigaud, C.; Elie, P. Evaluation of Visible Implant Elastomer as a Method for Tagging Small European Eels. *J. Fish Biol.* **2007**, *71*, 1546–1554. [CrossRef]
34. Archer, S.; Hope, A.; Partridge, J.C. The Molecular Basis for the Green-Blue Sensitivity Shift in the Rod Visual Pigments of the European Eel. *Proc. Biol. Sci.* **1995**, *262*, 289–295. [PubMed]
35. Hope, A.J.; Partridge, J.C.; Hayes, P.K. Switch in Rod Opsin Gene Expression in the European Eel, *Anguilla anguilla* (L.). *Proc. R. Soc. B Biol. Sci.* **1998**, *265*, 869. [CrossRef] [PubMed]
36. Akaike, H. Information Theory and an Extension of the Maximum Likelihood Principle. In *Selected Papers of Hirotugu Akaike*; Parzen, E., Tanabe, K., Kitagawa, G., Eds.; Springer: New York, NY, USA, 1998. [CrossRef]
37. Laffaille, P.; Acou, A.; Guillouet, J.; Legault, A. Temporal Changes in European Eel, *Anguilla anguilla*, Stocks in a Small Catchment after Installation of Fish Passes. *Fish Manag. Ecol.* **2005**, *12*, 123–129. [CrossRef]
38. Naismith, I.A.; Knights, B. Migrations of Elvers and Juvenile European Eels, *Anguilla anguilla* L., in the River Thames. *J. Fish Biol.* **1988**, *33*, 161–175. [CrossRef]
39. White, E.M.; Knights, B. Dynamics of Upstream Migration of the European Eel, *Anguilla anguilla* (L.), in the Rivers Severn and Avon, England, with Special Reference to the Effects of Man-Made Barriers. *Fish Manag. Ecol.* **1997**, *4*, 311–324. [CrossRef]
40. Briand, C.; Fatin, D.; Fontenelle, G.; Feunteun, E. Estuarine and Fluvial Recruitment of the European Glass Eel, *Anguilla anguilla*, in an Exploited Atlantic Estuary. *Fish Manag. Ecol.* **2003**, *10*, 377–384. [CrossRef]
41. Legault, A. Study of Some Selectivity Factors in Eel Ladders. *Bull. Fr. Peche Piscic.* **1992**, *325*, 83–91. [CrossRef]
42. Jellyman, D.J. Summer Upstream Migration of Juvenile Freshwater Eels in New Zealand. *N. Z. J. Mar. Freshw. Res.* **1977**, *11*, 61–71. [CrossRef]
43. Clough, S.C.; Turnpenny, A.W.H. *Swimming Speeds in Fish: Phase 1*; Environment Agency: Bristol, UK, 2001.
44. McCleave, J.D. Swimming Performance of European Eel (*Anguilla anguilla* (L.)) Elvers. *J. Fish Biol.* **1980**, *16*, 445–452. [CrossRef]

MDPI
St. Alban-Anlage 66
4052 Basel
Switzerland
www.mdpi.com

Fishes Editorial Office
E-mail: fishes@mdpi.com
www.mdpi.com/journal/fishes

www.ingramcontent.com/pod-product-compliance
Lightning Source LLC
Chambersburg PA
CBHW061615210926
51850CB00009B/2328
* 9 7 8 3 7 2 5 8 0 2 2 7 2 *